Clean Energy Production Technologies

The consumption of fossil fuels has been continuously increasing around the globe and simultaneously becoming the primary cause of global warming as well as environmental pollution. Due to limited life span of fossil fuels and limited alternate energy options, energy crises is important concern faced by the world. Amidst these complex environmental and economic scenarios, renewable energy alternates such as biodiesel, hydrogen, wind, solar and bioenergy sources, which can produce energy with zero carbon residue are emerging as excellent clean energy source. For maximizing the efficiency and productivity of clean fuels via green & renewable methods, it's crucial to understand the configuration, sustainability and techno-economic feasibility of these promising energy alternates. The book series presents a comprehensive coverage combining the domains of exploring clean sources of energy and ensuring its production in an economical as well as ecologically feasible fashion. Series involves renowned experts and academicians as volume-editors and authors, from all the regions of the world. Series brings forth latest research, approaches and perspectives on clean energy production from both developed and developing parts of world under one umbrella. It is curated and developed by authoritative institutions and experts to serves global readership on this theme.

Arindam Kuila
Editor

Biojet Fuel: Current Technology and Future Prospect

Springer

Editor
Arindam Kuila
Bioscience and Biotechnology
Banasthali Vidyapith
Newai, Rajasthan, India

ISSN 2662-6861 ISSN 2662-687X (electronic)
Clean Energy Production Technologies
ISBN 978-981-99-8782-5 ISBN 978-981-99-8783-2 (eBook)
https://doi.org/10.1007/978-981-99-8783-000

This Springer imprint is published by the registered company Springer Nature Singapore Pte Ltd.
The registered company address is: 152 Beach Road, #21-01/04 Gateway East, Singapore 189721, Singapore

Paper in this product is recyclable

Preface

Bio-aviation fuel (also known as biojet fuel, renewable jet fuel, or aviation biofuel in some literature), a type of biofuel for the air transport sector, is recognized as a short- to medium-term solution toward an overall reduction of the sector's GHG emissions. There are very few reports on biojet fuel production from different types of substrates. The present biojet fuel scenario demands improved low-cost technologies that can pay dividends in the long run. In lieu of these needs and demands, the present book seeks to explore the present status and future prospect of biojet fuel production. This book is divided into 12 chapters. The first and second chapters provide an overview of the aviation industry, feedstock, supply chain, and a basic introduction to biojet fuel. The production of biojet fuel from various substrates is compared in the third and fourth chapters. The fifth chapter discusses biojet fuel characteristics. The production of biojet fuel using catalytic cracking and hydrodeoxygenation is covered in the sixth and seventh chapters. The technoeconomic evaluation of producing biojet fuel is covered in the eighth chapter. The sustainability of biojet fuel and its many uses are covered in the ninth and tenth chapters. The last two chapters cover the life cycle evaluation, prospects for the future, and current state of biojet fuel production. Prominent scientists and researchers who have worked significantly in this field and have been actively involved for a number of years have contributed chapters to the book. We also thank the authors for their remarkable work and for accepting our invitation to participate to the book.

The book will be useful for students and researchers in the areas of various branches of life sciences like environmental biotechnology, bioprocess engineering, renewable energy, chemical engineering, nanotechnology, biotechnology, microbiology etc.

We are grateful to Springer Nature Publishing, especially Aishwarya Thyagarajan and Rhea Dadra for their complete cooperation and assistance in the timely publishing of this book. We would like to express our gratitude to the writers and the publication staff for their efforts for publishing this book.

Newai, Rajasthan, India Arindam Kuila

Contents

Editor and Contributors

About the Editor

Arindam Kuila is currently working as Assistant Professor in the Department of Bioscience and Biotechnology, Banasthali Vidyapith, Rajasthan, India. Previously, he worked as a research associate at Hindustan Petroleum Green R&D Centre, Bangalore, India. He did his PhD from Agricultural and Food Engineering Department, Indian Institute of Technology Kharagpur, India, in 2013 in lignocellulosic biofuel production. He was awarded Petrotech Research Fellowship from 2008. He has completed an Indo-Brazil collaborative project funded by DBT, India. He has guided six PhD students, and currently three PhD students are working under his guidance. He has published 10 edited books, 18 book chapters, 35 papers in peer-reviewed journals, and filed 5 patents (1 patent granted). He acted as guest editor for several international journals.

Contributors

Priyancka Arora Department of Biotechnology, Amity University, Mohali, Punjab, India

Yvan Jesus Olortiga Asencios Universidade Federal de São Paulo, UNIFESP, Instituto do Mar, Santos, São Paulo, Brazil

Nikita Bhati Department of Bioscience and Biotechnology, Banasthali Vidyapith, Newai, Rajasthan, India

Moumita Bishai Department of Botany, Gurudas College, Kolkata, India

Luiza Helena da Silva Martins Universidade Federal Rural da Amazônia, UFRA, Instituto de Saúde e Produção Animal, Belém, Pará, Brazil

Johnatt Allan Rocha De Oliveira Universidade Federal do Pará, UFPA, Faculdade de Nutrição, Belém, Pará, Brazil

Anthony Andrey Ramalho Diniz Universidade Federal de São Paulo, UNIFESP, Instituto do Mar, Santos, São Paulo, Brazil

Yatika Dixit Department of Bioscience and Biotechnology, Banasthali Vidyapith, Newai, Rajasthan, India

Y. Lalitha Kameswari Department of Electronics and Communication Engineering, Koneru Lakshmaiah Education Foundation (Deemed to be University), Vaddeswaram, Guntur, Andhra Pradesh, India

Andrea Komesu Universidade Federal de São Paulo, UNIFESP, Instituto do Mar, Santos, São Paulo, Brazil

Arindam Kuila Department of Bioscience and Biotechnology, Banasthali Vidyapith, Newai, Rajasthan, India

Depak Kumar Department of Chemical Engineering, Banasthali Vidyapith, Newai, Rajasthan, India

Sonu Kumar Department of Electronics and Communication Engineering, K.S.R.M. College of Engineering, Kadapa, Andhra Pradesh, India

Manipur International University, Imphal, Manipur, India

Sudesh Kumar Department of Chemistry, Banasthali Vidyapith, Newai, Rajasthan, India

Ankita Kumari Department of Chemistry, Banasthali Vidyapith, Newai, Rajasthan, India

Shubhankari Mishra Department of Biotechnology, Faculty of Engineering and Technology, Rama University, Kanpur, Uttar Pradesh, India

Shally Pandit Department of Bioscience and Biotechnology, Banasthali Vidyapith, Newai, Rajasthan, India

Eduardo Dellosso Penteado Universidade Federal de São Paulo, UNIFESP, Instituto do Mar, Santos, São Paulo, Brazil

Anand Prakash Department of Bioscience and Biotechnology, Banasthali Vidyapith, Newai, Rajasthan, India

Priyanka Sati Department of Chemistry, Banasthali Vidyapith, Newai, Rajasthan, India

Satyajit Saurabh DNA Fingerprinting Laboratory, Bihar State Seed and Organic Certification Agency, Mithapur, Patna, Bihar, India

Arun Kumar Sharma Department of Bioscience and Biotechnology, Banasthali Vidyapith, Newai, Rajasthan, India

Ajay Singh Verma Division of Research & Innovation, School of Applied and Life Sciences, Uttaranchal University, Dehradun, Uttarakhand, India

Samakshi Verma Department of Biotechnology, Faculty of Engineering and Technology, Rama University, Kanpur, Uttar Pradesh, India

Ashok Kumar Yadav Department of Chemical Engineering, Banasthali Vidyapith, Newai, Rajasthan, India

Preeti Yadav Department of Bioscience and Biotechnology, Banasthali Vidyapith, Newai, Rajasthan, India

Chapter 1
General Background and Introduction of Biojet Fuel

Satyajit Saurabh

Abstract Biojet fuel, sometimes referred to as aviation biofuel, is a substitute fuel produced mostly from plant-based renewable resources. It is made specifically to be used in aviation, primarily in jet engines, as a replacement for conventional jet fuel. In order to reduce greenhouse gas emissions and the aviation industry's reliance on fossil fuels, the idea of biojet fuel was developed. A variety of feedstocks, biomass, and algae are being used to make biojet fuel. The carbon footprint of biojet fuel is smaller than that of conventional jet fuel. In addition, biojet fuel has a higher energy density, which could lead to superior fuel economy and longer flight distances. The use of biojet fuel does not necessitate material modifications to aircraft engines because it can be blended seamlessly with regular jet fuel. As a result, airlines can use the fuel without experiencing any changes to their current infrastructure. In order to move toward a more environmentally friendly and sustainable aviation sector, the aviation industry, as well as governments and environmental organizations, continue to fund research, development, and implementation efforts to advance the production and use of biojet fuel.

Keywords Biojet fuel · Aviation turbine fuel · Avtur · AVGAS · Aviation · Feedstock · Lignocellulosic biomass · Algae · Pathways for biojet fuel production

1.1 Introduction

Global rise in fossil fuel prices and excessive greenhouse gas emission has drawn the attention of researchers from the aviation sector and biofuel firms for the development of alternative jet fuels, which must be environmentally sustainable and commercially suitable (Doliente et al. 2020; Faaij and van Dijk 2012). Jet fuel is a

---

S. Saurabh (✉)
DNA Fingerprinting Laboratory, Bihar State Seed and Organic Certification Agency, Mithapur, Patna, Bihar, India

A. Kuila (ed.), *Biojet Fuel: Current Technology and Future Prospect*, Clean Energy Production Technologies, https://doi.org/10.1007/978-981-99-8783-2_1

form of *av*iation *tur*bine fuel (ATF), also spelled as avtur, and is used in aircraft with turbine or gas engines. Biofuel is one of the most promising choices for commercially viable and sustainable fuel because of its renewable nature, reduced dependence on fossil fuel, and minimal greenhouse gas emission (Kubickova and Kubicka 2010). It can offer a more long-term and short-term environment-friendly solution to aviation sector than petroleum fuels. According to the International Air Transport Association's 2017 annual review, application of sustainable aviation fuel has the capacity to cut down CO_2 emission by 80% during their life cycle. However, some authors opined that they must possess characteristics like good thermostablility, cold flow properties, low freezing point, and low carbon output across their lifetimes (Mohammad et al. 2013 and Agusdinata et al. 2011). According to some workers, biojet fuel offers better properties like low tailpipe emissions and sulfur content, good cold flow property, and thermal stability, making them a suitable substitute for conventional fuel (Lokesh et al. 2015; Shankar and Khandelwal 2013; Timko et al. 2010). Also, biojet fuel has an advantage over other alternative fuels in that they are compatible with standard engine and fuel system and can be utilized without modifying the engine (Mofijur et al. 2015). The cooperation between national and international organizations, states, and nations can have an impact on the promising future of alternative jet fuel. The International Air Transport Association (IATA) anticipates that biojet fuel will contribute 30% of the jet fuel industry by 2030. Alternative biofuels have a number of benefits over traditional jet fuel, including lower costs and greater availability, in addition to having reduced greenhouse gas emissions. Cost fluctuations are the key factor affecting fossil fuels (Pandey 2011). A competent production process can lower the overall cost of fuel even while low-cost renewable sources are employed to produce biojet fuel.

Still there are concerns like the crops employed as a source of biojet fuel production should not affect food security, reforestation efforts, or environmental safety (Baljet 2010). Use of nonfood crops and useless land for the production of biojet fuel will be profitable and can support the supply chain process, particularly in developing nations. Although technologies for producing liquid fuel from biomass have advanced significantly, there is still much room for improvement (Tan et al. 2014; Elia et al. 2013; Jones et al. 2010). However, a trillion dollars would need to be spent to upgrade all of the flying engines so that they can run on a new kind of fuel, which is a major issue (Hileman and Stratton 2014).

1.2 Developments and Genesis of Biojet Fuel

Aviation simply cannot function without jet fuel as it is the sole power source for aircraft engines and the key to efficient air travel. This fuel has been vital since the earliest days of flight, especially after the introduction of jet engines in the 1930s. The importance of specialized fuel cannot be overstated as the use of AVGAS was initially attempted but quickly proved insufficient for the demands of jet engines.

Avtur, also known as aviation turbine fuel, is a petroleum-based fuel that is primarily utilized in aircraft turbine engines. This type of fuel is similar to AVGAS, which is specifically designed for use in piston engines. The distinction is that avtur serves to fuel aircraft powered by external combustion or turbine engines. Due to its widespread use in houses and small enterprises, this fuel type is very well known. Avtur is a type of aviation fuel designed for aircraft powered by gas turbine engines. It appears colorless and ranges in size from 8 to 16 carbon atoms (Balogun et al. 2022). It has an initial and final boiling point of approximately 125 °C and 290 °C, respectively, at atmospheric pressure.

A biojet fuel, also known as bio-aviation fuel (BAF) or avtur, is a type of fuel that is derived from renewable biomass sources, such as plants and algae, over a small period of time instead of fossil fuels like crude oil, which requires several years for its formation. The development of biojet fuel is driven by the need to reduce greenhouse gas emissions and reliance on finite fossil fuel resources in the aviation industry. The concept of biojet fuel has been researched and developed for several decades, but it gained significant attention in the early 2000s when concerns about climate change and energy security started to grow. The aviation industry is a major contributor to greenhouse gas emissions, and using biojet fuel could help reduce the carbon footprint of aircraft and mitigate climate change impacts. According to the International Air Transport Association (IATA), biojet fuel is a key element that can help reduce carbon footprint within the environmental impact of aviation (*"Developing Sustainable Aviation Fuel (SAF)")*. According to Bauen et al. 2009, biofuels can cut down CO_2 emissions by 20–98% compared to conventional jet fuel depending on the feedstock used. Biojet fuel is made from renewable resources such as vegetable oils, algae, or agricultural waste (Fig. 1.1). It holds immense potential in significantly reducing greenhouse gas emissions and dependence on fossil fuels. It serves as a sustainable alternative to traditional jet fuels. The production of biojet fuel involves converting biomass feedstocks into liquid hydrocarbon fuels that are similar to conventional jet fuels. These feedstocks can include various plant oils (such as soybean, camelina, and palm), animal fats, waste cooking oils, and algae. The conversion processes can include techniques such as transesterification, hydrotreating, and thermochemical processes like pyrolysis or gasification.

One of the primary advantages of biojet fuel is its potential to significantly reduce greenhouse gas emissions. While conventional jet fuels are derived from fossil fuels, biojet fuel is produced from renewable sources, which means that the carbon dioxide emitted during combustion is part of the natural carbon cycle. Also, biojet fuels have a lower sulfur content, reducing sulfur oxide emissions and the formation of particulate matter. Other advantages of biojet fuel are the availability of theoretically abundant feedstock, less risk in long-term use in case of fuel spoilage, and used as a "drop-in" alternative for existing engines (Bosch et al. 2017; de Jong et al. 2017). Biojet fuel also has the potential to enhance energy security by reducing dependence on volatile oil markets and geopolitical tensions associated with oil-producing regions. It can be produced domestically, diversifying the fuel sources for the aviation industry and reducing vulnerability to supply disruptions (Malode et al. 2021).

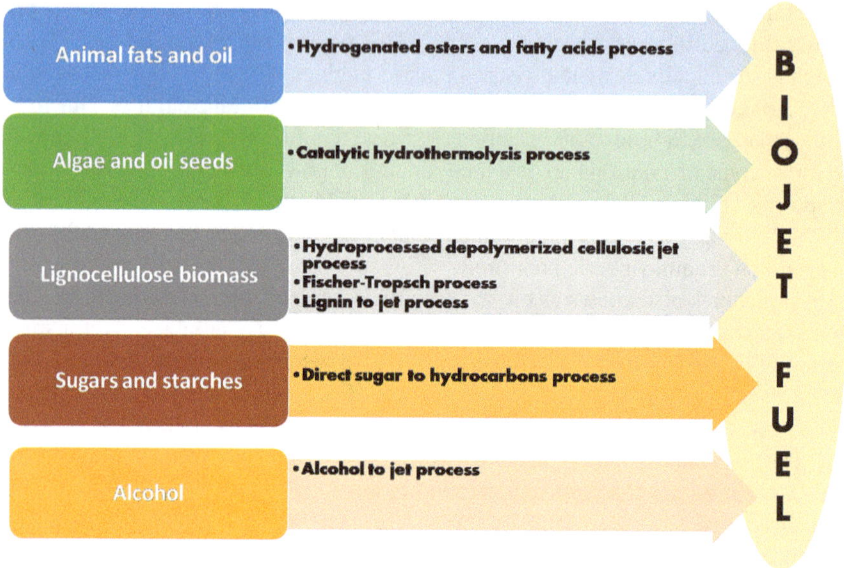

Fig. 1.1 Different renewable resources and routes for biojet fuel production

However, there are some challenges associated with the widespread adoption of biojet fuel. The production of biojet fuel requires a significant amount of land and resources, which could compete with food production if energy crops are more profitable or may contribute to deforestation and eutrophication from fertilizer use if not managed properly. Also, there may be cases of spatial or temporal boundaries, for example, feedstock may not be grown all year round or at all in some cases if specific conditions are required (Bosch et al. 2017; de Jong et al. 2017). Additionally, the cost of production is currently higher compared to conventional jet fuels, although technological advancements and economies of scale are expected to lower the cost in the future.

Overall, biojet fuel represents a promising alternative to conventional jet fuels, offering potential environmental benefits and reducing reliance on fossil fuels in the aviation industry. Continued research, development, and investment are necessary to overcome the challenges and enable the large-scale production and deployment of biojet fuel.

In commercial aviation, biojet fuels have already undergone successful testing. The first biofuel-powered test flight was conducted in 2008, and commercial flights were allowed to use blended fuels containing 50% biofuels in 2011. In 2009, IATA pledged to achieve carbon-neutral growth by 2020, while cutting down carbon emission to its one half by 2050 in a press statement titled "Carbon-Neutral Growth By 2020." Boeing claims that by 2015, 1% of the fuel used by airlines will come from biofuels (Bloomberg, July 22, 2010). In December 2011, the FAA awarded US$7.7 million to eight companies as part of its CAAFI and CLEEN projects for the development of drop-in sustainable fuels, primarily from biomass, alcohols, sugars, and organic matter. In 2015, production of fatty acid methyl esters and alkenones from *Isochrysis,* an alga, was being investigated as a potential feedstock for aviation biofuel (Reddy and O'Neil 2015). Thomas Brueck of Munich TU predicted that by 2050 algaculture might meet 3–5% of the world's demand for aviation fuel (Reuters 2016). International Airlines Group (IAG) is investing $400 million for developing sustainable biojet fuel in the coming 20 years. It is collaborating with Velocys and LanzaJet, two companies that provide sustainable aviation fuel, and is going to operate Europe's first domestic garbage-to-jet fuel plant in the United Kingdom in 2025. Additionally, it has made the historic commitment to power 10% of its firms' flights with sustainable aviation fuel by 2030, making it the first European airline group to do so. According to Boeing CEO Dave Calhoun, drop-in sustainable aviation fuels are "the only answer between now and 2050" to cut down carbon emissions (Norris 2021).

They have the benefit of being drop-in fuels, which enables their use without requiring large adjustments to current infrastructure or aircraft. They are thus a potential choice for the shift to more environmentally friendly flying. The use of biojet fuels is being investigated and promoted by a number of airlines, including British Airways and Virgin Australia, as well as industry players (Table 1.1). Initiatives for sustainable aviation fuel (SAF) have picked up steam globally, with goals set to raise the proportion of biojet fuels in aviation fuel mixes.

1.3 Feedstock Selection

Feedstock for biofuels production is renewable, biodegradable, sulfur-free, and nontoxic. An important factor in choosing a feedstock is its availability. For cultivated feedstocks, their availability and potential yield are interrelated (Doliente et al. 2020). There are different sources of biofuel that can be categorized into various generations, viz., first, second, third, and fourth generations (Ullah et al. 2018). Table 1.2 presents some examples of BAF production in each category (Doliente et al. 2020; Alalwan et al. 2019; Staples et al. 2018; Rödl 2018; Roth et al. 2018; Chiaramonti Horta and Nogueira 2017; Kandaramath Hari et al. 2015). Feedstock for first-generation biofuel are starch-, sugar-, or oil-based edible food crops such as sugarcane, maize, canola, sunflower. First, these biofuels appeared to have the potential to reduce the use of conventional fossil fuels and greenhouse gas emissions caused by their burning (Rodionova et al. 2017). But in reality it raised other global issues like food scarcity and the need for arable land, along with greenhouse gas emissions (Moodley 2021; Healey et al. 2015). Nonfood crops (jatropha and camelina), lignocellulose biomasss, and other biowastes make up the second generation of feedstock for aviation biofuels (Aguilar et al. 2018; Andree et al. 2017). These feedstocks have a high fatty acid content and can be hydroprocessed by esterification and isomorphism to produce biofuels (Azad et al. 2014). Some industrial co-products, such as crude tall oil from the paper industry, soapstocks, oil sediments, and acid oils from the edible oil refinery, can also be used as the feedstock to

Table 1.1 Compilation of biojet fuel production and aviation market (Wang et al. 2016)

Category	Pathways	Companies	US or international agencies	Airline companies/ manufacturers
Alcohol-to-jet (ATJ)	Ethanol-to-jet	Terrabon/MixAlco; Lanza Tech/ Swedish Biofuels: Coskata	Defense Advanced Research Projects Agency, FAA	Boeing, Virgin Atlantic
	Butanol-to-jet	Gevo; Byogy; Albemarle/Cobalt; Solazyme	US Navy/ NAWCWD, AFRL, DLA, USAF	Continental Airlines, United Airlines
Oil-to-jet (OTJ)	Hydroprocessed renewable jet (HRJ)	UOP; SG Biofuels; AltAir Fuels; Agrisoma Biosciences; Neste Oil; PetroChina; Sapphire Energy, Syntroleum/Tyson Food; PEMEX: ASA	US Navy, USAF, Netherland Air Force, NASA, Dutch Military, EADs	Boeing, Lufthansa, Virgin Atlantic, Virgin Blue, GE Aviation, Air New Zealand, Rolls-Royce, Continental, CFM, JAL, Airbus, KLM, Pratt & Whitney, Air China, TAM Airlines, Jet Blue Airways, IAE, United Airlines, Air France, Finnair, Air Mexico, Thomson Airways, Porter Airlines, Alaska Airlines, Horizon Air, Etihad Airways, Romanian Air, Bombardier
	Catalytic hydrothermolysis (CH)	Applied Research Assoc., Aemetis/ Chevron Lummus Global	FAA CLEEN, NRC Canada, AFRL	Rolls-Royce, Pratt & Whitney
	Hydrotreated depolymerized cellulosic jet (pyrolysis or HDCJ)	Kior/Hunt Refining/Petrotech Envergent, GTI, Dynamotive	FAA	N/A
Gas to jet (GTJ)	FT synthesis	Syntroleum; SynFuels; Rentech; Shell; Solena	US DOE, US DOD, USAF, Ontario government	Qatar Airways, United Airlines, Airbus, British Airways
	Gas fermentation	Coskata; INEOS Bio/Lanza Tech; Swedish Biofuels	N/A	Virgin Atlantic

(continued)

Table 1.1 (continued)

Category	Pathways	Companies	US or international agencies	Airline companies/ manufacturers
Sugar to jet (STJ)	Catalytic upgrading of sugar to jet	Virent/Shell, Virdia	AFRL, US DOE	N/A
	Direct sugar biological to hydrocarbons	Amyris/Total, Solazyme, LS9	US Navy, FAA	Boeing; Embraer; Azul Airlines; GE; Trip Airlines

Table 1.2 Feedstocks for bio-aviation fuel production (Doliente et al. 2020; Alalwan et al. 2019; Staples et al. 2018; Rödl 2018; Roth et al. 2018; Chiaramonti and Horta 2017; Suresh 2016; Kandaramath Hari et al. 2015)

First generation	Second generation	Third generation	Fourth generation
Oil-seed crops: camelina, oil palm, rapeseed, soybean, sunflower,	Oil-seed energy crops: jatropha, castor bean	Algae: Microalgae	Genetically modified organisms
salicornia	Grass energy crops: switch grass, miscanthus, Napier grass		Nonbiological feedstocks: CO_2, renewable electricity, water
Sugar and starchy crops: corn, wheat, sugarcane, sugar beets	Wood energy crops: poplar, willow, eucalyptus		
	Agricultural and forestry residues: corn stover, sugarcane bagasse, wood harvesting/ processing residues		
	Food and municipal waste: used cooking oil, animal fats, biogenic fraction of municipal solid waste		

be hydrogenated into jet fuel (Cvetkovic et al. 2016; Zhu et al. 2016). However, the need for expensive pretreatments and subsequent effluent treatment is a significant barrier to second-generation biofuel (Moodley 2021). Microalgae are used to make third-generation biofuels, which have gained a lot of attention due to their large-scale production, process optimization for high yield, ability to absorb CO_2, and ease of refinement (Moodley 2021). Fourth-generation biofuels are intended to use cyanobacteria that have been genetically modified to improve carbon dioxide capture (Sharma et al. 2020; Adeniyi et al. 2018). Fourth-generation biofuels are intended to use nonbiological resources and microbes that have been genetically modified to improve carbon dioxide capture (Sharma et al. 2020; Alalwan et al. 2019; Adeniyi et al. 2018). Genetically engineered organisms, most of which are still in the early stages of study, have been artificially engineered with increased sugar and oil yield and negative carbon capabilities (Alalwan et al. 2019). Such organism includes cyanobacteria, microalgae, yeast, and fungus. Despite the fact that their potential for biofuel production is quite promising, further research is required with respect to health and environmental concerns that these organisms might cause, as well as on containment and/or mitigation techniques that can be used when they are introduced into global supply chains (Abdullah et al. 2019).

Solar energy, CO2, water, and renewable power are examples of nonbiological feedstocks. When industrial plant flue gases are used, they may end up being the most ecologically friendly choice (Richter et al. 2018). In one method known as power-to-liquid (PtL), water is split into hydrogen and oxygen using an electrolyzer driven by renewable energy, and the hydrogen is then mixed with CO_2 and CO to generate BAF (Schmidt et al. 2018). The short-term costs of PtL fuels are higher than CJF, according to a recent techno-economic and environmental analysis by Schmidt et al. (2018), which was mostly driven by the cost of renewable energy. Another method is to split water and CO_2 using concentrated solar energy to create syngas, which is then used as a precursor for the manufacture of BAF (Richter et al. 2018). They found two European programs, Sunfire and SOLAR-JET, that proved the generation of jet fuel with CO_2, water, and solar energy even if both pathways are still in the early phases of study.

1.4 Conversion Technologies

The main technical approaches for biofuel production are represented by biochemical route and thermochemical pathway (Table 1.3). Enzymes and other microorganisms are often utilized in the biochemical approach to manufacture biofuel. The synthesis gas produced by pyrolysis or gasification technologies may be converted into biofuel via the thermochemical approach (Sims et al. 2010). The Fischer–Tropsch process, alcohol-to-jet, sugar-to-jet, and oil-to-jet are common examples (Wang and Ling 2016). Meanwhile, other researchers are concentrating on the synthesis of platform compounds generated from lignocellulose for the production of jet fuel range hydrocarbons (Wang et al. 2015; Sacia et al. 2015; Xia et al. 2014).

1.4.1 Conversion Process of Oil Feedstock

Depending on the type of oil feedstock used, hydrogenated esters and fatty acids (HEFA) and catalytic hydrothermolysis (CH) are two prevalent routes to convert oil feedstock into jet fuel. Vegetable oil, discarded cooking oil, and animal fats are the feedstocks for the former process, while oil plants or algal oils are the feedstocks for the latter. HEFA is a technique that hydrotreats triglycerides, saturated and/or unsaturated fatty acids, vegetable oils, discarded cooking oils, and animal fats. Typically, there are two steps in the procedure. First, unsaturated fatty acids and triglycerides

Table 1.3 Key conversion and catalysts of different jet fuel production pathways (Wei et al. 2019)

Pathway	Key conversion step	Catalyst
Hydrogenated esters and fatty acids (HEFA)	Catalytic hydrogenation Cracking and isomerization	Noble metals, transition metals Pt, Ni, or other precious metals
Catalytic hydrothermolysis (CH)	Catalytic hydrothermolysis Decarboxylation/hydrotreating	Zinc acetate Nickel
Hydroprocessed depolymerized cellulosic jet (HDCJ)	Hydrodeoxygenation	MoC/C, Pd–Mo
Fischer–Tropsch (FT)	FT process	Fe, Co, Ni, and Ru
Lignin to jet	Hydrodeoxygenation Hydrogenation	Transition metals, metal sulfides, metal phosphides, metal nitrides, carbides, and metal oxides Ru/C
Aqueous phase reforming (APR)	Acid condensation Hydrodeoxygenation	Acid catalysts Ru/C
Alcohol-to-jet (ATJ)	Ethanol dehydration Isobutanol dehydration Butanol dehydration	Al_2O_3, transition metal oxides, zeolite catalyst, and heteropolyacid catalysts Inorganic acids, metal oxides, zeolites, acidic resins Zeolite, zirconia, solid acid catalysts, HPW ($H_3PW_{12}O_{40}$), and mesoporous silica group

are initially converted into saturated fatty acids by catalytic hydrogenation; throughout the process, the triglycerides undergo a -hydrogen elimination reaction to generate a fatty acid. The saturated fatty acid is hydro-deoxygenated and decarboxylated to produce C15–C18 straight-chain alkanes. The deoxygenated straight-chain alkanes are then further selectively hydrocracked and deep isomerized to produce highly branched alkanes mixed liquid fuels. This is the second phase. As high-energy biofuels, the biojet fuels made by HEFA can be used in aircraft engines without mixing; this fuel has a high cetane number, good cold flow characteristics, great thermal stability, and low tailpipe emissions despite having a low aromatic content. CH, also known as hydrothermal liquefaction (HTL), is a different method for turning algae or oil plants into jet fuel. The conversion occurs in the presence of water at low pressures of 5–30 MPa and temperatures ranging from 250 to 380 °C. The moderate reaction conditions allow the method to use wet feedstock and have high-energy efficiency.

1.4.2 Conversion Process of Sugar Feedstock

Instead of first being converted into an ethanol intermediate, sugars can be fermented anaerobically to produce alkane-type fuels. This process is known as direct sugar to hydrocarbons (DSHC) or direct fermentation of sugar to jet (DF STJ). It can also be converted into jet fuel via a thermochemical process called aqueous phase reforming (APR), in addition to the biochemical approach.

1.4.3 Conversion Process of Alcohol Feedstock

Dehydration, oligomerization, hydroprocessing, and distillation are some of the reactions that can be utilized to convert alcohol into biofuels (Chiaramonti et al. 2014). To convert biomass into jet fuel, commercial manufacturing always uses ethanol, butanol, and isobutanol as the intermediary. There are several different ways to create jet fuel based on alcohols. Alcohols can be produced directly from sugars via yeast or microbial fermentation. Alcohols should be produced by fermenting starches after acidic or enzyme hydrolysis releases sugars. The conversion process for lignocellulosic feedstocks is more involved and may involve fermentation after hydrolysis, thermochemical conversion, or gasification after fermentation (Kennes et al. 2016). According to several studies, sugar can directly catalyze the conversion of alcohol (Carter 2017).

1.4.4 Conversion Process of Lignocellulosic Biomass

Lignocellulosic biomass constitutes lignin, cellulose, and hemicellulose. It has advantages over food supply technology in that it is less expensive, more readily available, and does not compete with food supply (Isikgor and Becer 2015; Somerville et al. 2010). Multiple processes, including Fischer–Tropsch process, lignin to jet, and hydroprocessed depolymerized cellulosic jet (HDCJ), can turn lignocellulose feedstock into fuel.

1.5 Limitations and Future Prospects

Although there is an overwhelming need for alternative biojet fuel, there are still a number of obstacles (Rekoske 2010). Environmental effects are one of the main drawbacks of producing biojet fuel. Major issues with the commercialization of alternative jet fuels include the price difference between conventional and biojet fuels, sustainability, and financial issues (Chiaramonti and Horta 2017; Nair and

Paulose 2014). One of the major challenges of biojet fuel production is its impact on the environment. Anthropogenic CO_2 emissions from the aviation industry were estimated to be 2.5% in 2005 and are projected to increase to 4–4.7% by 2050 (Lee et al. 2010). Additionally, as the demand for biojet fuels rises, the pattern of agricultural land usage will change from growing food to fuel crops, resulting in a shortage of food. Flexibility in terms of feedstock and process cost-effectiveness is another issue (Toro 2010). The development of feedstock should use the least amount of land, water, and nutrients, and the manufacturing method should be competent, consistent, and highly effective. This will inevitably result in the widespread development and use of sustainable alternative fuel (ICAO 2013). The production process must be seamless, expert, and extremely effective. For instance, process optimization, cost containment, property characterization, and certification are required for the production of biojet fuel from an algal source (Carlson and Lee 2010). Albrecht and Hallen (2011) claimed that the selection, production, and lipid extraction of beneficial algae species is a perplexing process.

The cost difference between conventional and biojet fuels is the following restriction, which is a significant barrier to the commercialization of alternative jet fuels (Nair and Paulose 2014). Factors contributing to the increased cost of biojet fuels include feedstock price, equipment and operating costs, conversion efficiency, and selling prices for distillate fuels and by-products (Bond et al. 2014). Reduced market prices are the only option to improve the biojet fuel market. This necessitates funding from the government and investments in research and development with an emphasis on feedstock productivity, inexpensive catalyst, equipment distribution and selection, ideal reaction conditions, factory scale, co-product recovery, and other factors (Jong et al. 2015). Airlines may be given incentives or compensation programs in exchange for acquiring biofuel and luring investors (ICAO 2013).

Given the aforementioned constraints, HEFA and Fischer–Tropsch synthesis are thought to be the most promising methods for the manufacture of biojet fuels in the future (Wei 2019). These technologies have a variety of feedstock resources, are technically advanced, emit less CO_2, and have reasonable purchase prices. Before the two technologies may be used for large-scale production, further work needs to be done. Additionally, there are numerous opportunities for the development of biojet fuels in the future, along with the cost reduction in emerging new technologies.

1.6 Conclusion

Production of jet fuel from renewable bioresources offers the aviation industry a possible means of reducing its dependence on fossil fuels and achieving its carbon emission reduction goals. There are many different feedstocks used in the production of biojet fuel, and different production processes call for different feedstocks. The main obstacles for commercialization of biojet fuel are the availability of feedstock, sustainability of process, and financial assistance. In order to compete with conventional jet fuels, efforts should be made to lower the market price for biojet fuels. These efforts should include scientific advancement and regulatory assistance while maintaining environmental sustainability. Future use of the abundant biomass feedstock has the potential to significantly reduce GHG emissions in light of the widespread use of biojet fuels.

References

Abdullah B, Syed Muhammad SAF, Shokravi Z, Ismail S, Kassim K, Mahmood AN et al (2019) Fourth generation biofuel: a review on risks and mitigation strategies. Renew Sust Energ Rev 107:37–50. https://doi.org/10.1016/j.rser.2019.02.018

Adeniyi OM, Azimov U, Burluka A (2018) Algae biofuel: current status and future applications. Renew Sust Energ Rev 90:316–335. https://doi.org/10.1016/j.rser.2018.03.067

Aguilar DL, Rodríguez-Jasso RM, Zanuso E, Lara-Flores AA, Aguilar CN, Sanchez A et al (2018) Operational strategies for enzymatic hydrolysis in a biorefinery. In: Kumar S, Sani RK (eds) Biorefining of biomass to biofuels, vol 4. Springer International Publishing, Cham, pp 223–248. https://doi.org/10.1007/978-3-319-67678-4_10

Agusdinata DB, Zhao F, Ileleji K, Dan DL (2011) Life cycle assessment of potential biojet fuel production in the United States. Environ Sci Technol 45(21):9133–9143. https://doi.org/10.1021/es202148g

Alalwan HA, Alminshid AH, Aljaafari HAS (2019) Promising evolution of biofuel generations. Subject review. Renew Energy Focus 28:127–139. https://doi.org/10.1016/j.ref.2018.12.006

Albrecht KO, Hallen RT (2011) A brief literature overview of various routes to biorenewable fuels from lipids for the National Alliance for advanced biofuels and bio-products (NAABB) Consortium

Andree BPJ, Diogo V, Koomen E (2017) Efficiency of second-generation biofuel crop subsidy schemes: spatial heterogeneity and policy design. Renew Sust Energ Rev 67:848–862. https://doi.org/10.1016/j.rser.2016.09.048

Azad AK, Rasul MG, Khan MMK, Sharma SC (2014) Review of non-edible biofuel resources in Australia for second generation (2G) biofuel conversion. In: International green energy conference, Tainjin, China, pp 867–878

Baljet M Aviation biofuel. Global Media Day 2010

Balogun SA, Ghazali I, Mohammed AT, Hermansyah D, Amanah A, Kurnia MT (2022) Renewable aviation fuel: review of bio-jet fuel for aviation industry. Eng Sci Lett 1(01):7–11. https://doi.org/10.56741/esl.v1i01.59

Bauen, A., Howes, J., Bertuccioli, L. and Chudziak, C., 2009. Review of the potential for biofuels in aviation. Final Report Prepared for the Committee on, Climate Change. E4tech, Switzerland.

Bond JQ, Upadhye AA, Olcay H, Tompsett GA, Jungho J, Xing R et al (2014) Production of renewable jet fuel range alkanes and commodity chemicals from integrated catalytic processing of biomass. Energy Environ Sci 7(4):1500–1523. https://doi.org/10.1039/c3ee43846e

Bosch J, de Jong S, Hoefnagels R, Slade R (2017) Aviation biofuels: strategically important, technically achievable, tough to deliver, in Grantham Briefing Papers (London: Imperial College London), 1–15. Available online at https://www.imperial.ac.uk/media/imperial-college/grantham-institute/public/publications/briefing-papers/BP-23-Aviation-Biofuels.pdf

Carlson LS, Lee MY (2010) Algae to alkanes. Department of Chemical & Biomolecular Engineering Senior Design Reports (CBE), University of Pennsylvania

Carter MK (2017) Direct catalytic conversion of sugars to ethanol. U.S. Patent 9,533,929, issued January 3

Chiaramonti D, Horta Nogueira LA (2017) Aviation biofuels: processes, opportunities, constraints, and perspectives. In: Singh R, Kumar S (eds) Biofuels production and processing technology. CRC Press, pp 295–318. https://doi.org/10.1201/9781315155067-11

Chiaramonti D, Prussi M, Buffi M, Tacconi D (2014) Sustainable bio kerosene: process routes and industrial demonstration activities in aviation biofuels. Appl Energy 136:767–774

Cvetkovic B, Cvetkovic I, Kojcin A, Trzin D, Bankovic-Ilic I, Stamenkovic O et al (2016) Side-stream products of edible oil refining as feedstocks in biodiesel production. Recycl Sustainable Dev 9(1):28–46

de Jong S, Antonissen K, Hoefnagels R, Lonza L, Wang M, Faaij A et al (2017) Life-cycle analysis of greenhouse gas emissions from renewable jet fuel production. Biotechnol Biofuels 10:64. https://doi.org/10.1186/s13068-017-0739-7

Doliente SS, Narayan A, Tapia JFD, Samsatli NJ, Zhao Y, Samsatli S (2020) Bio-aviation fuels, Sustainable feedstocks, energy crops, Waste biomass, Microalgae, production pathways, storage and transport, Supply Chains. Front Energy Res 8. https://doi.org/10.3389/fenrg.2020.00110

Elia JA, Baliban RC, Floudas CA, Gurau B, Weingarten MB, Klotz SD (2013) Hardwood biomass to gasoline, diesel, and jet fuel: 2. Supply chain optimization framework for a network of thermochemical refineries. Energy Fuel 27(8):4325–4352. https://doi.org/10.1021/ef400430x

Faaij A, van Dijk M (2012) White paper on sustainable jet fuel. SkyNRG (SkyNRG)

Hari TK, Yaakob Z, Binitha NN (2015) Aviation biofuel from renewable resources: routes, opportunities and challenges. Renew Sust Energ Rev 42:1234–1244. https://doi.org/10.1016/j.rser.2014.10.095

Healey AL, Lee DJ, Furtado A, Simmons BA, Henry RJ (2015) Efficient eucalypt cell wall deconstruction and conversion for sustainable lignocellulosic biofuels. Front Bioeng Biotechnol:3. https://doi.org/10.3389/fbioe.2015.00190

Hileman JI, Stratton RW (2014) Alternative jet fuel feasibility. Transp Policy 34:52–62. https://doi.org/10.1016/j.tranpol.2014.02.018

ICAO (2013) The challenges for the development and deployment of sustainable alternative fuels in aviation

International Air Transport Association (2017) Annual review

Isikgor F, Becer CR (2015) Lignocellulosic biomass: a sustainable platform for the production of bio-based chemicals and polymers. Polym Chem 6(25):4497–4559. https://doi.org/10.1039/C5PY00263J

Jones SB, Valkenburg C, Walkton CW, Elliott DC, Holladay JE, Stevens DJ et al (2010) Design case summary: production of gasoline and diesel from biomass via fast pyrolysis, Hydrotreating, and hydrocracking. EERE Publication and Product Library

Jong SD, Hoefnagels R, Faaij A, Slade R, Mawhood R, Junginger M (2015) The feasibility of short-term production strategies for renewable jet fuels – a comprehensive techno-economic comparison. Biofuels Bioprod Biorefin 9(6):778–800. https://doi.org/10.1002/bbb.1613

Kandaramath Hari T, Yaakob Z, Binitha NN (2015) Aviation biofuel from renewable resources: routes, opportunities and challenges. Renew Sust Energ Rev 42:1234–1244. https://doi.org/10.1016/j.rser.2014.10.095

Kennes D, Abubackar HN, Diaz M, Veiga MC, Kennes C (2016) Bioethanol production from biomass: carbohydrate vs syngas fermentation. J Chem Technol Biotechnol 91(2):304–317. https://doi.org/10.1002/jctb.4842

Kubickova I, Kubicka D (2010) Utilization of triglycerides and related feedstocks for production of clean hydrocarbon fuels and petrochemicals: a review. Waste Biomass Valor 1:293–308

Lee DS, Pitari G, Grewe V, Gierens K, Penner JE, Petzold A et al (2010) Transport impacts on atmosphere and climate: aviation. Atmos Environ 44:4678–4734

Lokesh K, Sethi V, Nikolaidis T, Goodger E, Nalianda D (2015) Life cycle greenhouse gas analysis of biojet fuels with a technical investigation into their impact on jet engine performance. Biomass Bioenergy 77:26–44. https://doi.org/10.1016/j.biombioe.2015.03.005

Malode SJ, Keerthi Prabhu K, Mascarenhas RJ, Shetti NP, Aminabhavi TM (2021) Recent advances and viability in biofuel production. Energy Convers Manage X 10:100070. https://doi.org/10.1016/j.ecmx.2020.100070. ISSN 2590-1745

Mofijur M, Hazrat MA, Rasul MG, Mahmudul HM (2015) Comparative evaluation of edible and non-edible oil methyl ester performance in a vehicular engine ☆. Energy Procedia 75:37–43. https://doi.org/10.1016/j.egypro.2015.07.134

Mohammad M, Hari TK, Yaakob Z, Sharma YC, Sopian K (2013) Overview on the production of paraffin based-biofuels via catalytic hydrodeoxygenation. Renew Sust Energ Rev 22:121–132

Moodley P (2021) 1 – Sustainable biofuels: opportunities and challenges, Editor(s): Ramesh C. Ray. In: Applied biotechnology reviews, sustainable biofuels. Academic Press, pp 1–20, ISBN 9780128202975. https://doi.org/10.1016/B978-0-12-820297-5.00003-7

Nair S, Paulose H (2014) Emergence of green business models: the case of algae biofuel for aviation. Energy Policy 65:175–184

Norris G (2021) Boeing moves forward with Airbus A321XLR-competitor plan. Aviation Week, February 4

Pandey PK (2011) Opportunities in bio-fuels for aviation. National Seminar on bioenergy Solutions

Reddy C, O'Neil G (2015) Jet fuel from algae? Scientists probe fuel potential in common ocean plant. Oceanus magazine. Woods Hole Oceanographic Institution, 28 January

Rekoske J (2010) Biofuels: challenges & opportunities. Asia Pacific Clean Energy Summit & Expo

Richter S, Braun-Unkhoff M, Naumann C, Riedel U (2018) Paths to alternative fuels for aviation. CEAS Aeronaut J 9:389–403. https://doi.org/10.1007/s13272-018-0296-1

Rodionova MV et al (2017) Biofuel production: challenges and opportunities. Int J Hydrog Energy 42:8450–8461. https://doi.org/10.1016/j.ijhydene.2016.11.125

Rödl A (2018) Lignocellulosic biomass. In: Kaltschmitt M, Neuling U (eds) Biokerosene: status and prospects. Springer, Berlin/Heidelberg, pp 189–220

Roth A, Riegel F, Batteiger V (2018) Potentials of biomass and renewable energy: the question of sustainable availability. In: Kaltschmitt M, Neuling U (eds) Biokerosene: status and prospects. Springer, Berlin/Heidelberg, pp 95–122

Sacia ER, Balakrishnan M, Deaner MH, Goulas KA, Toste FD, Bell AT (2015) Highly selective condensation of biomass-derived methyl ketones as a source of aviation fuel. ChemSusChem 8(10):1726. https://doi.org/10.1002/cssc.201500002

Schmidt P, Batteiger V, Roth A, Weindorf W, Raksha T (2018) Power-to-liquids as renewable fuel option for aviation: a review. Chem Ing Tech 90:127–140. https://doi.org/10.1002/cite.201700129

Shankar VSB, Khandelwal B (2013) Comparison of jet fuel produced by shale oil and biomass: manufacturing emission and performance. AIAA – JPC

Sharma S, Basu S, Shetti NP, Aminabhavi TM (2020) Waste-to-energy nexus for circular economy and environmental protection: recent trends in hydrogen energy. Sci Total Environ 713:136633. https://doi.org/10.1016/j.scitotenv.2020.136633

Sims RE, Mabee W, Saddler JN, Taylor M (2010) An overview of second generation biofuel technologies. Bioresour Technol 101(6):1570–1580. https://doi.org/10.1016/j.biortech.2009.11.046

Somerville C, Youngs H, Taylor C, Davis SC, Long SP (2010) Feedstocks for lignocellulosic biofuels. Science 329(5993):790–792. https://doi.org/10.1126/science.1189268

Staples MD, Malina R, Suresh P, Hileman JI, Barrett S (2018) Aviation CO_2 emissions reductions from the use of alternative jet fuels. Energy Policy 114:342–354. https://doi.org/10.1016/j.enpol.2017.12.007

Suresh P (2016) Environmental and economic assessment of transportation fuels from municipal solid waste. Diss. Massachusetts Institute of Technology

Tan ECD, Marker TL, Roberts MJ (2014) Direct production of gasoline and diesel fuels from biomass via integrated hydropyrolysis and hydroconversion process-a techno- economic analysis. Environ Prog Sustain Energy 33(2):609–617. https://doi.org/10.1002/ep.11791

Timko MT, Herndon SC, Blanco ER, Wood EC, Yu Z, Miake-Lye RC et al (2010) Combustion products of petroleum jet fuel, a fischer-tropsch synthetic fuel, and a biomass fatty acid methyl ester fuel for a gas turbine engine. Combust Sci Technol 183(10):1039–1068. https://doi.org/10.1080/00102202.2011.581717

Toro FA (2010) Biofuels in aviation major challenges, potentials major R&D activities and needs. Joint IEA-GHG-TransPoRD workshop

Ullah K, Sharma VK, Ahmad M, Lv P, Krahl J, Wang Z et al (2018) The insight views of advanced technologies and its application in bio-origin fuel synthesis from lignocellulose biomasses waste, a review. Renew Sust Energ Rev 82:3992–4008. https://doi.org/10.1016/j.rser.2017.10.074

Wang WC, Ling T (2016) Bio-jet fuel conversion technologies. Renew Sust Energ Rev 53:801–822. https://doi.org/10.1016/j.rser.2015.09.016

Wang Wei-Cheng, Tao Ling, Markham J, Zhang Y, Tan E, Batan L, Warner E, Biddy M (2016) Review of biojet fuel conversion technologies. National Renewable Energy Laboratory Technical Report NREL/TP-5100-66291 July 2016

Wang HL, Hao R, Pei HS, Wang HM, Chen XW, Tucker MP et al (2015) Biomass-derived lignin to jet fuel range hydrocarbons via aqueous phase hydrodeoxygenation. Green Chem 17(12):5131–5135. https://doi.org/10.1039/C5GC01534K

Wei H, Liu W, Chen X, Yang Q, Li J, Chen H (2019) Renewable bio-jet fuel production for aviation: a review. Fuel 254:115599

Xia QN, Cuan Q, Liu XH, Gong XQ, Lu GZ, Wang YQ (2014) Pd/$NbOPO_4$ multifunctional catalyst for the direct production of liquid alkanes from aldol adducts of furans. Angew Chem 53(37):9755. https://doi.org/10.1002/anie.201403440

Zhu LD, Xu ZB, Qin L, Wang ZM, Hiltunen E, Li ZH (2016) Oil production from pilot-scale microalgae cultivation: an economics evaluation. Energy Sources Part B 11(1):11–17. https://doi.org/10.1080/15567249.2015.1052594

Bibliography

BA begins offsetting domestic flight emissions. Flightglobal 3 January 2020

From green slime to jet fuel: algae offers airlines a cleaner future. Reuters 15 June 2016

Ways to make aviation fuel green. The Economist. 2022-08-17. ISSN 0013-0613

Chapter 2
Overview of Aviation Sector, Feedstock, and Supply Chain

Ankita Kumari, Depak Kumar, Priyanka Sati, Sudesh Kumar, Ashok Kumar Yadav, and Ajay Singh Verma

Abstract A refinery processes crude oils to manufacture a broad range of useful products, such as jet fuel, gasoline, petrochemicals, diesel, and asphalt constituents. In addition to being manufactured via hydro methods, kerosene can also be produced as a direct run product, particularly from heavier crude oil feedstocks. Hydrocarbon fuels like kerosene jet fuel are virtually exclusively made of carbon and hydrogen atoms. Aromatic compounds, cycloparaffins (naphthenes), and paraffins (iso and normal) make up the majority of the hydrocarbon composition. The ratios of these hydrocarbon components will vary in aviation jet fuel made from various feedstocks and processing methods. The burning of aviation turbine fuel has been linked to "global warming," which is why "biojet" has been suggested as a way to blend fuel to lessen carbon emissions. The maximum percentage of biojet to conventional jet fuel blend permitted by the standard is 50%. While bioethanol and biobutanol, which have been shown to be effective as automotive fuels, were determined to be inappropriate for use in aircraft because they did not meet ASTM D7566-09 requirements. Many technological solutions have surfaced as a result of extensive global R&D activities. These systems generated renewable hydrocarbon fraction as drop-in fuel known as "biojet" using feedstock like animal fat, leftover agricultural products, municipal solid waste (MSW), waste cooking oil, plant seed oil, and waste cooking oil. Reduced carbon emissions in aviation fuel are the main advantage of using feedstock made from plants or agricultural waste in place of crude oil. To ensure that biojet is a sustainable, affordable, and environmentally

A. Kumari · P. Sati · S. Kumar
Department of Chemistry, Banasthali Vidyapith, Newai, Rajasthan, India

D. Kumar (✉) · A. K. Yadav
Department of Chemical Engineering, Banasthali Vidyapith, Newai, Rajasthan, India
e-mail: kdepak@banasthali.in

A. S. Verma
Division of Research & Innovation, School of Applied and Life Sciences, Uttaranchal University, Dehradun, Uttarakhand, India

A. Kuila (ed.), *Biojet Fuel: Current Technology and Future Prospect*, Clean Energy Production Technologies, https://doi.org/10.1007/978-981-99-8783-2_2

friendly aviation fuel, a number of obstacles have to be overcome in order to meet the strict requirements of aviation fuel.

Keywords Jatropha oil · Aviation turbine fuel (ATF) · DSHC · Biojet · HEFA ASTM

2.1 Introduction

Our society has changed for the better because of air travel, which has improved community relations, fueled trade, and made it possible for things that were unavailable to earlier generations to be exchanged. Even though air travel makes the globe smaller and better, the energy necessary for flying comes at a significant cost. Highly specialized jet fuel (Jet-A-1), produced from crude oil, is used extensively in air transportation for people, cargo, and the military. The aviation industry has a significant need for 200 million tons of jet fuel annually due to the 5% annual increase in flight traffic worldwide (Davidson et al. 2014). Air carriers are a hazard to high-altitude propagation since they consume a lot of traditional fossil fuel (Jet-A), which is responsible for 2% of all greenhouse gas emissions from the transportation industry. To mitigate the effects of "global warming," certain nations have suggested levying a "carbon tax" on airlines that utilize aviation fuel derived from fossil fuels, such as Jet A-1. The following objectives have been committed to by International Air Transport Association (IATA) members:

- Increase fuel efficiency by 1.5% annually for the next 10 years
- Attain growth that is carbon-neutral for the aviation industry as a whole
- Achieve a 50% net reduction in CO_2 emissions by 2050

2.1.1 Aviation Industry Growth

India has one of the highest fuel prices and taxes in the world, if not the highest overall, which negatively impacts the bottom line of smaller airlines. Over the last 5 years, India's civil aviation sector has seen a 20% rise in both domestic and international air travel. Nitrogen oxides (NOX), carbon dioxide (CO_2), water vapor (H_2O), and particulate matter (PM) are the primary greenhouse gas (GHG) emissions from aviation during flight, and they all contribute to global air pollution (Martin et al. 2003). In order to replace fossil fuel-based Jet A-1 fuel, the aviation sector urgently needs green fuel. Second-generation biofuels could be a competitive alternative to Jet A-1, which is based on fossil fuels, and helps partially meet the demand for aviation fuel. This would help reduce the environmental impact of water vapor (H_2O), nitrogen oxides (NOX), particulate matter (PM), and net carbon dioxide (CO_2) (Özaydın et al. 2013).

The present refueling infrastructure may be used to partially mix these fuels with aviation gasoline, enabling the creation of a straightforward global supply chain system. Nevertheless, compared to other modes of transportation like cars, the use of biofuels in aviation is fraught with difficulties. One of the most difficult issues of using biofuels is in autos (Adhikari 2018).

The main prerequisites for environmentally friendly alternative jet fuels are as follows:

- It can be combined with regular jet fuel.
- Drop-in fuel is compatible with existing supply infrastructure and does not require engine or aircraft change.
- I can meet the same standards as traditional jet fuel, especially those pertaining to temperature resistance.
- It requires a lot of energy (minimum 42.8 MJ/kg).
- It can meet environmental criteria such as lowering freshwater requirements, minimizing lifecycle carbon emissions, without affecting food production, and preventing deforestation.

The current aircraft industry's strict requirements for aviation fuel must undoubtedly be met by any alternative fuel.

2.1.2 Biofuels That Are Sustainable for Aviation

The two most sustainable biofuels currently being produced for use in cars for surface transportation are bioethanol and biodiesel. While traditional automobiles in the EU, the United States, and India use a mixture of 5–10% (E5–E10) bioethanol in gasoline, Flexi cars in Brazil utilize an 85% bioethanol mixture in gasoline (E85) (Sinha et al. 2013). Table 2.1 shows compliance with aviation fuel specifications (Jet A-1). Diesel vehicles that use 20% biodiesel in place of petroleum diesel (B_{20}) perform well all around the world. Nonedible oils derived from plant seeds and lignocellulosic biomass are the most sustainable feedstocks for large-scale bioethanol and biodiesel production, respectively.

These two biofuels have a significantly lower carbon footprint than gasoline and diesel, according to their life cycle analysis (LCA) (Tirado et al. 2021). Consequently, the most environmentally friendly biofuels for cars are now bioethanol and biodiesel. The limitations of employing bioethanol and biodiesel as aviation biofuels are shown in Table 2.2.

Although biodiesel and bioethanol are utilized with great success in road transport vehicles, their intrinsic qualities have limited their usage as sustainable aviation fuel. The best solution is an alternative fuel that satisfies all necessary fuel criteria, ideally derived from renewable sources. As a result, the primary requirements for aviation biofuel, also known as biojet, are synthetic fuels that meet the fundamental requirements of aviation turbine fuel (ATF), whether they are hydrocarbon or non-hydrocarbon kinds and have "drop-in" qualities (Zhu et al. 2014). These biofuels have no net carbon footprint and need less capital for supply chain infrastructure management and refining operations.

Table 2.1 Compliance with aviation fuel specifications (Jet A-1)

Energy density	44 MJ/kg
Cold flow properties, pour point ° C	<44
Fuel composition	Proper ratio of n-alkanes, iso-alkanes, cyclo-alkanes, and aromatics (25 vol.%) Selective hydrocarbons
Viscosity at-20° C Cst max	8.0
Density at 15 °C	0.779–0.840
Smoke pt.mm min	19
Kinematic viscosity at 40 °C	1.2
Compatibility issues	Additives and materials used in jet engines
Existent gum	5.0–7.0
Extinction, ignition, and flammability	Within limits

Table 2.2 Biodiesel and bioethanol specification limitations as aviation fuel

	Bioethanol	Biodiesel
Energy density, MJ/kg	0.80	1.9–6.0
Kinematic viscosity at 40 °C	26.4	37.27
Flash point		130 °C higher
Pour pt. °C	Within limit	13–16
Impurities		Glycerol, acids
Compatibility	Corrosive with moisture	Noncompatible

1. They concentrated on creating a method that effectively creates jet propellant 8, or JP-8, an alternative to petroleum-based military jet fuel from oil-rich crops grown through aquaculture or agriculture. In the end, this method may provide an inexpensive substitute for JP-8 made from petroleum.
2. Alternative fuels in airplanes have unique challenges (safety, logistics, temperature, etc.). The only options available to consider, considering the amount of money invested in aviation, are drop-in fuels or fuels having kerosene-like qualities that do not require major modifications to infrastructure or equipment architecture.

2.1.3 Bio-Aviation Fuel

Bio-aviation fuel is a mixture of synthetic paraffinic kerosene (SPK) made from biomass and jet fuel derived from petroleum. The manufacturing platforms, as well as a brief process description, for these is SPK. Oil-to-jet production platform hydroprocessed esters and fatty acids production method (HEFA) generates HEFASPK by hydroprocessing deoxygenated oils and fats. Other oil-to-jet platforms include fast pyrolysis of cellulose and hydrothermal liquefaction of plant or algal oil. These platforms are then followed by jet fuel upgrading (Abdullah et al. 2019). The gas-to-jet platform converts biomass into syngas through the Fischer–Tropsch production pathway (FT), which is then hydroprocessed to create FTSPK. Alcohols are created by fermenting the fermentable sugars that are produced after the biomass is hydrolyzed (Behrendt et al. 2018). The process of converting fermentable sugars from biomass into farnesene through fermentation, hydroprocessing, and fractionation is known as the "sugar-to-jet production platform" or "direct sugar-to-hydrocarbon jet fuel synthesis" (DSCH). Aqueous-phase reforming is followed by direct sugar to hydrocarbons and sugar catalytic reforming as other sugar-to-jet platforms. Chemical or biochemical intermediates are another option (Anderson et al. 2012). Additional potential advantages include job creation, price stability, and energy security. The usage of bio-aviation fuel may lead to rural development, which may include a rise in employment in production and agriculture, as well as a higher productivity of marginal nonarable land. Deployment has not got enough support despite its economic advantages. The production pathways must receive investments in the form of subsidies and legislative support in order to become economically competitive with the output of oil refineries (Campbell 2018).

2.2 Feedstocks for Biomass-Derived

2.2.1 Synthetic Paraffinic Kerosene

The four generations that feedstocks are divided into are the first, second, third, and fourth generations. Table 2.3 shows feedstocks for bio-aviation fuel production (Azwan et al. 2016). When choosing a feedstock, availability is an important factor to take into account. There is a connection between the availability of cultivated feedstocks and their potential yield. With a production of 19.2 t/ha/year, oil palm was the most productive of these feedstocks. Since algae culture is mostly done at the lab- to pilot scale, the potential output for microalgae for 3-G feedstocks has been estimated to be substantially bigger at 91 t/ha/year; however, this number is questionable (Couto et al. 2017).

2.2.2 Renewable Feedstocks

Nonetheless, developing sustainable biojet fuel necessitates a continuous, quantitative, and qualitative supply of renewable feedstocks. To guarantee a consistent supply of fuel for international flights, such feedstocks must to be accessible on all continents of the planet (Domínguez-García et al. 2017). Commercial flights have already used biojet fuels made from plant seed oils in modest amounts (Dzięgielewski

Table 2.3 Feedstocks for bio-aviation fuel production

First generation	Second generation	Third generation	Fourth generation
Sugar and starchy crops: wheat, sugar beets, sugarcane, corn	Oil-seed energy crops: jatropha, castor bean	Algae, microalgae	Nonbiological feedstocks: CO_2, renewable electricity, water
Oil-seed crops: oil palm, camelina, rapeseed,	Wood energy crops: poplar, willow, eucalyptus		Genetically modified organisms
	Grass energy crops: Napier grass, miscanthus		
	Food and municipal waste		
	Forestry and agricultural residues		

et al. 2014). These flights, which were run by United, Lufthansa, JAL, and other airlines, all combined Jet A with biofuel made from the nonedible evergreen shrub *Jatropha* (Giudicianni et al. 2017). British Airways and Solena have partnered to begin manufacturing a synthetic paraffin product made from agricultural and urban waste (Fischer et al. 2018).

Three different kinds of raw materials could be used to produce carbon-free and affordable biojet fuel:

(a) Lipids – camelina, rapeseed, karanjia, maize (corn), palm oil, jatropha, and used cooking oil are all types of oils derived from seeds. Certain procedures seek to employ both algal and animal fats.

(b) Wood, agricultural wastes, and forest residues are examples of lignocellulosic biomass.

Energy crops include grasses and plants that grow quickly, such as giant reed, bamboo, and miscanthus.

(c) Sugars

Malaysia, Indonesia, China, India, the United States, Brazil, and Argentina account for a sizable portion of global plant oil production, rapeseed, palm, soybean, and jatropha, serving as the primary feedstock oils (Gutiérrez-Antonio et al. 2013). Plant oils have the potential to be sustainably available worldwide, but some nations, like India, have limited supply of some of them, especially edible oils (He et al. 2017). Table 2.4 shows a summary of supply chain model for bio-aviation fuel provision. Aquatic algae, which might be grown in offshore or marine water, is another alternative source. Yeast lipids, for example, are being researched as potential nonplant sources of microbial lipids. The future of bio-jet generation from fats and oils, on the other hand, is about creating sustainability in big quantities at a fair cost (Karmee 2017).

Table 2.4 Summary of supply chain model for bio-aviation fuel provision

Feedstock	Model	Model capability	Location
Forest residues	A mixed-integer linear programming model for the biomass-to-liquid supply chain at the national level	A framework for determining the best operating network in the supply chain.	United States
Cellulosic feedstock	Model of a biomass scenario	A model of system dynamics for simulating the complex incentive–production interaction	United States
Microalgae	Life cycle assessment with multiple actors integrated into a system dynamics model	Evaluating the potential of algal-based jet fuels to reduce GHG emissions	United States
Energy crops (miscanthus, willow) and waste biomass (waste wood)	Biomass value chain model (BVCM)	A thorough and adaptable whole-system optimization model	United Kingdom
Camelina, jatropha	Planning an aviation biofuel supply chain strategically	Lowering expenses and greenhouse gas emissions	Mexico
Wood residues	Total transportation cost model (TTCM)	Based on least-cost analysis, the biorefinery depot was chosen	United States
Rice husk, rice straw	A rice value chain multiobjective spatiotemporal mixed-integer linear programming model	Determine how to concurrently develop, create, and manage sustainable and successful rice value chains	Philippines
Waste biomass	Integrated biomass scenario model (BSM)	An efficient feedstock and fuel flow model combined with geospatial capabilities	United States
Oil crops, sugar crops	Techno-economic evaluation of biorefinery technologies	Developing scenarios for the combined synthesis of biochemicals and bio-aviation fuels	Brazil

Producing high-yielding nonfood grade oil feedstocks in areas with extremely wide production spaces, avoiding rivalry with (or displacement of) existing food production, and having enough water and other inputs are the only clear answers to these three key issues (Howe et al. 2015). Every year, India uses about 4.5 million tons of biojet fuel (Klein et al. 2018). Table 2.5 shows oil output from various oil plants per hectare (ha). Three million hectares of wasteland are available in India that might be planted with pongamia or jatropha, which would produce about nine million tons of oil annually, enough to produce biojets. Table 2.6 shows fuel properties and the role of fatty acid structure. Also, 70% of the world's biomass supply is produced by regions of tropical Asia and Africa.

Table 2.5 Oil output from various oil plants per hectare (ha)

Plant	Latin name	Productivity(l/ha/year)
Coconut	*Cocos nucifera*	2578
Castor bean	*Ricinus communis*	1354
Rapeseed	*Brassica napus*	1140
Cotton	*Gossypium hirsutum*	308
Palm	*Elaeis guineensis*	5698
Camelina	*Camelina sativa*	500
Corn	*Zea mays*	168
Jatropha	*Jatropha curcas*	1812
Karanja	*Pongamia pinnata*	1250

Table 2.6 Fuel properties and the role of fatty acid structure

Properties	Short chain	Long chain	Saturated	Unsaturated
Freezing point	Favorable	Unfavorable	Unfavorable	Favorable
Energy content	Low	High	High	Low
Flash point	Desirable	Undesirable	Undesirable	Favorable
Oxidation stability	Acceptable	Acceptable	Better	Unfavorable
Viscosity	Favorable	Unfavorable	Unfavorable	Favorable
NO_x	Increase	Reduced	Reduced	Increase
Combustion	Unfavorable	Favorable	Better	Unfavorable

2.3 Technologies for Biojet Fuel Processing

2.3.1 Hydroprocessed Ester and Fatty Acids (HEFA)

A company's renewable jet fuel process is an excellent illustration of cutting-edge technological techniques that could be used in the manufacturing of aviation biofuels (Jacobson et al. 2016). Tallow, algal oils, and a range of refined natural oils and fats, both edible and inedible, can all be converted using this procedure. Through the use of a selective cracking stage, the carbon chain lengths of natural oil, which range from C_{16}–C_{18}, are reduced to C_{10}–C_{14} for jet fuel in the renewable jet process (Doliente et al. 2020). The aim of this novel technique is to achieve a 50–70% Bio-SPK (bio-kerosene) yield (Molefe et al. 2019). The catalytic processes of deoxygenation, isomerization, and selective cracking of the hydrocarbons found in natural fats and oils must be optimized in order to produce high-quality, ultralow sulfur jet fuel that satisfies Jet A-1 specifications, such as a freeze point of −47 °C and a flash point of 38 °C (Fig. 2.1). This innovative technique yields diesel and naphtha range fuels as by-products (Newes et al. 2015).

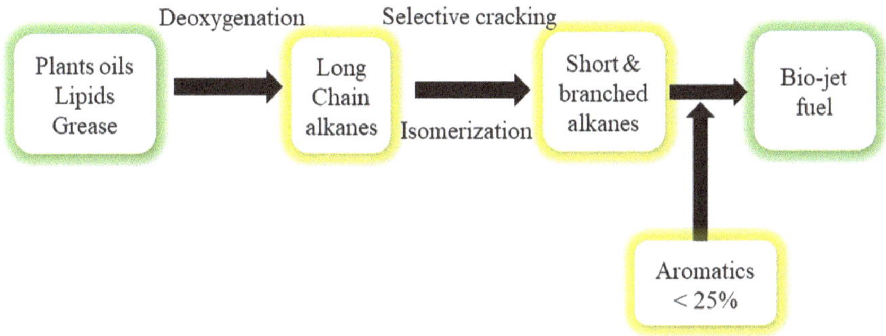

Fig. 2.1 Ecofining™ process scheme

The procedure can be changed to create the Bio-SPK at a particular freeze point or it can be run in diesel mode instead. The Indian Institute of Petroleum Dehradun, a part of CSIR, has developed another cutting-edge catalytic technique for producing biofuels for the aviation and transportation sectors by hydroprocessing nonedible vegetable oils like jatropha seed oil (Fontes and Freires 2018). A demonstration facility processing 100 kg of jatropha oil per day successfully proved the process, yielding 33–40% of biojet fuel and 99% conversion, with the remaining liquid products being used as diesel and gaseous fuel (Mohseni and Pishvaee 2016). The CSIR-IIP fuel processing plan for transportation and aviation is shown in Fig. 2.2. A 40% maximum yield and a superior conversion rate of 99% for biojet may be achieved by using hierarchical mesoporous zeolites, silica–alumina, and mesoporous alumina as selective catalysts for the hydroconversion of waste cooking oil, seed oil, and algal oil (Pham et al. 2010). Table 2.7 shows a comparison of the characteristics of the biojet generated by CSIR-IIP. Using oil with different FFA contents is one advantage of the technique. A scheme for synthesized bio- kerosene production is illustrated in Fig. 2.3 (Pirker et al. 2016).

2.3.2 Synthetic Bio-Kerosene Process

In order to create synthetic fuels like biojet fuel and biodiesel, the process is primarily based on the concept of converting biomass into liquid (BtL), employing feedstocks including lignocellulosic biomass, wheat straw wood debris, and forest leftovers. There are five steps in this process: (a) pretreatment of biomass, (b) gasification or pyrolysis, (c) syn-gas purification, (d) Fischer–Tropsch (FT) synthesis, and (e) hydroisomerization of FT-wax to biojet, naphtha, and biodiesel, depending on needs. Sulfur and other contaminants are absent from the very high-quality kerosene fraction that is produced by the BtL process (Sannan et al. 2017).

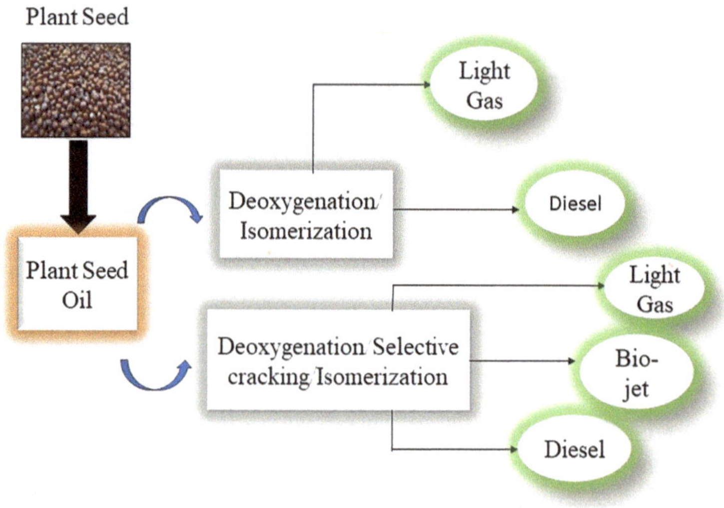

Fig. 2.2 CSIR-IIP fuel processing plan for transportation and aviation

Table 2.7 Comparison of the characteristics of the biojet generated by CSIR-IIP

Property	Units	Limit	Jet A-1	IIIP biojet
Viscosity (−20 °C)	Mm^2/S	8.00	3.72	3.45
Freezing pt.	0 C	Max.-47	−52.2	−63
Density	kg/m^3	775.84	793	780
Total aromatics	% v/v	Max.26.5	23	13
Smoke pt.	mm	25.0	26	34
Specific energy	MJ/kg	42.8	42.9	43.5
Sulfur	% m/m	0.3%	0.2	0.009

2.3.3 *Alcohol Oligomerization Process*

Numerous businesses have announced the discovery of methods to turn alcohols into jet fuel; however, the viability of these procedures depends on the source of the alcohols. According to the American renewable chemicals and biofuels business Gevo, fermentable sugars obtained from cellulosic biomass have been successfully used to produce isobutanol, which is then transformed into paraffinic kerosene (jet fuel) and isobutylene (Raje and Davis 1997). It is also claimed that Lanzatech, a different business, produces alcohol from "clean" carbon monoxide-containing industrial waste gases and uses oligomerization and hydrogenation to turn the alcohol into jet fuel as illustrated in Fig. 2.4.

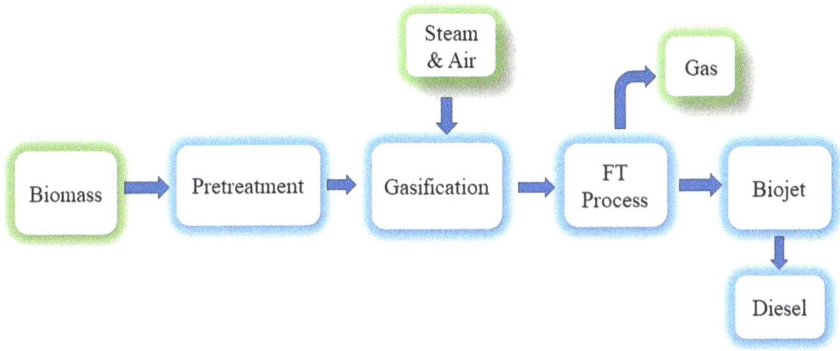

Fig. 2.3 Scheme for synthesized bio-kerosene production

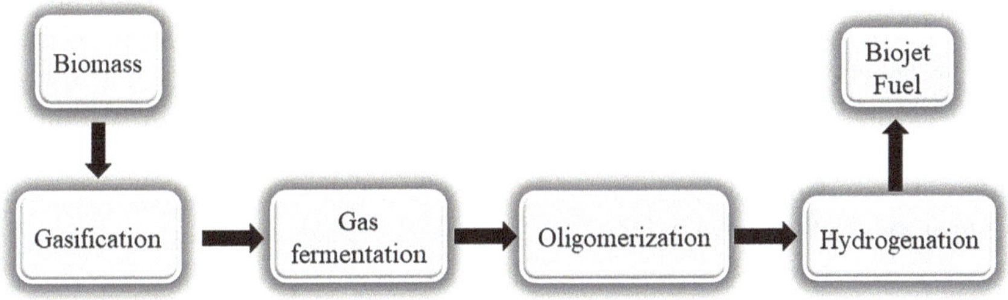

Fig. 2.4 Process plan for alcohol oligomerization to jet fuel (AJT)

2.3.4 *Direct Sugar to Hydrocarbons (DSHC)*

To complete conversion, a sophisticated fermentation technique is used in the procedure. This biological conversion, in contrast to "traditional" sugar-to-ethanol fermentation, occurs under aerobic circumstances (Schmitt et al. 2019). Then, this passes through one further conversion process, producing the saturated and hydrogenated hydrocarbon farnesane. The authorized route was created through an alliance between Amyris, an industrial bioscience business located in California, and Total, a French petroleum distribution and refining corporation (Seber et al. 2014). Most eukaryotes and higher bacteria include farnesene, a terpenoid olefin (1,6,10-dodecatrienes, $C_{15}H_{24}$), which is biochemically generated through the mevalonate or isoprenoid pathway (Samsatli and Samsatli 2019). It dissolves well in alcohols but not in water. Pour point is −76 °C, boiling point is 250 °C, and density is 0.83 (15 °C).

Because farnesene and its isomers are non-hygroscopic, have very low pour points, densities, and have flash points that meet aviation fuel criteria, they offer an advantage over butanol and ethanol as biofuels.

2.3.4.1 Mevalonate Pathway

Actyl-CoA + Acetoacetyl-CoA 3-Hydroxy-3-methylglutaryl CoA (HMG-CoA) Mavelonate Isopentyl-PP Geranyl-PP Farnasyl-PP Farnesene

The principal enzymes connected to the route of terpenoid biosynthesis include (a) acetyl-CoA acetyltransferase, (b) 3-hydroxy-3-methylglutarylcoenzyme A synthase, (c) HMG-CoA reductase, (d) mevalonate kinase, (e) phosphomevalonate kinase, (f) mevalonate pyrophosphate decarboxylase, (g) isopentenyl diphosphate (IPP) isomerase, (h) isoprene synthase, (i) farnesyl pyrophosphate (FPP) synthase, and (j) a-farnesene synthase.

Both aerobic and anaerobic microorganisms use glycolysis to produce the 2 mol of NADPH and the 3 mol of ATP needed for the process (Sinha et al. 2015). The ability to overproduce terpenoids through microbial fermentation instead of plant-based production has been made possible by recent advancements in synthetic biology and metabolic engineering. This has produced a number of significant discoveries, such as bulk chemicals and biofuels, as well as complicated natural products like artemisinin and taxol precursors. A scheme for the direct conversion of sugar to biojet is shown in Fig. 2.5.

2.3.5 *Hydroprocessing Bio-Oil to Produce Biojet Fuel*

The method is predicated on the hydrotreatment of bio-oil, which is made from wood and lignocellulosic biomass by fast and catalytic pyrolysis. Table 2.8 shows the status of development of the biojet fuel process. Co-processing heavy vacuum petrol oil (HVGO) and bio-oil in an oil refinery's hydrotreatment unit is an additional method of producing Jet-A1 fuel as shown in Fig. 2.6.

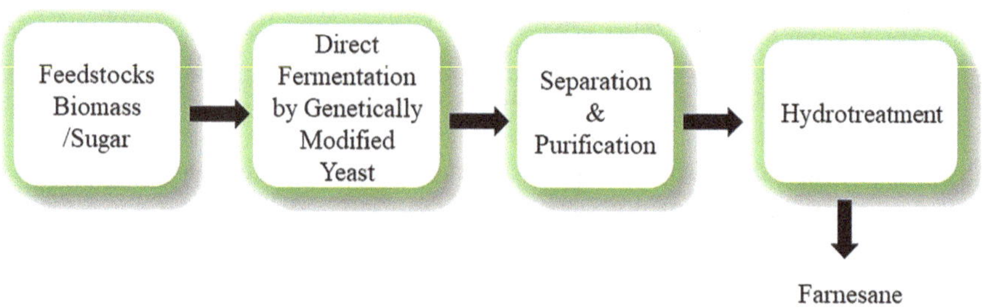

Fig. 2.5 Scheme for direct conversion of sugar to biojet (farnesane)

Table 2.8 Status of development of the biojet fuel process

Biojet fuel process	Certification	Feedstock type	Feedstock cost	Potential investment
Fischer–Tropsch (FT)	Maximum mix of 50% using Jet-A1	Lignocellulosic biomass and woody	Low	Very large
Alcohol oligomerization to jet fuel (ATJ)	Currently undergoing ASTM certification	Starches, sugars	Moderate but limited in certain nations on fuel vs. food	Medium
Hydrotreated pyrolysis oil	Currently undergoing ASTM certification	Starches, sugars	Medium	Very large
Direct sugar to hydrocarbons (DSHC)	Maximum 10% blend with fossil fuel jet	Sugars	Medium	Large
Hydro processed plant seed oil	ASTM 2011 maximum mix of 50% using Jet-A1	Plants oils	High for edible oils, medium for supplies of nonedible oils	Medium

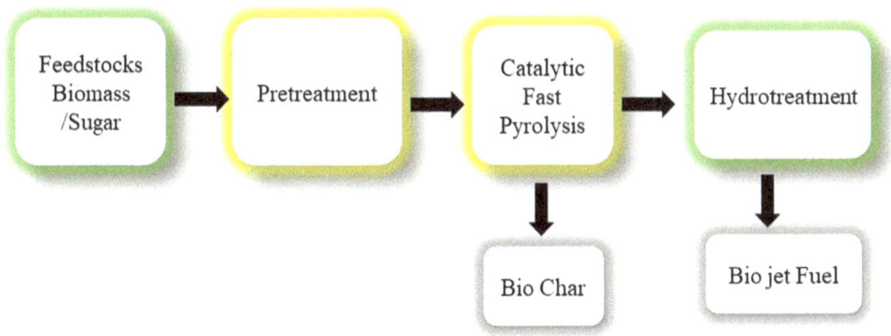

Fig. 2.6 Bio-oil hydrotreated to produce biojet fuel

A few large businesses, including BTG, Ensene, and UOP, are pushing the process, which is still in the developmental stage, to be demonstrated in a demo plant. Numerous chemicals, catalytic, and biocatalytic techniques have been demonstrated to be effective in converting various renewable feedstocks, including biomass, plant oilseeds, and sugars, into biojet fuel (Wang et al. 2021). However, in order to utilize such "biojet fuel" in commercial aircraft, it must be certified by the ASTM and be technologically and economically viable.

2.4 Feedstock Storage and Transportation, as well as Bio-Aviation Fuel

Planning and executing BAF provision present extra difficulties for the aviation sector due to the storage and supply chain mobility of intermediates, raw materials, and/or completed goods (Vasquez et al. 2017). Long-distance fuel and feedstock transportation dramatically raises supply chain expenses and greenhouse gas emissions. Therefore, it is necessary to reduce the associated effects in order to make BAF a more affordable and environmentally responsible substitute for CJF. The supply chain is generally not significantly affected by feedstock storage (Tao et al. 2017). However, across the supply chain, energy-intensive facilities that require a medium to long period to preserve and dry feedstocks may result in higher costs and emissions. Fortunately, storage becomes less of a problem after the final BAF products leave the biorefinery since sophisticated technologies, such as carrier tanks with particle settling and removal capabilities to preserve fuel, are already in place to support them during transport. However, the related effects of storage must be taken into account for its thorough planning, construction, and operation if they are to be taken into account within a supply chain for BAF provision (Wei et al. 2019).

A number of models have been put out to optimize the placement of industrial sites within BAF supply networks. The suggested models, however, only took the supply networks' transportation component into account. In order to account for the storage required to fulfill short-term future demand, as well as the consequences of biodiversity and the energy, food, water, and environment, supply chain models for BAF need to be more detailed (Taylor et al. 2010). To guarantee that the outcomes are trustworthy and pertinent, they should be conducted with as much recent data as feasible.

2.5 Economic and Environmental Analyses

The most popular ways to transport feedstocks are by truck, rail, and ship; pipelines are the least used method at the moment, but they could be important in the future. According to a recent analysis of feedstock logistics, the impact of transportation costs and distances on feedstock usage will lead to a rise in interest in multimodal movement, or the combination of modes. The cost of transportation is made up of a set cost and a variable cost that varies according to distance and is usually lower for both rail and ship than for vehicle. Geographical variations, feedstock composition and type, and transport capacities all affect how much transportation costs vary between nations (Woytiuk et al. 2017). The high cost of transportation makes biomass transport impractical for distances greater than 150–200 km. In a similar vein, long-distance feedstock transportation can result in higher emissions over the course of the life cycle.

In a BAF supply chain, feedstock storage serves primarily to handle the temporal unpredictability of demand, particularly in seasons of poor productivity (Trivedi et al. 2015). The problem of storing lignocellulosic feedstocks without experiencing large dry matter losses (DML) needs to be solved. The location and climate of storage have an impact on the quality of the lignocellulosic feedstocks being held. There are also reported costs for every storage infrastructure.

2.6 Issues Restricting the Use of Biojet Fuels Worldwide

Many obstacles prevent the widespread use of inexpensive biojet fuels, including lack of regulatory incentives, manufacturing capacity, high production costs, competing feedstock applications, technological restrictions, and the possibility of waste and residues (Yan et al. 2018).

2.6.1 High Production Costs

Biojet fuel is now more expensive than petro Jet-A1 fuel because of the unpredictable and poor feedstock supply chain on the global market, including pogoma, camellia, and jatropha oil. In general, the price of biojet fuel depends on (a) input cost and composition of feedstock, (b) process technologies, (c) conversion efficiency and product yield, (d) value-added coproducts, and (e) process energy efficiency (Zhang et al. 2018a). It was found that between 50% and 70% of the cost of biojet fuel is made up of feedstock and hydrogen (Zhang et al. 2018b). As a result, the important determinant for cost reduction is a steady supply of prospective feedstock.

2.6.2 Technology and Plant Capacity

Pilot or demonstration stages are reached by several biojet fuel processes. Some of them still require technological development (Zhu et al. 2018). The utilization of trash and residues as feedstock offers the most net greenhouse gas reduction when compared to other alternative feedstocks; nevertheless, supply and its availability chain strategy have not yet been recognized worldwide (Agusdinata and DeLaurentis 2015). High-capacity stand-alone production units require a large investment as well as a high running cost. Biojet fuel might be used as a drop-in fuel in the aviation sector in the near future. As a result, such plants should be built near a petroleum refinery or another biofuel facility (Agusdinata et al. 2011).

2.6.3 Absence of Policy Motivation

To bring fuels closer to parity with fossil fuels, a number of countries have set up incentives for mixing bioethanol and biodiesel with gasoline or diesel. On the other hand, the aviation industry has not implemented any similar measures concerning the utilization of biojet fuel (Bailis and Baka 2010). This circumstance may result in global pricing and availability of discrepancies for feedstock and biojet fuel (Ball et al. 2005).

2.7 Conclusion

Given that there will likely be a rise in demand for aviation services in the near future, the challenge will be to meet this request while also adhering to global initiatives to reduce emissions (Eller et al. 2016). A crucial step in decarbonizing the sector and separating it from the finite supply of fossil fuels is the use of alternate jet fuel. This chapter has examined the possibilities for the bio-aviation fuel industry by carefully examining the feedstocks, manufacturing methods, storage alternatives, and modes of transportation (Elsoragaby et al. 2019). The following are the main conclusions:

1. There are numerous feedstocks for the production of bio-aviation fuel that each have their own advantages in terms of the economy and the environment. Short- to medium-term inexpensive, high-yielding feedstocks with a high oil content might be a good interim option. While waste sources with uncertainty and variability, such used cooking oil and municipal solid waste, may limit their application, land-based crops with adverse environmental effects, like jatropha and oil palm, may also do the same.
2. There are production paths, but they are not all equally prepared. Due to its established nature, HEFA presents a potential solution for the quick and affordable usage of bio-aviation fuel. It is crucial to look at other production techniques more, especially FT, which has higher capital expenditures than other approaches but is closer to commercial maturity and produces superior greenhouse gas reductions.
3. Either centralized or dispersed, the transportation structure of refined petroleum products and biomass feedstock needs to be carefully designed to produce efficient supply chains. It has been found that using a variety of transportation modes throughout the chain can lower GHG emissions and transportation costs when traveling long distances.
4. Supply chains for biofuels may be planned and designed with the use of optimization models, which are helpful tools for decision-making. Decisions made in the supply chain are influenced by temporal and spatial factors.
5. These regulations must be simplified for every link in the supply chain in order to coordinate their growth and expansion and meet environmental and socioeco-

nomic sustainability goals. Since the aviation industry is international, certain regulations must be harmonized worldwide while also providing sufficient flexibility to meet the diverse national goals of different countries.

Acknowledgment It is my proud privilege and special appreciation toward the Department of Chemistry, Vigyan Mandir, Banasthali Vidyapith, Rajasthan.

References

Abdullah B, Muhammad SAFAS, Shokravi Z, Ismail S, Kassim KA, Mahmood AN, Aziz MMA (2019) Fourth generation biofuel: a review on risks and mitigation strategies. Renew Sust Energ Rev 107:37–50

Adhikari DK (2018) Bio-jet fuel. In: Biorefining of biomass to biofuels: opportunities and perception. Springer, pp 187–201

Agusdinata DB, DeLaurentis D (2015) Multi-actor life-cycle assessment of algal biofuels for the US airline industry. In: Algal biorefineries: volume 2: products and refinery design. Springer, pp 537–551

Agusdinata DB, Zhao F, Ileleji K, DeLaurentis D (2011) Life cycle assessment of potential biojet fuel production in the United States. Environ Sci Technol 45(21):9133–9143

Anderson S, Cooper J, Gudde N, Howes J (2012) Aviation renewable fuels: technical status and challenges for commercialisation. Aeronaut J 116(1185):1103–1122

Azwan MB, Norasikin AL, Abd Rahim S, Norman K, Salmah J (2016) Analysis of energy utilisation in Malaysian oil palm mechanisation operation. J Oil Palm Res 28(4):485–495

Bailis RE, Baka JE (2010) Greenhouse gas emissions and land use change from Jatropha curcas-based jet fuel in Brazil. Environ Sci Technol 44(22):8684–8691

Ball J, Carle J, Del Lungo A (2005) Contribution of poplars and willows to sustainable forestry and rural development. Unasylva 56(2):3

Behrendt D, Schreiber C, Pfaff C, Müller A, Grobbelaar J, Nedbal L (2018) Algae as a potential source of biokerosene and diesel–opportunities and challenges. In: Kaltschmitt M, Neuling U (eds) Biokerosene: status and prospects. Springer, Berlin/Heidelberg, pp 303–324

Campbell M (2018) Camelina–an alternative oil crop. In: Biokerosene: status and prospects. Springer, Berlin, pp 259–275

Couto ND, Silva VB, Monteiro E, Rouboa A, Brito P (2017) An experimental and numerical study on the Miscanthus gasification by using a pilot scale gasifier. Renew Energy 109:248–261

Davidson C, Newes E, Schwab A, Vimmerstedt L (2014) Overview of aviation fuel markets for biofuels stakeholders

Doliente SS, Narayan A, Tapia JFD, Samsatli NJ, Zhao Y, Samsatli S (2020) Bio-aviation fuel: a comprehensive review and analysis of the supply chain components. Front Energy Res 8:110

Domínguez-García S, Gutiérrez-Antonio C, De Lira-Flores JA, Ponce-Ortega JM, El-Halwagi MM (2017) Strategic planning for the supply chain of aviation biofuel with consideration of hydrogen production. Ind Eng Chem Res 56(46):13812–13830

Dzięgielewski W, Gawron W, Kaźmierczak U, Kulczycki A (2014) Butanol/biobutanol as a component of an aviation and diesel fuel. J Kones 21

Eller Z, Varga Z, Hancsók J (2016) Advanced production process of jet fuel components from technical grade coconut oil with special hydrocracking. Fuel 182:713–720

Elsoragaby S, Yahya A, Mahadi MR, Nawi NM, Mairghany M (2019) Energy utilization in major crop cultivation. Energy 173:1285–1303

Fischer M, Zenone T, Trnka M, Orság M, Montagnani L, Ward EJ, Ceulemans R (2018) Water requirements of short rotation poplar coppice: experimental and modelling analyses across Europe. Agric For Meteorol 250:343–360

Fontes CHDO, Freires FGM (2018) Sustainable and renewable energy supply chain: a system dynamics overview. Renew Sust Energ Rev 82:247–259

Giudicianni P, Pindozzi S, Grottola CM, Stanzione F, Faugno S, Fagnano M, Ragucci R (2017) Pyrolysis for exploitation of biomasses selected for soil phytoremediation: characterization of gaseous and solid products. Waste Manag 61:288–299

Gutiérrez-Antonio C, Gómez-Castro FI, Segovia-Hernández JG, Briones-Ramírez A (2013) Simulation and optimization of a biojet fuel production process. In: Computer aided chemical engineering, vol 32. Elsevier, pp 13–18

He CR, Kuo YY, Li SY (2017) Lignocellulosic butanol production from Napier grass using semi-simultaneous saccharification fermentation. Bioresour Technol 231:101–108

Howe D, Westover T, Carpenter D, Santosa D, Emerson R, Deutch S, Lukins C (2015) Field-to-fuel performance testing of lignocellulosic feedstocks: an integrated study of the fast pyrolysis–hydrotreating pathway. Energy Fuel 29(5):3188–3197

Jacobson RA, Keefe RF, Smith AM, Metlen S, Saul DA, Newman SM, Inman D (2016) Multi-spatial analysis of forest residue utilization for bioenergy. Biofuels Bioprod Biorefin 10(5):560–575

Karmee SK (2017) Fuel not food—towards sustainable utilization of gutter oil. Biofuels 8(3):339–346

Klein BC, Chagas MF, Junqueira TL, Rezende MCAF, de Fátima Cardoso T, Cavalett O, Bonomi A (2018) Techno-economic and environmental assessment of renewable jet fuel production in integrated Brazilian sugarcane biorefineries. Appl Energy 209:290–305

Martin VJ, Pitera DJ, Withers ST, Newman JD, Keasling JD (2003) Engineering a mevalonate pathway in Escherichia coli for production of terpenoids. Nat Biotechnol 21(7):796–802

Mohseni S, Pishvaee MS (2016) A robust programming approach towards design and optimization of microalgae-based biofuel supply chain. Comput Ind Eng 100:58–71

Molefe M, Nkazi D, Mukaya HE (2019) Method selection for biojet and biogasoline fuel production from castor oil: a review. Energy Fuel 33(7):5918–5932

Newes EK, Bush BW, Peck CT, Peterson SO (2015) Potential leverage points for development of the cellulosic ethanol industry supply chain. Biofuels 6(1–2):21–29

Özaydın B, Burd H, Lee TS, Keasling JD (2013) Carotenoid-based phenotypic screen of the yeast deletion collection reveals new genes with roles in isoprenoid production. Metab Eng 15:174–183

Pham V, Holtzapple M, El-Halwagi M (2010) Techno-economic analysis of biomass to fuel conversion via the MixAlco process. J Ind Microbiol Biotechnol 37(11):1157–1168

Pirker J, Mosnier A, Kraxner F, Havlík P, Obersteiner M (2016) What are the limits to oil palm expansion? Glob Environ Chang 40:73–81

Raje AP, Davis BH (1997) Fischer-Tropsch synthesis over iron-based catalysts in a slurry reactor. Reaction rates, selectivities and implications for improving hydrocarbon productivity. Catal Today 36(3):335–345

Samsatli S, Samsatli NJ (2019) The role of renewable hydrogen and inter-seasonal storage in decarbonising heat–comprehensive optimisation of future renewable energy value chains. Appl Energy 233:854–893

Sannan S, Jordal ABK, Roussanaly S, Giraldi C, Clapis A (2017) Understanding the cost of retrofitting CO2 capture in an integrated oil refinery: reference base case plants: economic evaluation. SINTEF Rapport

Schmitt N, Apfelbacher A, Jäger N, Daschner R, Stenzel F, Hornung A (2019) Thermo-chemical conversion of biomass and upgrading to biofuel: the Thermo-catalytic reforming process–a review. Biofuels Bioprod Biorefin 13(3):822–837

Seber G, Malina R, Pearlson MN, Olcay H, Hileman JI, Barrett SR (2014) Environmental and economic assessment of producing hydroprocessed jet and diesel fuel from waste oils and tallow. Biomass Bioenergy 67:108–118

Sinha AK, Anand M, Rana BS, Kumar R, Farooqui SA, Sibi MG, Joshi RK (2013) Development of hydroprocessing route to transportation fuels from non-edible plant-oils. Catal Surv Jpn 17:1–13

Sinha M, Sørensen A, Ahamed A, Ahring BK (2015) Production of hydrocarbons by aspergillus carbonarius ITEM 5010. Fungal Biol 119(4):274–282
Tao L, Milbrandt A, Zhang Y, Wang WC (2017) Techno-economic and resource analysis of hydroprocessed renewable jet fuel. Biotechnol Biofuels 10:1–16
Taylor JD, Jenni MM, Peters MW (2010) Dehydration of fermented isobutanol for the production of renewable chemicals and fuels. Top Catal 53:1224–1230
Tirado A, Trejo F, Ancheyta J (2021) Prediction of temperature profiles for catalytic hydrotreating of vegetable oil with a robust dynamic reactor model. Ind Eng Chem Res 60(38):13812–13821
Trivedi P, Malina R, Barrett SR (2015) Environmental and economic tradeoffs of using corn Stover for liquid fuels and power production. Energy Environ Sci 8(5):1428–1437
Vasquez MC, Silva EE, Castillo EF (2017) Hydrotreatment of vegetable oils: a review of the technologies and its developments for jet biofuel production. Biomass Bioenergy 105:197–206
Wang ZJ, Staples MD, Tyner WE, Zhao X, Malina R, Olcay H et al (2021) Quantitative policy analysis for sustainable aviation fuel production technologies. Front Energy Res 9:751722
Wei H, Liu W, Chen X, Yang Q, Li J, Chen H (2019) Renewable bio-jet fuel production for aviation: a review. Fuel 254:115599
Woytiuk K, Campbell W, Gerspacher R, Evitts RW, Phoenix A (2017) The effect of torrefaction on syngas quality metrics from fluidized bed gasification of SRC willow. Renew Energy 101:409–416
Yan X, Jiang D, Fu J, Hao M (2018) Assessment of sweet sorghum-based ethanol potential in China within the water–energy–food nexus framework. Sustainability 10(4):1046
Zhang Z, Wang Q, Chen H, Zhang X (2018a) Hydroconversion of waste cooking oil into bio-jet fuel over a hierarchical NiMo/USY@ Al-SBA-15 zeolite. Chem Eng Technol 41(3):590–597
Zhang B, Wu J, Yang C, Qiu Q, Yan Q, Li R et al (2018b) Recent developments in commercial processes for refining bio-feedstocks to renewable diesel. Bioenergy Res 11:689–702
Zhu F, Zhong X, Hu M, Lu L, Deng Z, Liu T (2014) In vitro reconstitution of mevalonate pathway and targeted engineering of farnesene overproduction in Escherichia coli. Biotechnol Bioeng 111(7):1396–1405
Zhu Z, Chu F, Dolgui A, Chu C, Zhou W, Piramuthu S (2018) Recent advances and opportunities in sustainable food supply chain: a model-oriented review. Int J Prod Res 56(17):5700–5722

Chapter 3
Production of Biojet Fuel

Eduardo Dellosso Penteado, Anthony Andrey Ramalho Diniz, Yvan Jesus Olortiga Asencios, Luiza Helena da Silva Martins, Johnatt Allan Rocha de Oliveira, and Andrea Komesu

Abstract The worldwide aviation industry emits almost 920 Mt. of CO_2, which is 2.5% of the global anthropogenic CO_2 emissions and 12% of the emissions from the transport sector. The aircraft also emits other gases that can cause changes in the atmosphere, such as nitrogen oxides (NOx), water vapor, and soot. Considering the emission of CO_2 and non-CO_2 (NOx, water vapor), the aviation sector is responsible for about 5% of the total global warming effect caused by humans. Because of the environmental impacts and the problems linked to the fossil fuel (price, availability, national security), it is necessary to reduce the fuel consumption and carbon footprint of the aviation industry. One possibility to achieve these aims is to use a sustainable aviation fuel based on biological source, the biojet fuels. The biojet fuel can be produced using a variety of processes, such as hydroprocessing of triglyceride feedstock, thermochemical processing of biomass, and alcohol-to-jet, direct sugar-to-hydrocarbon, and aqueous-phase reforming. Each one of these pathways has some advantages and disadvantages, and its technology readiness level is discussed in this chapter.

Keywords Aviation industry · Decarbonization · Fischer–Tropsch · Alcohol-to-jet · Hydroprocessed renewable jet · Direct sugar to hydrocarbon

---

E. D. Penteado (✉) · A. A. R. Diniz · Y. J. O. Asencios · A. Komesu
Universidade Federal de São Paulo, UNIFESP, Instituto do Mar, Santos, São Paulo, Brazil
e-mail: eduardo.penteado@unifesp.br

L. H. da Silva Martins
Universidade Federal Rural da Amazônia, UFRA, Instituto de Saúde e Produção Animal, Belém, Pará, Brazil

J. A. R. de Oliveira
Universidade Federal do Pará. UFPA, Faculdade de Nutrição, Belém, Pará, Brazil

A. Kuila (ed.), *Biojet Fuel: Current Technology and Future Prospect*, Clean Energy Production Technologies, https://doi.org/10.1007/978-981-99-8783-2_3

3.1 Introduction

The worldwide aviation industry uses kerosene produced from crude oil as more than 300 Mt. are consumed annually (Gutiérrez-Antonio et al. 2017; Ng et al. 2021; Wang and Tao 2016). Even with the unexpected COVID-19 pandemic that has caused a negative impact on the aviation industry (Cui et al. 2022), it is expected to raise the global demand for aviation fuel, rising around 5% annually until 2030 and doubling until 2050 (Ng et al. 2021). If this prediction is confirmed, the aviation industry would be affected by a serious energy crisis once the fossil fuel reserves are scarce and the fuel consumption rises every day (Gunerhan et al. 2023).

Moreover, the burning of kerosene produced from crude oil releases huge amounts of CO_2 emissions into the atmosphere, contributing to global warming and other severe environmental problems (Doliente et al. 2020; Gunerhan et al. 2023; Gutiérrez-Antonio et al. 2017; Wang and Tao 2016; Yang et al. 2019). In 2019, flights emitted almost 920 Mt. of CO_2, which is 2.5% of the global anthropogenic CO_2 emissions and 12% of the emissions from the transport sector (Su-ungkavatin et al. 2023). Before the COVID-19 pandemic, the forecasts estimated that these emissions could increase by 3.6% annually and double over the next decades (Sobieralski 2023; Wang et al. 2023). The aircraft also emitted other gases that can cause changes in the atmosphere, such as nitrogen oxides (NOx), water vapor, and soot (Okolie et al. 2023). Considering the emission of CO_2 and non-CO_2 (NOx, water vapor), the aviation sector is responsible for about 5% of the total global warming effect caused by humans (Okolie et al. 2023). Therefore, the worldwide aviation industry impacts the climate crisis using more fossil fuel and emitting more greenhouse gases (GHG) (Cui et al. 2022; Sobieralski 2023).

Because of the environmental impacts and the problems linked to the fossil fuel (price, availability, national security), it is necessary to reduce the fuel consumption and carbon footprint of the aviation industry (Gunerhan et al. 2023; Sobieralski 2023). The aviation sector recognizes the need to find sustainable alternatives to kerosene produced from crude oil. In 2016, the International Civil Aviation Organization (ICAO) launched a project called "Carbon Neutral Growth 2020" to reduce the carbon footprint in aviation from 2020 (Su-ungkavatin et al. 2023). Another organization of the aviation sector, the International Air Transport Association (IATA), is set to reduce the emission of GHG by 50% from 2005 to 2050 (Doliente et al. 2020; Yang et al. 2019).

One possibility to achieve these aims is to use a sustainable aviation fuel based on a biological source, the biojet fuels. Jet biofuels are hydrocarbons produced from renewable sources, also known as biokerosene, with the same boiling range of fossil jet fuels, which must be tailored for aircraft engines and that hold the same fuel properties and performance of fossil jet fuels (Monteiro et al. 2022). The biojet fuels are considered by IATA as a short- to medium-term solution toward the reduction of greenhouse gas (GHG) emissions from the aviation sector (Doliente et al. 2020; Ng et al. 2021; Yang et al. 2019). Up to 80% of CO_2 emissions can be reduced in all life cycles of the biojet fuel derived from renewable feedstock compared with the jet fuel from crude oil (Wei et al. 2019). The conventional jet fuel emits 87.5 g carbon dioxide equivalent per megajoule (gCO_2e/MJ), while the biojet fuel (from open pond algal oil) emits 1.5 gCO_2e/MJ, a 98% reduction (Wang and Tao 2016). It is worth mentioning that the biojet fuel is an important part to decrease the carbon footprint of the aviation industry but other approaches are required, such as improvement of the fuel efficiencies of aircraft (Su-ungkavatin et al. 2023).

A variety of feedstock can be used to produce biojet fuel, such as triglycerides, lignocellulosic biomass, and sugar and starch from plants, animals, wastes, and residues (Doliente et al. 2020; Gutiérrez-Antonio et al. 2017). This variety of renewable raw material allowed different biojet fuel pathways, such as hydroprocessing of triglyceride feedstock, thermochemical processing of biomass, alcohol-to-jet, direct sugar-to-hydrocarbon, and aqueous-phase reforming (Doliente et al. 2020; Gutiérrez-Antonio et al. 2017). Each one of these pathways has some advantages and disadvantages, and its technology readiness level. The American Society for Testing and Materials International (ASTM) certified seven pathways for the production of biojet fuel for commercial use, namely Fischer–Tropsch (FT) process, Fischer–Tropsch with increased aromatic content (FT-SPK/A), hydroprocessed esters and fatty acids (HEFA), direct sugar-to-hydrocarbon (DSHC), alcohol-to-jet (ATJ), catalytic hydrothermolysis jet (CHJ), and hydroprocessed hydrocarbons, esters, and fatty acids synthetic paraffinic kerosene (HHC-SPK or HC-HEFA-SPK) (ASTM D7566/2022 – Standard Specification for Aviation Turbine Fuel Containing Synthesized Hydrocarbons) (Gunerhan et al. 2023; Okolie et al. 2023; Su-ungkavatin et al. 2023).

Thereby, this chapter provides a deep discussion of the pathways to produce biojet fuel, showing the scientific and technological advances reported in the literature. To make it easier for the reader to understand the technologies used, the pathways to produce biojet fuel are separated into four sections according to the feedstocks and conversion processes: oil-to-jet (hydroprocessed esters and fatty acids, catalytic hydrothermolysis, hydroprocessed depolymerized cellulosic jet, and hydroprocessed hydrocarbons, esters, and fatty acids synthetic paraffinic kerosene) (Sect. 3.2), alcohol-to-jet (ATJ) (ethanol-to-jet and butyl alcohols-to-jet) (Sect. 3.3), gas-to-jet (GTJ) (Fischer–Tropsch and biomass-to-fuel) (Sect. 3.4), and sugar-to-jet (STJ) (direct sugar-to-hydrocarbon and aqueous-phase reforming) (Sect. 3.5).

3.2 Oil-to-Jet

3.2.1 Hydroprocessed Esters and Fatty Acids

The hydroprocessed esters and fatty acids (HEFA) is a catalytic chemical process whose purpose is to convert animal fats and vegetable oils into liquid transportation fuels, which are chemically equivalent to the transportation fuels produced from fossil resources. It is a commercially deployed technology that converts the triglycerides from animal fats and vegetable oils into hydrocarbons suitable for being used in jet fuels and diesel (Han et al. 2021; Pearlson et al. 2013).

HEFA is one of the seven technology pathways certified by the American Society for Testing Materials (ASTM) to produce biojet fuels under ASTM D7566. For HEFA, the certification was issued by the ASTM in 2011, and it is required before any commercial airline decides to use fuel for an international flight (IRENA 2017; Julio et al. 2021; Misra et al. 2023; Tiwari et al. 2023).

Triglycerides (TG, triacylglycerol, TAG, or triacylglyceride) compose the structure of all edible and nonedible, vegetable oils, and fats found in nature, whose general structure is shown in Fig. 3.1, where R1, R2, and R3 represent three different free fatty acids attached to a glycerol unit, and the chemical and physical characteristics of the triglycerides depend on the types and length of those radicals (fatty acids) (Caltzontzin-Rabell et al. 2022; Han et al. 2021).

The HEFA process can use waste cooking oil, animal fats, vegetable oils, pyrolysis oil, and algal oil to formulate jet fuel with hydroprocessing. Figure 3.2 shows the process, which involves a set of reactions to extract, from the biomass, its free fatty acids. After this extraction, there is a reaction of isomerization (rearrangement of molecules), followed by a hydrocracking reaction (reduction of the length of carbon chain of molecules) to obtain the jet fuels in accordance with the legal specification (Ng et al. 2021).

Even though, as the triglyceride molecule contains oxygen as shown in Fig. 3.1, it has low thermal stability, it could cause operational problems if used directly at high concentration as fuel for some internal combustion engine. To overcome this problem, there are three mechanisms of converting it to a paraffin component by hydrodeoxygenation reaction, which makes it possible to remove the double bonds and oxygen components contained in those vegetable oils, where two of them are used in the process shown in Fig. 3.2: (1) hydrodeoxygenation (HDO) converted only to water, (2) carbon monoxide and decarbonylation (DCO) converted to water, and (3) decarboxylation (DCO2) converted to carbon dioxide (Han et al. 2021; Ng et al. 2021).

HEFA is a process limited by the catalysts and reaction conditions, which can lead to the production of light gases (liquified petroleum gases and propane), diesel, naphtha, and jet fuel (Dolah et al. 2022). It produces fuels from renewable

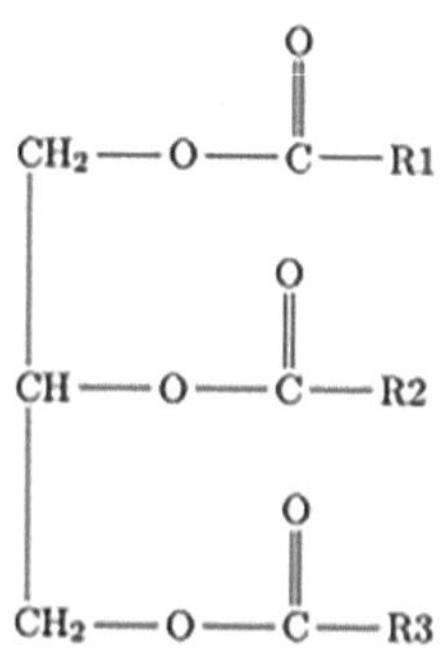

Fig. 3.1 Representation of a general triglyceride molecule (Caltzontzin-Rabell et al. 2022)

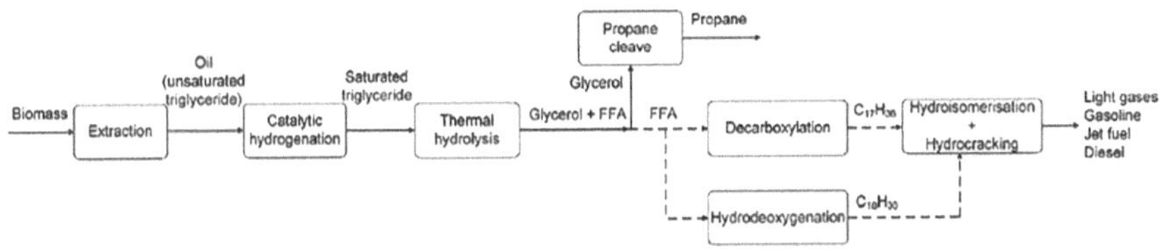

Fig. 3.2 HEFA process for the production of fuels from oily biomass (Ng et al. 2021)

feedstocks, with a performance comparable to conventional petroleum-based fuels, with high cetane numbers, low sulfur and aromatic content, and reduced greenhouse gas emission in a promising conversion pathway for industrial applications (Goh et al. 2022).

From the technology readiness perspective, HEFA can be considered the most mature pathway among those certified by the ASTM. Its feedstocks consist of oleic and linoleic fatty acids, from vegetable oils and animal fats, which are well known to have similar physicochemical properties, and are well-defined feedstocks (Goh et al. 2022; Misra et al. 2023).

In terms of economic aspects, HEFA is by far the most competitive technology to produce sustainable aviation fuels in the market today because its conversion cost is relatively low, with a high fuel yield, even with the price of triglycerides being several times the price of lignocellulosic material (Tanzil et al. 2021). Among the seven certified pathways certified under the ASTM D7566, HEFA biojet is regarded, from the commercial point of view, as the most advanced and the only one that produces high volumes of sustainable aviation fuel (SAF) (Misra et al. 2023).

Till now, the HEFA pathway has been used to produce the vast majority of SAF, with various organizations around the world having already used biofuel produced from this pathway for their test flights, even though there are still some barriers to limit this pathway as also the availability and cost of its feedstock and the inadequate knowledge of the conversion process mechanism of this pathway (Okolie et al. 2023).

3.2.2 *Catalytic Hydrothermolysis*

Catalytic hydrothermolysis (CH) is one of the routes used to convert oil into jets using oils, fatty acids, triglycerides, and biomass sources such as lignocellulosic as raw material. Here, we will discuss how this path occurs for materials considered sustainable, such as waste or by-products of the agroindustry (Chong and Ng 2021a).

CH is also commonly called hydrothermal liquefaction. This process makes use of a technique that uses subcritical water, which is capable of converting wet biomass into a biological material rich in carbon. This material is designated as bio-oil, which is the difference between the CH process and pyrolysis that needs dry biomass to be carried out. Accordingly, CH is a promising technique as there is no need to carry out the drying step, which is a unitary operation that consumes energy, and could be more profitable for the company in case this step is skipped (Chong and Ng 2021a).

The main material used to produce the biojet using CH technology is algae; this material is promising because it is an excellent source of carbon, and the lipid content tends to increase from 5% to 30%, which would be feasible for this material (Chong and Ng 2021a).

CH technology was designed to convert triglycerides into renewable fuels in supercritical water conditions at high pressures and temperatures, and when a catalyst is used this reaction can be accelerated (Eswaran et al. 2021).

According to the literature, the biojet fuels produced from materials such as tung oil, jatropha oil, and camelina oil using the CH process have very similar properties to those fuels obtained conventionally from oil. Carinata oil has already been used and has shown high yields relative to high-density aromatics as well as energy with good low-temperature properties (Eswaran et al. 2021).

The procedure is as follows: the oil is fed into a supercritical hydrothermal reactor, then several reactions take place, such as decarboxylation, cracking, isomerization, recombination, and cracking. But this oil must enter this process in a preconditioned way because this will reduce the consumption of hydrogen since this raw material has a considerable amount of triglycerides that will react in the CH-type reactor with supercritical water and will undergo hydrolysis catalysis and also the cyclization reaction in the aqueous medium; therefore, there is so much reduction in the consumption of hydrogen in the form of gas that it ends up being more viable compared with other processes (Wen et al. 2009).

An interesting study that showed the application of CH was that of Li et al. (2010), in which these researchers used the residual biomass of tung, soy, and jatropha oils for application in a biojet fuel, the JP-8. The results of these researchers showed that these materials were able to meet the navy's specifications. The study conditions were a temperature of 450–475 °C at a pressure of 210 bar, and the experiment was conducted using a catalyst in one condition and in the other without a catalyst. The organic or biocrude phase was treated with the CH process, obtaining fuels with yields of 60% in aromatics, which are the main ingredients desired by the biojet fuel; hence, the results obtained in this study demonstrated that this CH technique can be applied to other materials that have triglycerides in their composition (Li et al. 2010).

A patent number US20080071125A1 (Lixiong Li 2008) was created to make it possible to convert triglycerides into biofuels; this process made use of CH for jet application, performing steps that, when described, are extremely important for the process as a whole, as mentioned next.

The first stage includes what we call pretreatment; here, the unsaturated triglycerides undergo important reactions called catalytic conjugation, cyclization, and cross-linking.

In the second stage, since the triglycerides are already modified in the previous stage because they have undergone a series of reactions, they are now able to be modified with supercritical water at high temperatures, so what will happen now?

Well, it also follows a sequence of reactions such as cracking, hydrolysis, decarboxylation, dehydration, aromatization, and even the combination of all of these, from which a crude hydrocarbon oil will be produced in addition to a phase of glycerol in water and small molecular weight molecules.

In the third stage, the product described in the second stage is refined; here, it will be transformed into various biofuels through different reactions, such as chemical, fermentative, etc.

The last stage aims at obtaining a biofuel that presents linear, branched, and cyclic and aromatic chain paraffins, which are derived from the conversion of triglycerides. As for aromatic compounds, we can say that they come from the conversion of triglycerides, oil, or coal.

3.2.3 Hydroprocessed Depolymerized Cellulosic Jet (HDCJ)

Hydroprocessed depolymerized cellulosic jet (HDCJ) consists of a very innovative approach for the production of jet fuels using renewable energy sources, such as lignocellulosic biomass. Throughout this chapter, we have seen many promising materials categorized as lignocellulosic and certainly those that are considered waste are more interesting to create a green technology. Processes that decompose lignocellulosic biomass will include depolymerization reactions and certainly involve several hydrolysis processes; such steps are extremely important for the release of sugars, which, in their monomeric form, are excellent substrates to lead to fermentations that give rise to biofuels such as ethanol or butanol (Demirbas 2009).

In this context, when there is a jet-oriented approach, the HDCJ process transforms these biomasses into promising fuels for aviation. Chemical compounds such as alcohols (ethanol and butanol) are subjected to the hydroprocessing process, which consists of subjecting the material to high temperatures and pressures in the presence of a catalyst. When converted to hydrocarbons, they will resemble the traditional fuels used in the jet. This product, however, is still mixed with conventional fuel to comply with the legislation required in aviation, but it is a renewable technology and very interesting for us, who are living in an era of environmental concerns (Kazi et al. 2010).

The HDCJ procedure has various advantages for the environment. It does this in two ways. First, it employs biomass sources other than food, lessening the rivalry between food and fuel production. Cellulosic biomass is also widely available and may be obtained responsibly. Second, compared to fossil jet fuels, the resultant jet fuel has the potential to dramatically lower greenhouse gas emissions. This is so because the process generally uses less carbon-intensive processes and the raw material comes from renewable sources (Biddy et al. 2016).

When compared to other processes such as post-pyrolysis, the HDJC process turns out to be quite promising; it has two stages, where the first stage can be described as a hydrothermal pretreatment that has a catalyst as an auxiliary for the hydrodeoxygenation of the bio-oil. The second stage, on the other hand, must be more controlled and uses catalysts for hydrogenation under conditions of high temperatures in order to obtain fuels rich in hydrocarbons. In the HDJC process, ZSM-5 and Raney nickel catalysts are most often used for these fins (Gutiérrez-Antonio et al. 2017).

3.2.4 Hydroprocessed Hydrocarbons, Esters, and Fatty Acids Synthetic Paraffinic Kerosene (HHC-SPK or HC-HEFA-SPK)

The bio-derived hydrocarbons and lipids from *Botryococcus braunii* (algae species) are processed, then cracked and isomerized to form the biojet fuel in hydroprocessed hydrocarbons, esters, and fatty acids synthetic paraffinic kerosene process (HHC-SPK) (Gunerhan et al. 2023). This process is almost the same as the HEFA pathway, only differing in the feedstock used. In 2020, the ASTM approved the HHC-SPK to produce biojet fuel, with the restriction to blend with jet fuel up to 10% by volume (Gunerhan et al. 2023).

3.3 Alcohol-to-Jet

Alcohol-to-jet synthetic paraffinic kerosene (ATJ-SPK), also known as alcohol oligomerization, was approved in 2016 (Claudia Gutiérrez-Antonio et al. 2021). ATJ-SPK is a fuel produced from alcohols, such as methanol, ethanol, butanol, and long-chain fatty alcohols (Wang and Tao 2016). According to the ASTM D7566 standard, the produced biofuel can be utilized up to 30% by volume in blends with fossil jet fuel (ASTM 2016).

Alcohols as feedstock can be obtained in a variety of methods such as fermentation of sugars obtained from sugar-containing crops, the liquefaction and saccharification of starch-containing crops, or via hydrolysis of lignocellulose. The alcohol is then subjected to four steps, including dehydration, oligomerization, hydrogenation, and separation (Achinas et al. 2021).

In spite of the fact that various alcohols or intermediate pathways are feasible for manufacturing jet fuel, the most common ATJ production pathways are through ethanol or butanol (Chong and Ng 2021a). These alcohols are discussed in this chapter.

3.3.1 Ethanol-to-Jet

Bioethanol derived from biomass is readily available as a feedstock for ATJ fuel because it is primarily used as a transportation fuel to replace gasoline (Chong and Ng 2021a). The overall process diagram for ethanol to jet fuel is shown in Fig. 3.3.

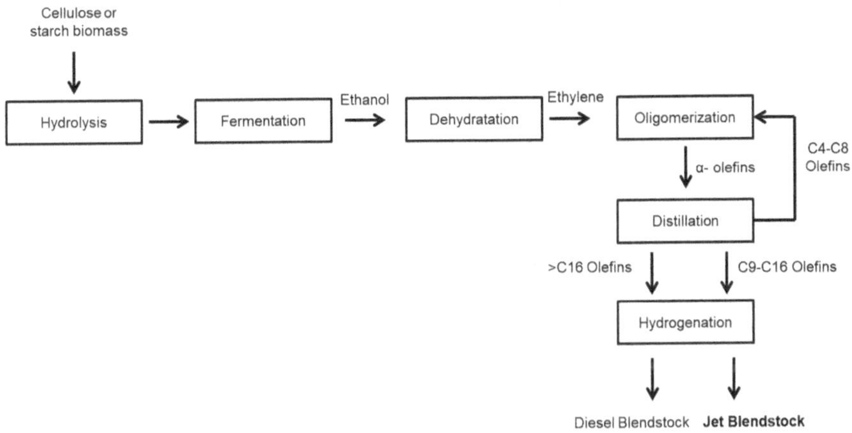

Fig. 3.3 Process of converting cellulose and starch biomass into biojet fuel via the ethanol-to-jet pathway. (Adapted from Chong and Ng 2021a; Wang and Tao 2016)

Alcohols have matured in technology, so each of their stages has undergone extensive research.

First, ethanol is dehydrated to produce ethylene. This process is well developed and commercially practiced. In 2010, Braskem opened a bio-based ethanol dehydration plant that produces green polyethylene in Brazil. Ethylene monomer is produced by dehydration of ethanol obtained from sugarcane (Brooks et al. 2016).

There are two possible dehydration pathways for ethanol: direct dehydration into ethylene or formation of diethyl ether, followed by cracking into ethylene. In the presence of a strong acid, diethyl ether can be formed at temperatures as low as 300 °C and then cracked into ethylene and water (Bokade and Yadav 2011; Chong and Ng 2021a). Studies on dehydration catalysis started with alumina and transition metal oxides (TiO_2, Fe_2O_3, Mn_2O_3, Cr_2O_3) and moved to silicoaluminophosphates (SAPO), H-ZSM-5 zeolite catalyst, and heteropolyacid catalysts (Wang and Tao 2016). Recently, a novel gamma alumina was synthesized by solvent protection and a hydrothermal procedure (Lv et al. 2023). According to Lv et al. (2023), the novel gamma alumina showed higher activity, higher selectivity of ethylene, and higher reaction stability. Under the optimal conditions, both the conversion of ethanol and selectivity of ethylene were higher than 99% (Lv et al. 2023).

As reported by Styskalik et al. (2020), mesoporous aluminosilicate catalysts prepared by non-hydrolytic sol-gel (NHSG) feature intermediate levels of acidity (both in strength and nature), resulting in intermediate catalytic activity. It is important to note that the best NHSG-made samples remained highly stable over time, showed no trace of ethylene oligomers, and exhibited no signs of coke formation (Styskalik et al. 2020).

Masih et al. (2019) demonstrated an efficient and sustainable catalytic conversion of ethanol-to-ethylene using Rho zeolite. The steady-state selectivity to ethylene remained above 99% for the dehydration reactions carried out in the temperature range of 250–400 °C. The catalytic properties of Rho zeolite were superior compared with another zeolite, ZSM-5 and γ-Al_2O_3 materials. In another study using zeolite, Soh et al. (2017) modified zeolite-Y (80) with phosphoric acid. Despite a drop in ethanol conversion with phosphorus-modified catalysts, a modified zeolite-Y with 10 wt% H_3PO_4 can achieve 99% ethylene selectivity when operating at 723 K and ethanol partial pressure of 16 kPa (Soh et al. 2017).

Gamma alumina (γ-Al_2O_3) and zeolites (microporous HZSM-5) are commonly used in industrial ethanol dehydration (Chong and Ng 2021a). Specialized ethanol-to-ethylene heterogeneous catalysts such as Syndol have become commercially available with high selectivity and conversion (Geleynse et al. 2018). Therefore, many catalysts have been studied because the ethanol dehydration step is strongly dependent on the acid sites in the chosen catalyst.

In the oligomerization step, alkenes produced by alcohol dehydration must be oligomerized into the desired hydrocarbon chain length distribution; typically 8–16 carbons for kerosene. To achieve an appreciable yield and meet fuel specifications, the oligomerization process must be carefully designed (Geleynse et al. 2018). There are a variety of pathways to producing jet fuel from ethylene such as direct oligomerization and conversion of ethylene to intermediate olefins for oligomerization. The first pathway is the most difficult.

In order to improve the sustainability of jet fuel production, Villareal-Hernández et al. (2023) applied intensification ideas using reactive distillation. In the results, the return-on-investment value improved by approximately 50% and the annual cost decreased by approximately 90%. Additionally, efficiency, mass intensity, and thermodynamic efficiency improved by close to 30% and the global warming potential value improved by 70% (Villareal-Hernández et al. 2023).

Homogeneous and heterogeneous catalysts have long been used to produce long-chain alpha olefins. Panpian et al. (2021) investigated ethylene oligomerization using NiAlKIT-6 as the catalyst. NiAlKIT-6 catalysts converted ethylene to >95% with up to 55% selectivity for C8+. In a 30 h period, the catalyst maintained good stability while maintaining C8+ selectivity. It was also possible to regenerate the spent catalyst so that its catalytic activity could be maintained.

Many companies have developed different oligomerization processes, depending on the feedstock used, like Chevron Phillip's "Ziegler" process. In Ziegler's "one-step" process, the catalyst cannot be recycled, but must be disposed of. However, the catalyst can be reused in the "two-step" reaction (Doliente et al. 2020; Weissermel and Arpe 2008).

After oligomerization, the resulting olefins are distilled to diesel- and jet-range fuels and light olefins. As shown in Fig. 3.3, light olefins (C4–C8) separated through distillation return to the oligomerization step. Jet fuel-range products (C9–C16) can be subjected to hydrogenation at 370 °C and WHSV of 3 h^{-1} using palladium or platinum catalysts over 5% by weight of hydrogen. The C9–C16 alkanes produced from the hydrogenation step are suitable for renewable jet fuels (Wang and Tao 2016).

3.3.2 Butyl Alcohols-to-Jet

In addition to ethanol, other alcohols are further along in the development of jet fuel production processes. *n*-Butanol is primarily produced from petrochemicals using propylene feedstock and hydroformylation. Also, it is possible to produce it by fermenting sugars using *Clostridia* bacteria, which is widely known for its acetone–butanol–ethanol (ABE) process. The production of isobutanol using genetically modified yeast cells also uses alcoholic fermentation (Brooks et al. 2016). Several companies have developed technologies to produce alternative jet fuel based on butanol, including UOP, Gevo, and Cobalt/US Navy (Chong and Ng 2021a). The overall process diagram for *n*-butanol and isobutanol to jet fuel is shown in Fig. 3.4.

After fermentation, dehydration of butyl alcohol can be divided into *n*-butanol and isobutanol. Using *n*-butanol, 1-butene and 2-butane are produced, while using isobutanol, olefins are produced, such as 1-butene, *cis*-2-butene, *trans*-2-butene, and isobutene.

In the dehydration of *n*-butanol over an acid catalyst, butene is produced at lower temperatures, but higher temperatures are required for skeletal isomerization of *n*-butanol (Chong and Ng 2021b). Conesa et al. (2023) studied the graphite-supported heteropolyacid ($H_4SiW_{12}O_{40}$ and $H_3PW_{12}O_{40}$) as a regenerable catalyst in

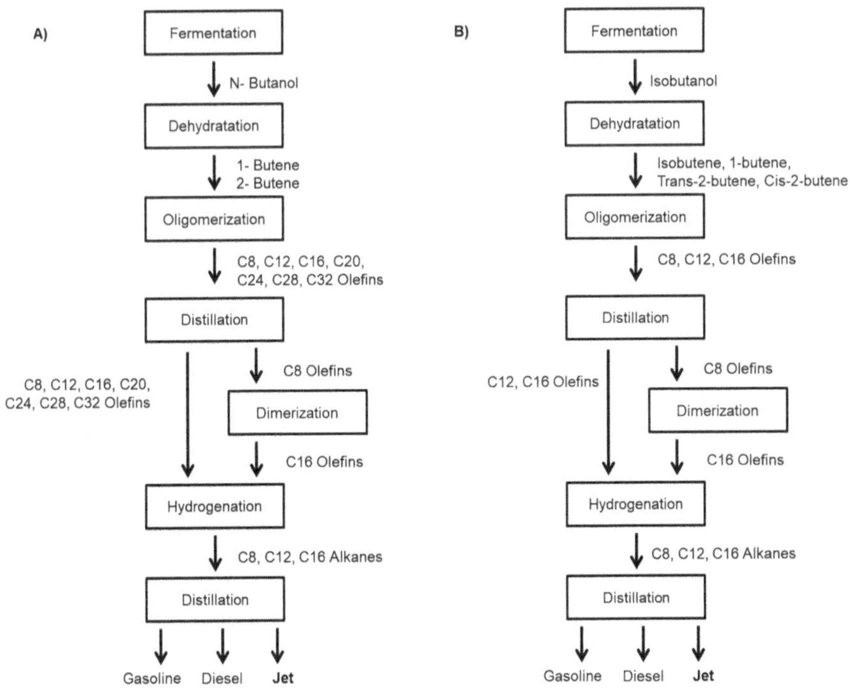

Fig. 3.4 Process of converting *n*-butanol (**a**) and isobutanol (**b**) into biojet fuel. (Adapted from Chong and Ng 2021b; Wang and Tao 2016)

the dehydration of *n*-butanol to butenes. The catalyst activity was compared against a zeolite HZSM-5 with a Si/Al ratio = 23. The results evidence that graphite-$H_4SiW_{12}O_{40}$ interactions efficiently tailored the acidity, resulting in an active regenerable catalyst and *n*-butenes selective (>98%).

In another study, Zhang et al. (2023b) studied the selective dehydration of ABE (or low-carbon alcohols) into light olefins over the Ce@Fe@SAPO-34 catalyst and olefin polymerization into jet fuels over the ionic liquid catalyst ([bmim]Cl–2$AlCl_3$). By coupling the two-step process, high ABE conversion (89.3%) and high jet fuel yield (71.5%) were achieved under atmospheric pressure.

Following the process, as described by (Wang and Tao 2016), 1-butene is oligomerized to produce olefins ranging from C8 to C32 with 97% conversion. The reaction is carried out at ambient temperature for 16 h with methylaluminoxane and transition-metal catalysts from group 4 (Cp2ZrCl_2/MAO). The unreacted olefins of 2-butene, including *cis*-2-butenes and *trans*-2-butenes, are distilled at a temperature-controlled rate. The C8 olefin, 2-ethyl-1-hexene, is distilled and sent to the dimerization reactor. During dimerization, the C8 olefin is completely converted, yielding 90% of $C_{16}H_{32}$ under Nafion catalyst at 116 °C for 2 h. The products from 1-butene oligomerization, ranging from C12 to C32, together with C16 olefins produced from dimerization, are sent to hydrogenation process over 0.08 wt% PtO_2 catalyst. As a result, the C12–C16 paraffins can be blended with jet fuel, while the C20–C32 alkanes can be separated and sold as lubricants (Wang and Tao 2016).

Isobutanol is mostly dehydrated over mildly acidic α-Al_2O_3 catalysts, but other catalysts such as inorganic acids, metal oxides, zeolites, and acidic resins, among others, have been reported to be feasible (Chong and Ng 2021b). In their review of scientific and patent literature on selective dehydration of isobutanol to isobutene, Dubois et al. (2023) summarized the state of the art of this reaction. The selective dehydration of isobutanol into isobutene can be performed under low or high pressures. This choice impacts not only the downstream recovery and purification of isobutene, but also the stability of the catalyst and the reactor volume (Dubois et al. 2023).

Guo et al. (2021) studied the catalytic oligomerization of isobutyl alcohol using dealuminated zeolite beta. The results showed that isobutyl alcohol can be quantitatively oligomerized over dealuminated zeolite beta with the selectivity of C8–C16 exceeding 50% at a conversion of 98%. In another study about oligomerization of isobutene, Al-Kinany et al. (2019) showed that the conversion of isobutene into distillates ranges between 97% and 100% using phosphoric acid on H-zeolite-Y zeolite by acid impregnation and ultrasonic vibration technique.

After oligomerization, the jet and diesel yields can be improved by distilling the C8 olefins and initiating another dimerization at 116 °C over a Nafion catalyst. In addition, C8 olefins can either be dimerized or reacted with butenes to form C12 olefins, allowing jet-range chemicals to contain more C12 and C16.

Generally, higher alcohols have higher energy content and lower water solubility than ethanol, but they are not used as widely as ethanol. When comparing *n*-butanol, isobutanol, and ethanol, *n*-butanol has the highest greenhouse gas (GHG) emissions, while ethanol has the lowest. However, butanol has a lower heat of vaporization and less corrosivity, making it a more attractive feedstock than ethanol (Doliente et al. 2020).

3.4 Gas-to-Jet

Among several technologies available for biojet fuel production, the Fischer–Tropsch (FT) synthesis and biomass-to-biojet processes offer environmental benefits such as reduced life cycle greenhouse gas (GHG) emissions, and economic benefits associated with availability and costs, in comparison with conventional jet fuels (Liu et al. 2013; Tiwari et al. 2023; Zhao et al. 2023). Both processes are in strong relation since both need syngas (H_2/CO) as a feedstock, and the only difference between them is the source to produce syngas; some recent studies are discussed.

3.4.1 Fischer–Tropsch

In the mid-1920s, German scientists Franz Fischer and Hans Tropsch discovered that it is possible to synthesize liquid hydrocarbons from syngas (a gaseous mixture composed of CO and H_2) obtained from coal gasification. At that time, this synthesis was carried out over catalysts of alkalized iron chips at 673 K and under high-pressure conditions (>100 bar). This synthesis received the name of the respective German scientists, today known as Fischer–Tropsch (FT) synthesis (Liu et al. 2013; Suppes and Storvick 2016).

The FT synthesis allows the possibility to synthesize liquid hydrocarbons from almost all carbon-based matter feedstocks, which is a major advantage since among coal and natural gas, it is possible to use renewable feedstock, for example, biogas, bioethanol, bio-oil, lignocellulosic materials, and many other types of biomass. The carbon-based matter feedstocks need to be transformed into clean syngas (H_2/CO) first before going to FT synthesis.

FT fuels as jet fuel have many characteristics that make them very attractive, for example, FT fuels are clean of sulfur, oxygen, and nitrogen, among other contaminants commonly found in fossil fuels. It is aromatic-free. Also, the combustion of FT fuels is free from aerosol emissions, thus extending the combustor and turbine life, all these characteristics reducing the deposits on fuel lines and engine components (Bermúdez et al. 2011; Meurer and Kern 2021).

The FT jet fuel needs to possess superior properties to conventional jet fuel available today, such as higher cetane number, lower cloud point, and lower emissions (de Klerk et al. 2023). This makes FT fuels very attractive to be used as jet fuel because it is possible to handle the hydrocarbon properties by managing the FT synthesis. For instance, synthetic paraffinic kerosene (SPK), a jet fuel, can be produced from coal by FT synthesis, also known as FT-SPK. The FT-SPK fuels are a mixture of normal and iso-paraffins with a small percentage of cyclo-paraffins. The iso- and *n*-type paraffins and carbon number of FT-SPKs (from C9 to C15) are typical in jet fuel obtained from fossil fuels (de Klerk et al. 2023; Meurer and Kern 2021).

On the other hand, some disadvantages are mentioned in the literature regarding the absence of aromatics in the FT fuels. For instance, FT kerosene does meet the minimum density requirement for conventional kerosene; Dahal et al. (2021) and Yang et al. (2019) reported that the absence of aromatics in FT jet fuels, such as synthetic paraffinic kerosene (SPK) and hydroprocessed esters and fatty acids (HEFA), typically has lower specific energy density than conventional jet fuels (Dahal et al. 2021; Yang et al. 2019). Furthermore, the absence of aromatics can favor leaks in certain types of fuel systems. All these authors mentioned that the disadvantages can be solved by blending with conventional jet fuels and other fuel additives (Liu et al. 2013).

The FT process can be organized into three principal steps:

1. Syngas production
2. Removal of CO_2 and undesired compounds as well as impurities from syngas
3. FT synthesis indeed.

FT synthesis is strongly exothermic, whose simplified reaction can be described as follows (Schulz et al. 2021):

$$n\,\mathrm{CO} + 2n\,\mathrm{H_2} \rightarrow (-\mathrm{CH_2}-)_n + n\,\mathrm{H_2O} \quad \Delta_\mathrm{R} H_0 (250°\mathrm{C}) = -158.5\,\mathrm{kJ} \tag{3.1}$$

where n can be 1, 2, 3..; $(-CH_2-)$ represents the main products, which can be straight-chain paraffins, olefins, and alcohols. The reaction product variety is more complex. Oxygenates and branched hydrocarbons can also be produced in lower amounts. The oxygen molecules are rejected as water according to Eq. 3.1, or, in some cases, as CO_2 through water gas shift (WGS) reaction; the latter is usually promoted by some active catalysts (Schulz et al. 2021).

WGS has some regulatory roles to FT synthesis; methanation, the coking reaction (by Boudouard reaction), also occurs in parallel.

FT synthesis makes mainly straight-chain hydrocarbons. The composition of the product varies depending on the H_2:CO ratio, the type of catalyst, and the process conditions (pressure and temperature). These straight-chain hydrocarbons need to be further processed to be considered an acceptable jet fuel. After the FT synthesis stage, the hydrocarbon products are upgraded to liquid fuels using well-known and well-established processes such as cracking, isomerization, and, in some cases, distillation (to separate middle distillates and naphtha), all these processes are common in petroleum refineries. It is important to mention that all products have a near-zero level of sulfur, nitrogen, nickel, vanadium, aromatics, and asphaltenes (Carvalho et al. 2019).

FT synthesis can be carried out in low-temperature (LTFT process) or high temperatures (HTFT process), with temperature ranges between 200–240 °C and 300–350 °C, respectively (Teimouri et al. 2021). The main difference between LTFT and HTFT is that no liquid phase is present outside the catalyst particles in the HTFT reactors (Carvalho et al. 2019). Table 3.1 summarizes the principal characteristics of the LTFT and HTFT processes.

Table 3.1 Principal characteristics of the LTFT and HTFT processes

Process	Reaction temperature range (°C)	Type of catalyst	Products
Low-temperature Fischer–Tropsch (LTFT)	220–240	Co-based	Hydrocarbon of long chains/waxes, paraffins
High-temperature Fischer–Tropsch (HTFT)	300–350	Fe-based	Hydrocarbons of short chains, olefins, and gasoline, among other chemicals

Source: Adapted from Carvalho et al. (2019), Teimouri et al. (2021)

3.4.1.1 The Catalysts for FT Synthesis

Ni, Fe, Co, and Ru-based catalysts have been employed for FT synthesis (Schulz et al. 2021), Ni generally favoring CH_4 formation, an undesirable product; on the other hand, Ru shows very good activity and selectivity. However, it is very expensive. The industry uses Fe and Co-based catalysts, which are considered large-scale viable catalysts (Liu et al. 2013). As usual in heterogeneous catalysis, the catalytic support should be a porous material with a large surface area, thus favoring finely dispersed active centers for catalysis. A moderate metal–support interaction is strongly desired. Typical porous materials have been used as catalytic support in FT synthesis: SiO_2, Al_2O_3, TiO_2, zeolites (H-ZSM-5, H-BETA, and H-Y), active carbon, and pillared clays, among others (Boymans et al. 2022; Valero-Romero et al. 2021).

Co-based catalysts are usually preferred for the synthesis of long-chain paraffins as they are more resistant to deactivation by water. When Fe-based catalysts are used, the water gas shift (WGS) reaction occurs simultaneously as Fe is active for WGS (WGS consumes CO and H_2O to produce additional H_2 and CO_2). On the other hand, Fe is cheaper than Co, has lower methane selectivity, and low sensitivity to poisoning, among other properties. The industry uses both Co- and Fe-based catalysts according to the requirements and specifications of the desired final products.

3.4.2 Biomass-to-Fuel

The feedstock for this process includes lignocellulosic biomass, for example, woody energy crops, residues generated by forestry and agriculture, and other organic wastes. Although a wide variety of feedstock can be used, their characteristics and properties will affect the efficiency of the process, the type of gasifier, and, principally the quality of syngas.

The typical process starts with the pretreatment of biomass. This pretreatment has the objective of increasing the density of the respective biomass; during the pretreatment, the particle size is reduced and the moisture is reduced as well, thus facilitating the subsequent chemical processes and its transport and logistics (Carvalho et al. 2019).

After this pretreatment, the gasification step takes place; this is the first fundamental step since it produces syngas, which will be used as feedstock for FT synthesis. Fluidized bed gasifiers are usually used for this step.

Once syngas is obtained, it is then conditioned to remove CO_2 and impurities inherent to biomass. The H_2:CO ratio of syngas is also adjusted at this stage. After these processes, syngas is ready for FT synthesis, which follows the previously described processes.

According to the ASTM D7566, these FT biofuels can only be used mixed with conventional aviation fuel to ensure the minimum amount of all components in the final fuel composition. This also guarantees the compatibility of the fuel with the aircraft engine (Chuck 2016). The first route approved by the ASTM is the synthesis of FT in 2009, in which coal, natural gas, or biomass can be used as raw material; the adapted process is shown in Fig. 3.5. Due to the low content of aromatics present in the biofuel obtained, it is necessary to be mixed with at least 50% conventional aviation fuel for its use.

The sustainable supply of biomass for FT synthesis allows possible competition for the raw biomass from other sectors (Dahal et al. 2021). For instance, some forest products have an established market and destination; thus, the production of biojet fuels should come from alternative low-cost and abundant agricultural residues. The heat generated by gasification and FT synthesis can be used for another process; gasification of biomass also produces chemicals, and these characteristics may increase the economic performance and thermal efficiency of the process.

Some studies have been reported in the literature. Hanaoka et al. (2015), Li et al. (2016), Yan et al. (2013), Zhang et al. (2023a) studied biojet synthesis through the FT synthesis, where the conceptual bases and technical aspects are described, and they used woody biomass and different types of catalysts to produce biojet fuels.

The FT biojet fuels are promising; however, some disadvantages of slow large-scale applications are difficulties in the logistics of raw materials and the cleaning of syngas, and high capital investment.

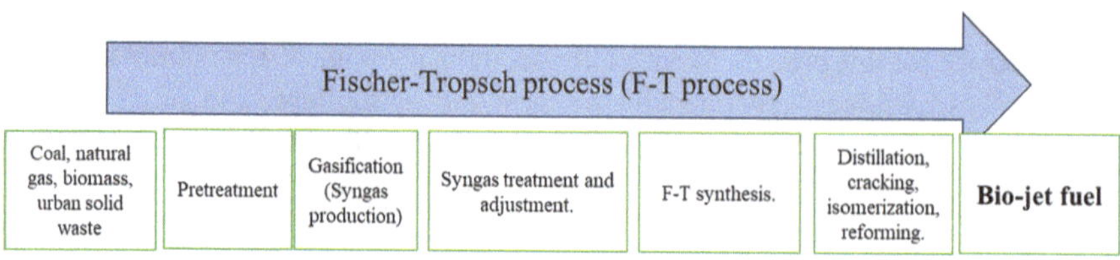

Fig. 3.5 ASTM-approved Fischer–Tropsch technology route for biojet production. (Adapted from Chuck 2016)

3.5 Sugar-to-Jet

3.5.1 Direct Sugar-to-Hydrocarbon

Aviation has its share of the blame for the emission of polluting gases; hence, biotechnological processes that use the sustainable conversion of sugars into fuel can contribute enormously to reduce carbon emissions in the environment.

The US Navy carried out experiments with a new fuel with characteristics comparable to diesel, which was obtained from the direct conversion of sugar, via transformation by yeasts, into hydrocarbons, a process known as direct sugar-to-hydrocarbon (DSH), in which the yeast produced farnesene, a branched-chain hydrocarbon with multiple double bonds, which is processed into a simple moderately branched alkane molecule, much more interesting than conventional diesel, due to the higher amount of cetane in its composition (Hamilton et al. 2014).

The production of hydrocarbons is similar to other processes found for other sustainable fuels. It follows a production path consisting of steps such as pretreatment and conditioning of biomass, enzymatic hydrolysis, separation of solids, biotechnological conversion, and product recovery and purification, which, in the American case, is obtained for corn. Still, it can be carried out for other raw materials, such as sugarcane, in the case of large producers, such as Brazil.

It is possible to describe this as a basic process for obtaining through Fig. 3.6.

To improve this type of fuel, more in-depth studies have been carried out to analyze more improved combustion and modify the fuel composition through the development and use of single-molecule, binary blends, and synthetic substitute studies (Carr et al. 2012; Caton et al. 2011; Mathes et al. 2010).

A solid study was carried out to evaluate the thermal behavior and the influence of farsene in the developed biojet, but the data are still very limited, even so it was possible to obtain information through pressure diagnostics of diesel engines based on farsene, which provide the same, to a limited extent, combustion metrics (Conconi and Crnkovic 2013).

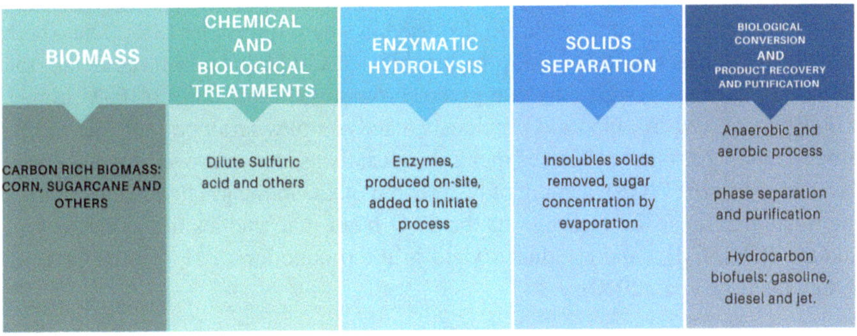

Fig. 3.6 A basic process for obtaining biojet from biomass

3.5.2 Aqueous-Phase Reforming

This technology is an alternative capable of generating the production of green hydrogen that can be used for processes to obtain hydrocarbons through the hydrogenation of unsaturated and saturated fatty acids and then deoxygenated. This process can be carried out through the use of vegetable oils. In general, this process uses a catalytic reaction to produce hydrogen through catalytic reactions of reagents that are generally in the liquid state (water and glycerol) and in which high pressures and temperatures are used (Oliveira 2014).

In this process, glycerin is an essential material with great potential in hydrogen production via aqueous-phase reform (APR). For this, the energy supply in the APR is much lower than that in the reform process (SR) since the APR, as already mentioned, is carried out at low temperatures. This fact is essential as this temperature can also promote the displacement of water gas and thus have lower CO concentrations in a reactor.

Another critical advantage of APR is producing 7 mol of hydrogen for every 1 mol of glycerol, four from glycerol and three from water (Davda et al. 2005). In this way, the hydrogen produced can be used as fuel in internal combustion engines or even in fuel cells and can be used to produce chemical products such as methanol and ammonia.

The APR allows several oxygenated hydrocarbon compounds obtained from biomass, such as ethanol, ethylene glycol, polyols, cellulose, among others, to be converted into hydrogen. This process is undoubtedly a technology of great interest for the production of fuels, and of course of great attention to the aviation sector, which uses large volumes of fuel. Within this technology, several factors can influence the selectivity of the process, such as the nature of the active catalytic metal, pH of the solution, and feed and process conditions (He et al. 2013).

The aqueous-phase process occurs at reduced temperatures, at values between 220 °C and 270 °C and high pressure, which requires the need for catalysts such as platinum (Pt) and nickel (Ni), which increases hydrogen productivity (Eloffy et al. 2022).

As mentioned, the breakdown processes of biomass molecules rich in cellulose and hemicellulose generate monomers that can serve as inputs for hydrogen production processes in a system similar to that used in steam reforming (SR). Thus, both of these processes create H2 from the breaking of C–C bonds. In more detail, the contact of a catalyst with ethylene glycol promotes the breaking of C–C bonds, which release CO. As reported previously, catalysts play an important role in this process. As in the other processes mentioned, catalysts also improve processes such as water gas shift (WGS), leaving only 300 ppm of CO in the gas flow. Several other intermediates favor the cleavage of the C–O bond that enables the generation of alkanes (CH_4, C_2H_6) and a reduced yield in the production of H_2 (Cortright et al. 2002; Tanksale et al. 2010).

3.6 Economic and Environmental Analysis of Biojet Fuel Pathways

Table 3.2 summarizes the economic and environmental issue characteristics of different jet fuel production pathways and the year approved by the ASTM. To choose the best pathway to be employed to biojet fuel production, it is desirable to have a lower minimum selling price (MSP), cost of production, and greenhouse gas (GHG) emissions. In contrast, the technology readiness level must be higher, showing that this technology is available to be commercialized (Okolie et al. 2023; Wei et al. 2019). In this sense, the technology readiness level controls the production cost, reducing the minimum selling price (Wang and Tao 2016). HEFA and Fischer–Tropsch were first proved by the ASTM and are the most mature technology to produce biojet fuel.

Table 3.2 Economic and environmental analyses of pathways of biojet fuel production

Pathway	Year of approval by the ASTM	Minimum selling price (US$/L)	Production cost (US$/L)	GHG emissions (g CO2-eq/MJ)	Technology readiness level
HEFA	2008	0.43–1.51	0.36	3.06–53.10	8–9
CH	2020	0.65–1.34	1.53	21.20–39.30	4–5
HDCJ	Not approved	1.38–1.89	–	−2.70–49.50	6
Hydroprocessed HHC-SPK or HC-HEFA-SPK	2020	–	–	–	6–7
ATJ-SPK	2018 (ethanol) 2016 (butanol)	0.96–2.88	0.55	−27.00–117.50	6–7
Fischer–Tropsch	2009	0.65	0.37	−1.60–18.20	8–9
Biomass-to-fuel	Not approved	1.65–2.00	–	–	6
Direct sugar-to-hydrocarbon	2014	1.89–6.45	1.54	22.00–80.00	6
Aqueous-phase reforming	Not approved	1.23–1.25	–	–	5

Source: Adapted from Okolie et al. (2023), Wei et al. (2019)

3.7 Final Remarks

Biojet fuel is a possible way to reduce the environmental impacts and the problems linked to the fossil fuel (price, availability, national security) in the aviation industry. Several pathways to produce the biojet fuel have been developed, and some of them are being used to produce it commercially. In Table 3.3, the advantages and disadvantage of the main technologies discussed in this work are shown. However, HEFA and FT were first proved by the ASTM and are the most mature technology to produce biojet fuel; several efforts should be taken to make biojet fuels more attractive than conventional jet fuel, especially to reduce the market price for biojet fuels.

Table 3.3 Summary of the pros and cons of each biojet fuel pathway

Pathway	Pros	Cons
HEFA	Exothermic reaction, reducing the energy cost and the environmental impact Biojet fuel with high quality (ignites fast and has a great heating value – 44 MJ/kg) Production of less reactive soot	Availability of resources/ feedstock is limited relative to the projected industrial demand High demand of hydrogen for the cracking of triglyceride (10–15 mole per mole of triglyceride)
Catalytic hydrothermolysis (CH)	Low capital costs and good energy efficiency when likened to other processes Transportability and storability of liquid fuels	Low quality and stability of biojet fuel
Hydroprocessed depolymerized cellulosic jet (HDCJ)	Variety of lignocellulosic feedstocks	Low yields Complexity process
Alcohol-to-jet (ATJ-SPK)	Permissible aromatic content Well-established structure for ethanol production, decreasing transportation cost Reduced cost of production due to lower temperature and pressure required Low demand of hydrogen	Issues with feedstock availability Low yield Long processing routes involved
Fischer–Tropsch	High feedstock options (including not food varieties) High-energy efficiency Permitted range of aromatic content Sulfur-free leading to less emission during engine combustion	Expensive technology
Direct sugar-to-hydrocarbon	No energy intensive High feedstock options (including not food varieties) Hydrocarbon yields up to 92%	Only 10% can be used according ASTM Special enzymes needed Market immaturity
Aqueous-phase reforming	Use renewable supplies	Expensive catalyzes used Pretreatment needed High demand of chemical to pretreatment

References

Achinas S, Margry S, Euverink GJW (2021) A technological outlook of biokerosene production. In: Sustainable biofuels. Elsevier, pp 225–246

Al-Kinany MC, Al-Drees SA, Al-Megren HA, Alshihri SM, Alghilan EA, Al-Shehri FA et al (2019) High-quality fuel distillates produced from oligomerization of light olefin over supported phosphoric acid on H-Zeolite-Y. Appl Petrochem Res 9(1):35–45

ASTM A (2016). Standard specification for aviation turbine fuel containing synthesized hydrocarbons. ASTM International West Conshohocken

Bermúdez V, Lujan JM, Pla B, Linares WG (2011) Comparative study of regulated and unregulated gaseous emissions during NEDC in a light-duty diesel engine fuelled with Fischer Tropsch and biodiesel fuels. Biomass Bioenergy 35(2):789–798

Biddy MJ, Davis R, Humbird D, Tao L, Dowe N, Guarnieri MT et al (2016) The techno-economic basis for coproduct manufacturing to enable hydrocarbon fuel production from Lignocellulosic biomass. ACS Sustain Chem Eng 4(6):3196–3211

Bokade VV, Yadav GD (2011) Heteropolyacid supported on montmorillonite catalyst for dehydration of dilute bio-ethanol. Appl Clay Sci 53(2):263–271

Boymans E, Nijbacker T, Slort D, Grootjes S, Vreugdenhil B (2022) Jet fuel synthesis from syngas using bifunctional cobalt-based catalysts. Catalysts 12(3):288

Brooks KP, Snowden-Swan LJ, Jones SB, Butcher MG, Lee G-SJ, Anderson DM et al (2016) Low-carbon aviation fuel through the alcohol to jet pathway. In: Biofuels for aviation. Elsevier, pp 109–150

Caltzontzin-Rabell V, Romero-Izquierdo AG, Moreno-Gómez AL, Martínez-Guido SI, Gómez-Castro FI, Gutiérrez-Antonio C (2022) Raw materials for a biomass-based industry. In: Biofuels and biorefining. Elsevier, pp 25–52

Carr MA, Caton PA, Hamilton LJ, Cowart JS, Mehl M, Pitz WJ (2012) An experimental and modeling-based study into the ignition delay characteristics of diesel surrogate binary blend fuels. J Eng Gas Turbines Power 134(7)

Carvalho F, Silva FTF, Szklo A, Portugal-Pereira J (2019) Potential for biojet production from different biomass feedstocks and consolidated technological routes: a georeferencing and spatial analysis in Brazil. Biofuels Bioprod Biorefin 13(6):1454–1475

Caton PA, Hamilton LJ, Cowart JS (2011) Understanding ignition delay effects with pure component fuels in a single-cylinder diesel engine. J Eng Gas Turbines Power 133(3)

Chong CT, Ng J-H (2021a) Biojet fuel in aviation applications: production, usage and impact of biofuels. Elsevier

Chong CT, Ng J-H (2021b) Biojet fuel production pathways. In: Biojet fuel in aviation applications. Elsevier, pp 81–141

Chuck C (2016) Biofuels for aviation: feedstocks, technology and implementation. Academic Press

Conconi CC, Crnkovic PM (2013) Thermal behavior of renewable diesel from sugar cane, biodiesel, fossil diesel and their blends. Fuel Process Technol 114:6–11

Conesa JM, Morales MV, García-Bosch N, Ramos IR, Guerrero-Ruiz A (2023) Graphite supported heteropolyacid as a regenerable catalyst in the dehydration of 1-butanol to butenes. Catal Today 420:114017

Cortright RD, Davda RR, Dumesic JA (2002) Hydrogen from catalytic reforming of biomass-derived hydrocarbons in liquid water. Nature 418(6901):964–967

Cui Q, Hu Y, Yu L (2022) Can the aviation industry achieve carbon emission reduction and revenue growth simultaneously under the CNG2020 strategy? An empirical study with 25 benchmarking airlines. Energy 245:123272

Dahal K, Brynolf S, Xisto C, Hansson J, Grahn M, Grönstedt T, Lehtveer M (2021) Techno-economic review of alternative fuels and propulsion systems for the aviation sector. Renew Sust Energ Rev 151:111564

Davda RR, Shabaker JW, Huber GW, Cortright RD, Dumesic JA (2005) A review of catalytic issues and process conditions for renewable hydrogen and alkanes by aqueous-phase reform-

ing of oxygenated hydrocarbons over supported metal catalysts. Appl Catal B Environ 56(1–2):171–186
de Klerk A, Chauhan G, Halmenschlager C, Link F, Montoya Sánchez N, Gartley B et al (2023) Sustainable aviation fuel: pathways to fully formulated synthetic jet fuel via Fischer–Tropsch synthesis. Energy Sci Eng 91. https://doi.org/10.1002/ese3.1379
Demirbas A (2009) Progress and recent trends in biodiesel fuels. Energy Convers Manag 50(1):14–34
Dolah R, Zafar S, Hassan MZ (2022) Alternative jet fuels: biojet fuels' challenges and opportunities. In: Value-chain of biofuels. Elsevier, pp 181–194
Doliente SS, Narayan A, Tapia JFD, Samsatli NJ, Zhao Y, Samsatli S (2020) Bio-aviation fuel: a comprehensive review and analysis of the supply chain components. Front Energy Res 8
Dubois J-L, Segondy S, Postole G, Auroux A (2023) Isobutanol to isobutene: processes and catalysts. Catal Today 418:114126
Eloffy MG, Elgarahy AM, Saber AN, Hammad A, El-Sherif DM, Shehata M et al (2022) Biomass-to-sustainable biohydrogen: insights into the production routes, and technical challenges. Chem Eng J Adv 12:100410
Eswaran S, Subramaniam S, Geleynse S, Brandt K, Wolcott M, Zhang X (2021) Techno-economic analysis of catalytic hydrothermolysis pathway for jet fuel production. Renew Sust Energ Rev 151:111516
Geleynse S, Brandt K, Garcia-Perez M, Wolcott M, Zhang X (2018) The alcohol-to-jet conversion pathway for drop-in biofuels: techno-economic evaluation. ChemSusChem 11(21):3728–3741
Goh BHH, Chong CT, Ong HC, Seljak T, Katrašnik T, Józsa V et al (2022) Recent advancements in catalytic conversion pathways for synthetic jet fuel produced from bioresources. Energy Convers Manag 251:114974
Gunerhan A, Altuntas O, Caliskan H (2023) Utilization of renewable and sustainable aviation biofuels from waste tyres for sustainable aviation transport sector. Energy 276:127566
Guo X, Guo L, Zeng Y, Kosol R, Gao X, Yoneyama Y et al (2021) Catalytic oligomerization of isobutyl alcohol to jet fuels over dealuminated zeolite Beta. Catal Today 368:196–203
Gutiérrez-Antonio C, Gómez-Castro FI, de Lira-Flores JA, Hernández S (2017) A review on the production processes of renewable jet fuel. Renew Sust Energ Rev 79:709–729
Gutiérrez-Antonio C, Romero-Izquierdo AG, Gómez-Castro FI, Hernández S (2021) Renewable feedstock and its conversion routes to biojet fuel. In: Production processes of renewable aviation fuel. Elsevier, pp 33–54
Hamilton L, Luning-Prak D, Cowart J, McDaniel A, Williams S, Leung R (2014) Direct Sugar to Hydrocarbon (DSH) fuel performance evaluation in multiple diesel engines. SAE Int J Fuels Lubr 7(1):2014-01–1472
Han GB, Jang JH, Ahn MH, Suh Y-W, Choi M, Park N-K et al (2021) Operation of bio-aviation fuel manufacturing facility via hydroprocessed esters and fatty acids process and optimization of fuel property for turbine engine test. Korean J Chem Eng 38(6):1205–1223
Hanaoka T, Miyazawa T, Shimura K, Hirata S (2015) Jet fuel synthesis from Fischer–Tropsch product under mild hydrocracking conditions using Pt-loaded catalysts. Chem Eng J 263:178–185
He L, Yang J, Chen D (2013) Hydrogen from biomass. In: Renewable hydrogen technologies. Elsevier, pp 111–133
IRENA (2017) Biofuels for aviation: technology brief. International Renewable Energy Agency, Abu Dhabi. https://www.irena.org/-/media/Files/IRENA/Agency/Publication/2017/IRENA_Biofuels_for_Aviation_2017.pdf
Julio AAV, Batlle EAO, Trindade AB, Nebra SA, Reyes AMM, Palacio JCE (2021) Energy, exergy, exergoeconomic, and environmental assessment of different technologies in the production of bio-jet fuel by palm oil biorefineries. Energy Convers Manag 243:114393
Kazi FK, Fortman JA, Anex RP, Hsu DD, Aden A, Dutta A, Kothandaraman G (2010) Techno-economic comparison of process technologies for biochemical ethanol production from corn Stover. Fuel 89:S20–S28

Li L, Coppola E, Rine J, Miller JL, Walker D (2010) Catalytic hydrothermal conversion of triglycerides to non-ester biofuels. Energy Fuel 24(2):1305–1315
Li J, Yang G, Yoneyama Y, Vitidsant T, Tsubaki N (2016) Jet fuel synthesis via Fischer–Tropsch synthesis with varied 1-olefins as additives using CO/ZrO2–SiO2 bimodal catalyst. Fuel 171:159–166
Liu G, Yan B, Chen G (2013) Technical review on jet fuel production. Renew Sust Energ Rev 25:59–70
Lixiong Li (2008) US20080071125A1
Lv J, Wang D, Peng L, Guo X, Ding W, Yang W (2023) Ethanol dehydration to ethylene over high-energy facets exposed gamma alumina. Catalysts 13(6):994
Masih D, Rohani S, Kondo JN, Tatsumi T (2019) Catalytic dehydration of ethanol-to-ethylene over rho zeolite under mild reaction conditions. Microporous Mesoporous Mater 282:91–99
Mathes A, Ries J, Caton P, Cowart J, Luning Prak D, Hamilton L (2010) Binary mixtures of branched and aromatic pure component fuels as surrogates for future diesel fuels. SAE Int J Fuels Lubr 3(2):794–809
Meurer A, Kern J (2021) Fischer–Tropsch synthesis as the key for decentralized sustainable kerosene production. Energies 14(7):1836
Misra P, Alvarez-Majmutov A, Chen J (2023) Isomerization catalysts and technologies for biorefining: opportunities for producing sustainable aviation fuels. Fuel 351:128994
Monteiro RRC, dos Santos IA, Arcanjo MRA, Cavalcante CL, de Luna FMT, Fernandez-Lafuente R, Vieira RS (2022) Production of jet biofuels by catalytic hydroprocessing of esters and fatty acids: a review. Catalysts 12(2):237
Ng KS, Farooq D, Yang A (2021) Global biorenewable development strategies for sustainable aviation fuel production. Renew Sust Energ Rev 150:111502
Okolie JA, Awotoye D, Tabat ME, Okoye PU, Epelle EI, Ogbaga CC et al (2023) Multi-criteria decision analysis for the evaluation and screening of sustainable aviation fuel production pathways. iScience 26(6):106944
Oliveira EV (2014) Evaluation of hydrogen production from alumina-supported catalysts (master in chemical engineering). State University of Campinas, Campinas
Panpian P, Tran TTV, Kongparakul S, Attanatho L, Thanmongkhon Y, Wang P et al (2021) Production of bio-jet fuel through ethylene oligomerization using NiAlKIT-6 as a highly efficient catalyst. Fuel 287:119831
Pearlson M, Wollersheim C, Hileman J (2013) A techno-economic review of hydroprocessed renewable esters and fatty acids for jet fuel production. Biofuels Bioprod Biorefin 7(1):89–96
Schulz C, Kolb P, Krupp D, Ritter L, Haas A, Soorholtz M et al (2021) Preparation and high-throughput testing of TiO2-supported co catalysts for Fischer–Tropsch synthesis. Catalysts 11(3):352
Sobieralski JB (2023) Sustainable air transportation through the operational use of a social cost index. J Clean Prod 385:135663
Soh JC, Chong SL, Hossain SS, Cheng CK (2017) Catalytic ethylene production from ethanol dehydration over non-modified and phosphoric acid modified Zeolite H-Y (80) catalysts. Fuel Process Technol 158:85–95
Styskalik A, Vykoukal V, Fusaro L, Aprile C, Debecker DP (2020) Mildly acidic aluminosilicate catalysts for stable performance in ethanol dehydration. Appl Catal B Environ 271:118926
Suppes GJ, Storvick TS (2016) Energy and sustainability. In: Sustainable power technologies and infrastructure. Elsevier, pp 75–119
Su-ungkavatin P, Tiruta-Barna L, Hamelin L (2023) Biofuels, electrofuels, electric or hydrogen?: a review of current and emerging sustainable aviation systems. Prog Energy Combust Sci 96:101073
Tanksale A, Beltramini JN, Lu GM (2010) A review of catalytic hydrogen production processes from biomass. Renew Sust Energ Rev 14(1):166–182

Tanzil AH, Brandt K, Wolcott M, Zhang X, Garcia-Perez M (2021) Strategic assessment of sustainable aviation fuel production technologies: yield improvement and cost reduction opportunities. Biomass Bioenergy 145:105942
Teimouri Z, Abatzoglou N, Dalai AK (2021) Kinetics and selectivity study of Fischer–Tropsch synthesis to C5+ hydrocarbons: a review. Catalysts 11(3):330
Tiwari R, Mishra R, Choubey A, Kumar S, Atabani AE, Badruddin IA, Khan TMY (2023) Environmental and economic issues for renewable production of bio-jet fuel: a global prospective. Fuel 332:125978
Valero-Romero MJ, Rodríguez-Cano MÁ, Palomo J, Rodríguez-Mirasol J, Cordero T (2021) Carbon-based materials as catalyst supports for Fischer–Tropsch synthesis: a review. Front Mater 7
Villareal-Hernández AC, Ramírez-Mendiola MD, Quiroz-Ramírez JJ, Sánchez-Ramírez E, Huerta-Rosas B, Contreras-Zarazua G, Segovia-Hernández JG (2023) Intensification of the oligomerization and hydrogenation stage for biojet fuel production: preliminary outlines. Ind Eng Chem Res 62(22):8820–8833
Wang W-C, Tao L (2016) Bio-jet fuel conversion technologies. Renew Sust Energ Rev 53:801–822
Wang Y, Zhang Q, Hao J, Luo Z, Xiao J (2023) Global warming potential of bio-jet fuel produced by biomass aqueous-phase conversion. Sustain Prod Consumption 37:284–293
Wei H, Liu W, Chen X, Yang Q, Li J, Chen H (2019) Renewable bio-jet fuel production for aviation: a review. Fuel 254:115599
Weissermel K, Arpe H-J (2008) Industrial organic chemistry. John Wiley & Sons
Wen D, Jiang H, Zhang K (2009) Supercritical fluids technology for clean biofuel production. Prog Nat Sci 19(3):273–284
Yan Q, Yu F, Liu J, Street J, Gao J, Cai Z, Zhang J (2013) Catalytic conversion wood syngas to synthetic aviation turbine fuels over a multifunctional catalyst. Bioresour Technol 127:281–290
Yang J, Xin Z, He Q(S), Corscadden K, Niu H (2019) An overview on performance characteristics of bio-jet fuels. Fuel 237:916–936
Zhang Q, Xiao J, Hao J (2023a) Cumulative exergy analysis of lignocellulosic biomass to bio-jet fuel through aqueous-phase conversion with different lignin conversion pathways. Energy 265:126301
Zhang R, He Y, Luo Y, Lou D, Zhu R, Zhu C, Li Q (2023b) Selective preparation of jet fuels from low carbon alcohols and ABE at atmospheric pressure. Energy 281:128246
Zhao Z, Li Y, Zhu H, Lyu Y, Ding Y (2023) A review of CO/CO_2 C-based catalysts in Fischer–Tropsch synthesis: from fundamental understanding to industrial applications. Chem Commun 59(26):3827–3837

Chapter 4
Comparative Analysis of Biojet Fuel Production from Different Potential Substrates

Eduardo Dellosso Penteado, Anthony Andrey Ramalho Diniz, Yvan Jesus Olortiga Asencios, Luiza Helena da Silva Martins, Johnatt Allan Rocha de Oliveira, and Andrea Komesu

Abstract The aviation sector emits 2.5% of the global anthropogenic CO_2 emissions, which can rise by 3.6% annually and double over the next decades. Therefore, the decarbonization of the aviation sector is necessary. One possible solution to reduce the environmental impacts and the problems linked to the fossil fuel (price, availability, national security) in the aviation industry is the biojet fuel. The biojet fuel can be produced using a variety of biomass feedstocks that can be classified considering its chemical nature in triglyceride (soybeans, sunflower, castor bean, palm, rapeseed/canola, jatropha, camelina, algae), sugar and starch (sugarcane, corn, cassava), lignocellulosic (sugarcane bagasse, grasses, plant residues, wood), and wastes (municipal solid waste, sewage, gad flue). The choice of feedstock for biojet fuel production is a complex decision influenced by economic, environmental, and technological considerations. Each feedstock presents unique advantages and challenges, and the optimal choice may vary depending on regional conditions and priorities. Thereby, this chapter discusses each one of these raw materials, showing the scientific and technological advances reported in the literature. Moreover, a comparative analysis is done using environmental and economic aspects.

Keywords Aviation industry · Decarbonization · Raw material · Lignocellulose · Waste · Triglyceride · Sugar and starchy

E. D. Penteado (✉) · A. A. R. Diniz · Y. J. O. Asencios · A. Komesu
Universidade Federal de São Paulo, UNIFESP, Instituto do Mar, Santos, São Paulo, Brazil
e-mail: eduardo.penteado@unifesp.br

L. H. da Silva Martins
Universidade Federal Rural da Amazônia, UFRA, Instituto de Saúde e Produção Animal, Belém, Pará, Brazil

J. A. R. de Oliveira
Universidade Federal do Pará. UFPA, Faculdade de Nutrição, Belém, Pará, Brazil

A. Kuila (ed.), *Biojet Fuel: Current Technology and Future Prospect*, Clean Energy Production Technologies, https://doi.org/10.1007/978-981-99-8783-2_4

4.1 Introduction

The aviation sector will increase by around 5% annually until 2030 and double until 2050 (Ng et al. 2021). Moreover, this kind of industry emits 2.5% of the global anthropogenic CO_2 emissions (almost 1000 Mt. CO_2 per year) and 5% of the global greenhouse gases (CO_2 and non-CO_2 – NOx, water vapor) (Okolie et al. 2023; Su-ungkavatin et al. 2023). In the future, these emissions can rise by 3.6% annually and double over the next decades (Sobieralski 2023; Wang et al. 2023). Thereby, the decarbonization of the aviation sector is necessary (Cui et al. 2022; Sobieralski 2023).

One option to reduce the environmental impacts and the problems linked to the fossil fuel (price, availability, national security) in the aviation industry is the biojet fuel (Su-ungkavatin et al. 2023). According to the International Air Transport Association (IATA), the biojet fuel is the short- to medium-term solution to decrease the emissions of greenhouse gases (Doliente et al. 2020; Ng et al. 2021; Yang et al. 2019). Biojet fuel is a complex mixture of organic compounds (*n*-paraffins, iso-paraffins, olefins, naphthenes, and aromatics), whose carbon range distribution is from C8 to C16 (Emmanouilidou et al. 2023). The D7566 standards of the American Society for Testing and Materials (ASTM) establish a few technical specifications that biojet fuel must meet to be used due to the high safety standards and compatibility with aircraft fleet and refueling infrastructure (Emmanouilidou et al. 2023; Panoutsou et al. 2021).

The biojet fuel can be produced using a variety of biomass feedstocks that can be classified considering its chemical nature in triglyceride, lignocellulosic, sugar and starch, and wastes (Doliente et al. 2020; Gutierrez-Antonio et al. 2017). Nowadays, the triglyceride feedstocks, which include oil and fats that are edible, nonedible, and wastes, are the main raw material to produce biojet fuel (Emmanouilidou et al. 2023). However, with the increase in future demand, triglyceride feedstocks will not be enough to supply the market and lignocellulosic residues can contribute to sustainable biojet fuel production (Romero et al. 2022). Wood, husks, straws, grasses, and kernels are the most important lignocellulosic feedstock used to produce biojet fuel. Furthermore, sugarcane, corn, and cassava are classified as sugar and starch feedstocks, while municipal solid wastes, sewage, and flue gas are wastes. This variety of renewable raw material allowed different biojet fuel pathways, such as hydroprocessing of triglyceride feedstock, thermochemical processing of biomass, and alcohol-to-jet, direct sugar-to-hydrocarbon, and aqueous-phase reforming (Doliente et al. 2020; Gutierrez-Antonio et al. 2017).

Thereby, this chapter provides a deep discussion of the potential feedstock to produce biojet fuel, showing the scientific and technological advances reported in the literature. To make it easier for the reader to understand, raw material to produce biojet fuel was separated into sugar and starch feedstock (Sect. 4.2), triglyceride feedstock (Sect. 4.3), lignocellulosic feedstock (Sect. 4.4), and wastes (Sect. 4.5).

4.2 Sugars and Starch

4.2.1 Sugarcane

Around the world, in tropical regions of many countries, sugarcane *(Saccharum officinarum*) is grown over larger areas, mainly for being used for human consumption, through its transformation in sugar (sucrose). There is also the possibility of its application for the production of many other products, such as electric energy, organic chemicals, fuels, and paper, derived directly from it or its resulting products (Renouf et al. 2010).

The jet fuel demand has been increasing as a natural result of the aeronautical industry being fast growing, which leads to concerns about the demand for fossil fuels and their atmospheric emissions. In this scenario, the use of biofuels produced from sugarcane feedstock can be considered a part of the solution as one of the main advantages of using it can be the possibility of growing it on a large scale (Escalante et al. 2022).

Alkane-type fuels can be produced directly from sugars, instead of first converting them to ethanol intermediate, called direct sugar-to-hydrocarbons (DSHC), and it was named by the ASTM as synthetic iso-paraffin from fermented hydroprocessed sugar (SIP). It can be also found in the literature as direct fermentation of sugar-to-jet (DFSTJ). DSHC is a technology, from the economic point of view, mainly affected by feedstock. Hence, the studies indicate the best feedstock for this pathway is sugarcane (Wei et al. 2019).

Among the pathways certified by the ASTM, there is also another one known called alcohol-to-jet (ATJ), which involves converting the biomass into alcohol that can be processed into long-chain hydrocarbons (SPK), which will be used as the desired aviation fuel. For this technology, numerous types of feedstocks can be applied, sugarcane being one of those options (Okolie et al. 2023). Yao et al. (2017) conducted a techno-economic study of the ATJ pathway, where they tested three different feedstocks, with sugarcane among their options. Finally, they concluded that sugarcane was the most cost-effective and environmentally sustainable feedstock for that technology.

A derived potential feedstock is the sugarcane bagasse (SCB) due to its chemical composition, which is typically rich in cellulose (44%), but it also contains hemicellulose (28%), lignin (21%), ashes (5%), and extractive (2%). However, before it is used in the production of biochemicals and biofuels, its de-lignification is necessary to make it more susceptible to enzyme attacks, followed by pretreatment and hydrolysis (Ajala et al. 2021). Luo et al. (2023) demonstrated the possibility of producing biojet fuels from bagasse through the integration of bio- and chemical catalysis reaction processes. Finally, they obtained high selectivity of jet-range fuels (83.0%) and high conversion of acetone/butanol/ethanol (95.3%).

4.2.2 Corn

Corn (*Zea mays*) is an important agricultural crop grown worldwide as one of the major cereal crops (Ruan et al. 2019). In 2017, its production was around 1.03 billion tons, with more than 80% being produced by eight countries/regions, the United States leading (37% of total), followed by China (21%), Brazil (8%), the European Union (6%), Argentina (4%), Ukraine (3%), India (2%), and Mexico (2%), which means that just three countries (the United States, China, and Brazil) are responsible for about two-thirds of the world's production (Ruan et al. 2019; da Silva et al. 2022).

The literature indicates that one of the major challenges for the wide adoption of the biojet fuels is their competitiveness with the price of conventional jet fuels, which implicates a demand for strategies to reduce their production costs. An alternative to this question is the integration of the production of these sustainable aviation fuels with existing industrial plants, aiming at cost reduction, in general. In this scenario, corn ethanol is one of the industries that make possible this integration for producing sustainable aviation fuel (Tanzil et al. 2021).

Staples et al. (2014) evaluated the greenhouse gas (GHG) emissions in the life cycle of some raw materials that could be a candidate for the production of biojet fuel, involving the corn grain. Considering seven stages and a reference baseline, they verified that the cultivation and production of biojet fuels were the stages with the greatest contribution to the life cycle of greenhouse gas emissions. Finally, those authors concluded that, in the assumption of the baseline corn, its contribution to the life cycle GHG emissions decreased by 30% compared to the conventional jet fuel (Escalante et al. 2022; Staples et al. 2014).

According to Yoo et al. (2022), the life cycle GHG emission of sustainable aviation fuel (SAF) derived from corn grain ethanol is 26% lower than the emission of the jet fuel produced from petroleum, and they also indicate that this feedstock is the dominant and maturely used to develop biofuel in commercialized biorefineries. Hence, they indicate that it is worth paying attention to the application of this feedstock in green technologies for lowering the life cycle of GHG emission.

4.2.3 Cassava

In over 100 tropical and subtropical countries, cassava (*Manihot esculenta*) is mainly cultivated by family farmers, who use it as a major source of energy as its roots are rich in carbohydrates (Tiago et al. 2020). It has many local varieties, which can be adapted to different regions, and has been used for decades as a source of starch that can be applied as raw material in food, paper, textile, and many other industries (Ogundari et al. 2012; Tiago et al. 2020).

In Thailand and southern Brazil, cassava is widely cultivated, as a single crop (Olusola Sanusi et al. 2023). Escalante et al. (2022) evaluated the potential feedstock for the biojet production, focusing on the Brazilian context. In their research, they did not focus, specifically, on the application of cassava as feedstock. However, they mentioned that EMBRAPA – Agroenergy identified approximately nine potential feedstocks that can be used in biojet fuel, including cassava, which indicates that it can be a promising option for future research, in Brazil, aiming at increasing its biojet production.

4.3 Triglyceride

Nowadays, agriculture is facing big challenges such as increasing productivity without environmental footprints, facing climate change, and providing renewable fuels.

To avoid climate change, a deep change in the energy sources that move the current society is needed; in this scenario, many countries signed international agreements, thus one of these actions is replacing fossil fuel-derived jet fuels to sustainable fuels (Ortiz et al. 2020). Plant oils are potential candidates to replace fossil oil due to the similarities in their chemical structures and physicochemical properties.

The production of biojet fuels through the conversion of vegetable oils and animal fats is known as hydroprocessed esters and fatty acids (HEFA) route; the HEFA received approval from the ASTM in 2011 (Chuck 2016). Its use is limited to a 50% mixing ratio with conventional jet fuel because of the lack of aromatics in the composition of this biofuel. HEFA is the technology with the greatest development, being the most widely used by the industry (Vasquez et al. 2017). Figure 4.1 summarizes the HEFA process.

The biojet fuels obtained through this process have properties that are advantageous for higher altitude flight, characterized by low temperatures. HEFA fuels possess low lubricity due to the absence of O and S. As its cetane number is lower than conventional jet fuels, the ignition can be affected; however, these problems can be addressed by blending with conventional jet fuel (Escalante et al. 2022). The presence of aromatics is the main difference between conventional jet fuel and biojet fuels as the latter lacks them in their composition; this is the principal reason for restricting their use in certain mixing ratios to ensure fuel compatibility with the aircraft engine.

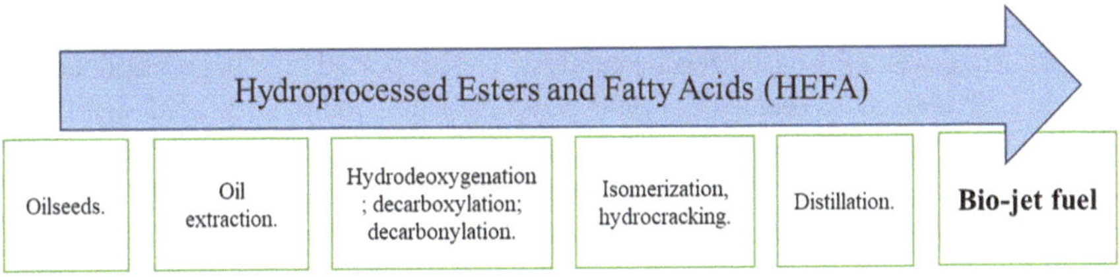

Fig. 4.1 ASTM-approved HEFA technology route for biojet production. (Adapted from Chuck 2016)

Some of the oilseed crops currently used for commercial HEFA production include camelina, sunflower, soybean, jatropha, palm, canola, coconut, cotton, castor bean, among others. It is important to consider, among the sustainability of its oil extraction, that there should be an economic advantage to using oilseed crops instead of fossil fuels. This chapter is dedicated to some of these sources such as soybeans, palm, sunflower, and castor bean.

Chu et al. (2017), Eller et al. (2016), Wu et al. (2017) reported the biojet fuel process through the hydroprocessing of various oilseeds using various catalysts. The raw materials used by Chu et al. (2017) were camelina, carinata (nonedible oil), and used cooking oil (UCO); they used a nickel–molybdenum-based catalyst. Wu et al. (2017) studied the catalytic cracking of vegetable oils (composed principally of palmitic acid, stearic acid, oleic acid, linoleic acid, and linolenic acid) using zeolite (HZSM-5) as a catalyst. Similarly, Eller et al. (2016) used coconut oil as a starting material, with sulfided $NiMo/Al_2O_3$ as a catalyst. The important aspects studied in these works include feedstock composition, hydrogen consumption, energy demand, and process selectivity. In general, their yields achieved higher than 60%.

Hydroprocessing means a treatment for the removal of oxygen from fats and oil in the presence of hydrogen as ASTM D7566 and ASTM D1655 stated that the composition of final biofuel must contain at least 99.5% of carbon and hydrogen (Why et al. 2019). Hydroprocessing is a key process for the HEFA (Fig. 4.1). This process includes several catalytic reaction mechanisms. This technology is more advantageous compared to other conversion pathways since it is flexible and can use low-quality feedstock. Furthermore, the separation stage of the by-products is not so complex compared to, for example, transesterification technology, which needs a further purification process (Why et al. 2019).

4.3.1 Soybeans

The oil extracted from soybean has a huge potential for the HEFA process to produce biojet fuel because of its low cost, developed agriculture, and available facilities. However, it is important to consider that the oil extracted from soybeans is the main feedstock for the production biodiesel (e.g., in Brazil); thus, it is very important to evaluate alternative potential crops (de Souza et al. 2020).

Choi et al. (2015) studied the conversion of waste soybean oil and palm fatty acid distillates in a single-step reaction, without the addition of hydrogen, and the reaction occurred over a Pd-based zeolite catalyst. They reported high degrees of deoxygenation of wasted soybean oil (yield = 95.5%) and palm fatty acid distillates (yield = 94.3%).

Scaldaferri and Pasa (2019) studied the production of biofuel jet from oil extracted from soybean; the reaction occurred over $NbOPO_4$ catalyst, and in their research they used mild experimental conditions in a one-step process. The yield for biojet fuel was 62%. Among biojet fuel, biogasoline and green diesel were obtained.

4.3.2 Palm

Palm oil is the second largest source of edible oil. It is a multipurpose vegetable oil, with products ranging from food to biodiesel. Currently, vegetable oils such as palm oil are used by some industries to produce biodiesel. Although this diesel-oriented production represents only a small part of the total world production, its use for this purpose is very attractive as it can be considered an alternative to soybean (Arunachalam 2012; Tan et al. 2009).

Palm oil is rich in long-chain saturated fatty acids (such as palmitic and stearic acids, in addition to oleic acid). Palm oil can be a promising feedstock to produce biojet fuel as biojet fuel is compatible with Jet A-1 commercial fuel standard (Why et al. 2022).

Lin et al. (2020) produced biojet fuel from palm oil through hydroprocessing and hydrocracking/isomerization. Palm oil was first hydroprocessed given mainly alkanes. The latter was transformed into jet fuel-range products through further hydrocracking/isomerization processes, and the reaction occurred over the Ni–Ag supported on silico-aluminio-phosphates (SAPO-11) catalyst. The effects of the reaction parameters on the product distributions were investigated. Lin et al. (2020) reported that at high temperatures due to the occurrence of cracking, the contents of C15 to C18 decreased, and the ones of C8 to C14 increased.

Why et al. (2022) produced biojet fuel from different types of palm kernel oil (PKO). The deoxygenation process was carried out over Pd/C catalysts at 400 °C for 2 h. They concluded that at 8 wt% loading it achieved the highest selectivity of jet paraffins (96% liquid product containing *n*- and iso-paraffins, olefins, naphthenes, and aromatics; and 73% of jet paraffins selectivity). The physicochemical properties of produced jet fuel such as density, kinematic viscosity, cloud point, smoke point, pour point, flash point, and final boiling point obey the standard Jet A-1 fuel established in the ASTM standards.

4.3.3 Sunflower

Sunflower oils are rich in oleic acid content. Sunflower is the third most produced oilseed in the world and is one of the most important oilseed meal feed sources. The sunflower oil industry maintained its competitiveness in oilseeds markets during the last decades influenced by its continuous innovation (Zhao et al. 2016).

In the study reported by Zhao et al. (2015), nonedible sunflower oils that were catalytically cracked in a fixed-bed reactor, the reaction was carried out at three different reaction temperatures: 450 °C, 500 °C, and 550 °C, and over a ZSM-5 catalyst. They concluded that the reaction temperature influences the yield and quality of liquid products. The highest conversion efficiency was 30.1% at 550 °C. They observed that the reaction temperature affected the composition of the noncondensable gases.

4.3.4 Castor Bean

Castor bean is a perennial plant found in subtropical and tropical regions. The best-known commercial source of hydroxy fatty acid (HFA) is ricinoleic acid (12(R)-hydroxy-octadec-*cis*-9-enoic acid; 12-OH 9c-18:1), which is obtained from castor oil (*Ricinus communis L.*; Euphorbiaceae). This makes castor oil essential to the chemical industry (Kenar et al. 2017). Castor oil is used in coatings, paints, lubricants, inks, and a wide variety of products.

Castor bean is a promising feedstock for biojet production given its relatively high yield and because it is a nonfood oil source. However, because it is the principal commercial source of hydroxy fatty acid, castor bean oil is a valuable feedstock for the industry; therefore, a higher price than other seed oils is expected (Tao et al. 2017).

Liu et al. 2015 synthesized biojet fuel by hydroprocessing of castor oil in a continuous-flow fixed-bed reactor, obtaining high yields. The highest yield reached 91.6 wt% of alkane, with a high isomer/*n*-alkane ratio, and they used a Ni/acidic zeolites catalyst. They also observed that the degree of hydrodeoxygenation (HDO) and hydrocracking influences the content of alkanes in the final product.

4.3.5 Rapeseed/Canola Seed

Rapeseed (*Brassica napus*) is the third largest oilseed crop in the world, considered healthy for cooking, especially for its beneficial balance of fatty acids (Cisneros-Yupanqui et al. 2021). It has been traditionally grown for the animal feed and production of vegetable oil for human consumption (Diaz et al. 2010). However, in the last years, rapeseed has been considered a promising raw material for biodiesel production and biojet fuel due to the high oil content (40%) (Diaz et al. 2010). Moreover, rapeseed oil has a high energy content and relatively low viscosity (Shi et al. 2017). Therefore, rapeseed oil stands out as an ideal feedstock for biojet fuel due to its wide availability and favorable properties.

The production of biojet fuel from rapeseed oil involves a process known as hydroprocessing (HEFA), removing impurities and adjusting its molecular structure (Cheng et al. 2014). The resulting biojet fuel can be used as a drop-in replacement for conventional jet fuel, requiring no modifications to existing aircraft engines or infrastructure. During the use in the turbine, the biojet fuel made from rapeseed emits harmful particulates and sulfur compounds, which could improve the air quality around airports (Labeckas and Slavinskas 2015).

Numerous researchers have highlighted the comparatively low carbon footprint and comprehensive eco-friendliness exhibited by biojet fuel from rapeseed oil (Obnamia et al. 2020; Shi et al. 2017; Ukaew et al. 2016). The greenhouse gas emission of biojet fuel produced from rapeseed using the HEFA can vary from 36–51 g

CO_2e/MJ (Ukaew et al. 2016) and −55 to −107 g CO_2e/MJ (Shi et al. 2019) depending on the agricultural methodologies and geographical settings.

One concern is the competition between biofuel production and food production (Tao et al. 2017). As rapeseed oil is also used for culinary purposes, there is a need to balance its use in both industries to ensure food security. Moreover, the relatively high cost of this raw material is another drawback (IRENA 2021).

4.3.6 Jatropha

Jatropha curcas L. is a member of the Euphorbiaceae family and is categorized as an inedible oilseed. Its tree can withstand challenging conditions like high temperatures and low humidity, and its seeds are extremely rich in oil (about 36 g/100 g). This culture has grown all over the world, even in regions with challenging climates, so it adapts quickly to these environments. *Jatropha* cultivation does not compete with the area used for growing food crops because it only grows in arid regions (Arockiasamy et al. 2021).

This plant can be used as a source of biofuel because it is rich in oils and does not produce a significant amount of greenhouse gases (Escalante et al. 2022). In fact, there is scientific proof that using these kinds of biofuels actually lowers the emission of these gases. Based on this culture, the aviation industry is one of the applications for these biofuels, making it a sustainable and environmentally friendly method to use these fuels (Arockiasamy et al. 2021). dos Santos et al. (2017) observed that *Jatropha* is the more viable feedstock for biojet fuel production in Brazil.

The minimum selling price stands around U$ 1.78 per liter for biojet fuel produced using *Jatropha* as feedstock (Wang 2016). This price could vary depending on feedstock costs, refinery capital cost, co-products credits, and energy cost (Escalante et al. 2022).

4.3.7 Camelina

Camelina is an oleaginous plant, similar to *Jatropha curcas L*, that is neither useful for food production nor competes with land that will be utilized for it. It is a fairly widespread plant in Canada and is known to withstand harsh weather conditions like cold and drought. It is also intriguing since it requires no costly agricultural inputs (Li et al. 2018).

When compared to conventional fuels generated from petroleum, its raw material potential for the use of fuels that may be employed in the aviation sector showed encouraging results with regard to greenhouse gas emissions, revealing that there was a reduction in the range of 65–97% (Li et al. 2018).

As they conformed with the law, many businesses have already started utilizing camelina fuel, which is a potential raw material for aviation that may be used with drop-in aviation using hydroprocessed esters and fatty acids (Li et al. 2018).

4.3.8 Algae

When compared to plant cultures, algae are even superior as a source of raw materials for the production of biofuels (Doliente et al. 2020; Lim et al. 2021). Among their advantages are their ease of cultivation and high oil yields, as well as the fact that they can be grown without needing any land at all, including using wastewater (Lim et al. 2021). They can also help reduce greenhouse gas emissions (O'Neil et al. 2019). Algae farming is a recognized and developing technique that does not compete with food culture and is highly promising for use in biojet (O'Neil et al. 2019).

Proteins, carbohydrates, and lipids make up algae, and the latter is particularly promising for use in biojets. Both dry and wet seaweed may be treated, and this oil can be recovered by separating it from the other molecules that will be applied. Aviation fuel hydrocarbons can be separated from their product or manufactured using a variety of methods, each with pros and cons (O'Neil et al. 2019). For instance, when algae are transformed into lipids, they undergo hydrocracking (isomerization) through the HEFA to create the biojet fuel (Gutierrez-Antonio et al. 2017; Tao et al. 2017). On the other hand, if algae are converted into bio-oil through liquefaction, this bio-oil is subsequently subjected to a hydrotreatment process to generate the biojet fuel (Chiaramonti et al. 2017). When the algae feedstock is converted into syngas via gasification, the resulting syngas is employed in the Fischer–Tropsch technique to produce the biojet fuel (Elkelawy et al. 2022). Lastly, in cases where algae are transformed into sugars through a pretreatment process, these sugars are then subjected to fermentation to produce the biojet fuel via alcohol to jet (Elkelawy et al. 2022).

An economic evaluation of biojet fuels derived from microalgae projected a minimum selling price of U$ 8.46 per liter (in 2011) (Wei et al. 2019). This price can decrease to U$ 2.43 per liter (in 2011) once the technology advances and the market evolves (Doliente et al. 2020; Wei et al. 2019). There are several issues in cultivation, harvesting, and oil extraction technologies, which are still inefficient and/or capital- and resource-intensive (Doliente et al. 2020).

Several studies reported the impact of using algae biomass as feedstock to biojet production (Agusdinata et al. 2011; Bennion et al. 2015; Fortier et al. 2014; Lim et al. 2021; Wei et al. 2019). The amount of the GHG emission can be reduced by as much as 90% by using algae-based biojet fuel compared to fossil jet fuel (Bennion et al. 2015; Fortier et al. 2014).

Genetically modifying algae and microalgae species is again interesting due to their potential to directly synthesize biojet fuel (Lim et al. 2021). Another interesting point of investigation is the adjustment in the cultivation conditions of these species leading to enhanced biomass, lipids, or targeted compounds improving the biojet fuel yields and reducing the production costs (Doliente et al. 2020). Concurrently, efforts must also address the challenges of optimizing the oil extraction process to curtail energy consumption. Presently, this factor constitutes a substantial proportion of overall processing expenses (Wei et al. 2019).

4.4 Cellulosic Materials

4.4.1 Sugarcane Bagasse and Trash

The sugarcane industry produces in great excess by-products such as straw and sugarcane bagasse; these materials are considered lignocellulosic and have great potential for application in sustainable processes such as the production of biofuels. The bagasse comes from the cane after its grinding, consisting of a fibrous residue, whereas the straw comes from the leaves, tips, and other parts of the plant that are not harvested. The composition of this lignocellulosic material is what makes it promising since it consists of a structure composed of cellulose, hemicellulose, and lignin, the first and second are polysaccharides that can be fermented and/or transformed into biofuels, while the third is a structure of phenolic nature that can be applied for the production of phenolics (Carvalheiro et al. 2008).

The use of this material includes the collection and preparation of both bagasse and sugarcane straw; hence, there is a need for treatments that can deconstruct the recalcitrance of this material so that it can be applied for the purpose of biofuels, and thus, we have several examples of pretreatment such as chemical, physical, and biological (Dias et al. 2012).

After the pretreatment stage, saccharification techniques can be used to convert polysaccharide sugars into fermentable monomers, so that enzymatic or acid hydrolysis techniques can be used, and the sugars obtained will be used to produce ethanol through fermentation through yeast. Thus, we will have biologically based products, such as biofuels (ethanol) or biochemicals (organic acids) (Galbe and Zacchi 2012).

Straw and sugarcane bagasse are renewable resources; thus, using them in these processes has various advantages in terms of waste reduction and environmental effects. Furthermore, by converting trash into useful resources, this strategy adheres to the ideals of a circular economy (Galbe and Zacchi 2012).

Many businesses are working on the production of these fuels using biomass, and the products obtained meet the ASTM D7566 specifications and can be used in aircraft turbines; hence, the jet fuel/ATJ mixture is already being used in commercial airlines. The alcohol obtained from biomass is promising for use in aviation through transformation pathways where these alcohols will form a product rich in hydrocarbons that are used by airplanes (Escalante et al. 2022).

4.4.2 Grasses

Like many other biomasses, grasses are also materials rich in complex carbohydrates, which, after transformation, can give rise to high-quality fuel for aircraft. The current production of sustainable aviation fuel still needs to be more efficient, much more for land use than for large scale. Thus, Uludere Aragon et al. (2022) carried out a study seeking to evaluate the use of grasses and land use, seeking more efficient help of the land for the production of biomass used in the production of the biojet. In that same study, good results were observed for crops such as grasses, miscanthus, and switchgrass, given the possible harmful effects on the surrounding climate or soil moisture, which showed interesting results regarding the final production cost of the fuel.

As in many other processes, the use of biomass has an almost always standard sequence for use, which includes pretreatments, enzymatic hydrolysis, and fermentation, recovery, and purification stages that make obtaining fuels viable or not from the point of view of production (de Oliveira et al. 2021). This fact is no different for the production of biojet from such biomasses, as is the case of grasses.

There is much interest in the use of grasses for the production of fuel, including biojet, since this material has exciting characteristics from a productive point of view as they grow quickly without significant challenges in terms of cultivation and harvesting. In addition, this material can be cultivated in several areas worldwide in the most different types of soil (Tye et al. 2016).

Producing biojet fuel from this type of material can enable more efficient and sustainable aviation, given that the sector requires high fuel demands (Cervi et al. 2021). Despite the production of biojet from biomass such as grasses still being very recent, good initiatives are already found, with policies suggesting using such biomass for this production. This fact can be proven from the indications of the use of this biomass for the production of biojet by the International Civil Aviation Organization (ICAO), in which the body suggests the culture of grasses due to the noncompetition for its use as a food crop (ICAO 2013). In addition, the Sustainable Aviation Fuel Users Group (SAFUG) declared that not compromising the availability of water or biodiversity from the use of this biomass is an additional boost for the use of this biomass and emphasized the importance of using sources of biomass without compromising water availability or biodiversity (SAFUG 2018).

Aviation biofuel is seen as an emerging bioenergy supply chain, which may require large amounts of biomass resources in the coming years. Although, globally, biojet fuel production is currently at a development stage, many dedicated initiatives and policy statements have already suggested the conditions for using biomass for this purpose. For example, the International Civil Aviation Organization (ICAO) has indicated that biomass crops for biojet fuel should not compete with food crops. The Sustainable Aviation Fuel Users Group (SAFUG) has emphasized using biomass sources without compromising water availability or biodiversity (SAFUG 2018).

Biojet production must use biomass always available at sufficient production levels to meet a large production volume, which is precisely the cause of the change in supply throughout the year seasonality (Richter et al. 2018). In the case of grasses, if they are not edible, the biggest problem with their use would be their availability in the current scenario. In addition, there would be a lack of production to support the pasture and fuel industries during periods of drought (Herr et al. 2016). Climate change can cause changes in rainfall patterns, promoting changes and increasing the complexity of the possible use of this material for fuel production (Perkis and Tyner 2018).

4.4.3 Plant Residues

Plant residues, in general, are rich in carbohydrates, mainly cellulose and hemicellulose, which can be converted into fermentable monosaccharides and thus produce fuels. Based on this, new technologies have emerged, focused on producing fuels for sectors with high demand and specific properties, as is the case of biojet fuel.

Around the world, many airlines have successfully tested biofuels to make the sector more sustainable (Gutierrez-Antonio et al. 2017; Wei et al. 2019). In 2008, Virgin Atlantic Airlines became the first to carry out a flight using the biojet; since then, several companies have carried out the same type of test, as was the case of Turkish Airlines in 2022, which used biojet produced from algae (Sharno and Hiloidhari 2022).

There are many plant residues with properties that qualify for the production of biofuels (Hao et al. 2021). These materials can be divided into four groups. First, the materials directed to the first-generation production (sugarcane, beet, oil palm). Second, materials used for production called second-generation. In this group, we can find biomass residues, agricultural and forestry residues, the so-called plant wastes. Furthermore, the material for the third generation is found in the last two groups, such as algae and microalgae, and for the fourth generation, which includes fungi and bacteria, in addition to genetically modified microorganisms (Doliente et al. 2020).

The use of biomasses, such as plant wastes, starts already in the cultivation stage and is followed by the transport of the material from the field to the conversion sites, the plants, and ends in the biojet conversion stage, in the conversion plants (Sharno and Hiloidhari 2022). For this large volume of material to be used, most of them (biomass and forest) need to have their seasonality evaluated, as this is what will determine the availability of such materials (Caputo et al. 2005; Madlener and Bachhiesl 2007; Nilsson and Hansson 2001). This phenomenon is limited to harvest periods, climatic conditions, and the need to plant and replant fields and forests (Wood and Layzell 2003).

4.4.4 *Wood and Wood By-Products*

The biojet fuel is considered in the long and medium term as a possible alternative to using fossil fuels in aircraft. However, the production volume capable of meeting the great demand for the sector still needs to be increased and corresponds to only 0.5% of the identified demand (van Dyk and Saddler 2021). These limitations occur since many production processes occur by processing oils into esters and hydrotreated fatty acids (HEFA). However, the oil required for this production is limited. Another fact is due to the lack of policies that encourage the development of technology and the different characteristics of the biojet (Bjornsson and Ericsson 2022).

The production of fuel from it occurs through residues from the silviculture and forestry industries, mainly crowns and branches, after removing the main parts to produce cellulose and paper. The second most abundant source of wood fuel is saw-mill residues, which move a total volume of around 2.3 Mt. DM, or 43 PJ, annually (De Jong et al. 2018).

The use of silviculture residues for the production of biofuels is still limited. However, the wide availability of this material encourages its use for these purposes in Sweden, including the production of biojet, and the study of this use on a pilot scale needs more stimuli and on a priority basis (Statens Offentliga Utredningar 2019).

4.5 Wastes

4.5.1 *Municipal Solid Wastes*

The term municipal solid waste (MSW) is used to describe recyclables, compostable materials, and garbage from houses, businesses, institutions, construction, and demolition sites (Rahman et al. 2022). The composition of MSW varies greatly between regions and includes a wide variety of organic and inorganic waste (Dornau et al. 2020). The organic content of MSW in developed nations is about 32%, but it is over 50% in developing nations. It raises important questions about waste management strategies (Bhattacharjee et al. 2023).

Every year, just over 2 billion tons of MSW are produced worldwide. By 2050, MSW volumes are expected to reach 3.4 billion tons per annum due to population growth, industrialization, and urbanization (Dornau et al. 2020; Kaza et al. 2018). The accumulation of MSW leads to air pollution, water pollution, and greenhouse gas emissions (GHG), and takes up valuable space in landfills. Hence, properly managing MSW is essential for protecting the environment and public health.

Converting complex feedstocks like MSW into transportation fuels like gasoline, diesel, and jet fuel is becoming increasingly attractive as conversion technologies and processes advance (Rahman et al. 2022). The American Society for Testing and Materials (ASTM) has approved the use of renewable feedstocks (agro-waste and MSW) to produce jet fuel (Morgan et al. 2019; Sajid et al. 2022).

MSW can be converted into jet fuel by gasification through the Fischer–Tropsch process and/or some other thermochemistry/biochemistry routes (Wei et al. 2019). Emmanouilidou et al. 2023 performed a systematic review using PRISMA. For bio-jet fuel conversion, catalytic hydroprocessing of waste lipid feedstocks is the most commonly used. The catalytic pyrolysis of waste plastics and co-pyrolysis with solid biomass residues can also contribute to the development of cost-effective technologies and effective policy support. In addition, gasification, coupled with Fischer–Tropsch and alcohol-to-jet processes, proved to be an excellent pathway for developing sustainable aviation fuel (Emmanouilidou et al. 2023).

Harisankar and Vinu (2023) reviewed the feasibility of hydrothermal treatment of heterogeneous and co-mingled waste feedstocks, and the scale-up challenges associated with it. It has been shown in existing studies that the hydrothermal liquefaction treatment of lignocellulosic biomass, as well as algal biomass, is a sustainable way to produce biofuels. However, the literature on the hydrothermal treatment of waste feedstocks such as MSW is relatively lacking (Harisankar and Vinu 2023).

A study by Suresh (2016) examined the life cycle greenhouse gas emissions and economic feasibility of middle distillate fuels derived from MSW, including diesel and jet fuel, through three thermochemical conversion pathways: conventional gasification and Fischer–Tropsch (FT MD); plasma gasification and Fischer–Tropsch (plasma FT MD); and conventional gasification, catalytic alcohol synthesis, and alcohol-to-jet upgrading (ATJ MD). According to this analysis, diesel and jet fuels produced from MSW can reduce the GHG emissions intensity of transportation, but policy mechanisms may be necessary to ensure economic viability (Suresh 2016).

Despite the potential reported in the literature, the challenges of MSW jet fuel production are the highly inconsistent and heterogeneous composition of the raw material that affects the product quality and yield.

4.5.2 Sewage

Municipal and industrial wastewater treatment plants produce sewage sludge, a solid by-product that is lipid-rich, widely available, and has a negative cost (Hari et al. 2015). Municipalities incur a disposal cost of US$200–600 per dry ton (or US$40–120 per wet ton, given an assumed solids content of 20%), which means the cost of fuels derived from this material can potentially be offset by US$1.5–7/gallon if this material is used as a feedstock (Cronin et al. 2022). In addition, the use of waste materials for biofuel production can overcome a number of difficulties, for example, needing fertilizer, irrigation, land, and labor (Hari et al. 2015; Saynor et al. 2003).

Farooq et al. (2020) studied the feasibility of an integrated hydrothermal liquefaction (HTL) plant in the United Kingdom. Using the Aspen Plus simulation approach, an integrated HTL plant with a feed throughput of 10 t h^{-1} was modeled. Three HTL configurations are considered for the assessment of technical–economical, regional resource, and carbon footprint, such as a base case without energy and resource recovery, an HTL with heat integration, and an HTL with energy and resource recovery. Feedstocks such as algae, food waste, and sewage sludge were investigated, with sewage sludge having the lowest minimum fuel selling price. As a result of heat integration, heating and cooling utilities are reduced by 96.4% and 77.8%, respectively. Additionally, heat integration and resource recovery reduce the minimum fuel selling price by 10.5%. This technology is capable of meeting 22.8% of the UK jet fuel demand, according to the regional resource assessment. According to the carbon footprint assessment, the technology can reduce CO_2 emissions by 18.3% compared to current aviation emissions at maximum production (Farooq et al. 2020).

In another study with hydrothermal liquefaction, Chiaberge et al. (2021) explored the possibility of co-distilling a blend of fossil crude with hydrothermal liquefaction (HTL) biocrude from primary sewage sludge. It was observed that biocrude contributes mainly to high boiling point fractions, particularly diesel and residue, and that kerosene also has a significant contribution. The distilled fractions, however, contained significant amounts of nitrogen, which corresponded to compounds that were resistant to hydrotreating with a different carbon number and double-bond equivalent (DBE). This issue could be controlled by reducing the blending ratio or with specific upgrading treatments. The co-distillation of HTL biocrude with fossil fuels is, therefore, an attractive method for introducing renewables into existing refineries.

Cronin et al. (2022) studied the hydrothermal liquefaction from food waste, sewage sludge, and fats, oils, and grease. According to the findings of this work, the upgraded HTL biocrude material shows key fuel properties, such as carbon number distribution, distillation profile, surface tension, density, viscosity, heat of combustion, and flash point, which all fall within the range required for aviation fuels.

Bashir et al. (2022) converted sewage sludge into sustainable jet fuel-range hydrocarbons (C8–C16). The biocrude oil was produced from sewage sludge in a thermo-catalytic reforming (TCR) system (2 kg/h) at 450 °C pyrolysis and 700 °C post-reforming. In a subsequent two-step hydroprocessing process, which was carried out in a bench-scale batch high-pressure reactor, the biocrude oil was hydrodeoxygenated and hydrocracked separately. H_2 pressure was varied while temperature, feed volume, catalyst loading, and batch time remained constant. Thus, at 60 bar H_2, about 25% by weight of hydroprocessed oil was recovered via atmospheric distillation, including normal, cycle, and iso-paraffins and aromatic oils between C8 and C16. Most of the specifications for jet fuel were met by sewage sludge-derived range fractions.

4.5.3 *Flue Gas*

The flue gas (also called exhaust gas or stack gas) is the gas that is emitted from a combustion plant. It is composed of reaction products of fuel and combustion air, as well as residual substances like particulate matter (dust), sulfur oxides, nitrogen oxides, and carbon monoxide (Speight 2019).

In recent years, the conversion of CO_2 into fuels and high-value chemicals has attracted widespread attention as it contributes to mitigating greenhouse gas emissions while also producing valuable chemicals (Yao et al. 2020).

Using novel, inexpensive iron-based catalysts, Yao et al. (2020) developed a synthetic protocol for fixing carbon dioxide directly into aviation jet fuel. Fe–Mn–K catalyst was prepared using the organic combustion method. It converted carbon dioxide into hydrocarbons at a rate of 38.2%, yielded 17.2%, and showed a 47.8% selectivity, along with a low selectivity for carbon monoxide (5.6%) and methane (10.4%). Additionally, a conversion reaction produces ethylene, propylene, and butenes, totaling 8.7% of the yield, which are raw materials for petrochemicals and are currently derived from fossil crude oil (Yao et al. 2020).

Using layered double-hydroxide precursors, Zhang et al. 2021 produced a high-selectivity Na-modified CoFe alloy catalyst capable of converting CO_2 directly into a jet fuel containing C8–C16 hydrocarbons. With a temperature of 240 °C and pressure of 3 MPa, this catalyst achieves 63.5% selectivity in C8–C16 with 10.2% CO_2 conversion and a combined selectivity of fewer than 22% toward undesirable CO and CH_4.

4.6 Comparative Analysis of Feedstock

As previously discussed, there are several sources to produced biojet fuel and can be classified into three distinct categories: first generation (1-G), second generation (2-G), and third generation (3-G) (Doliente et al. 2020; Wei et al. 2019).

The 1-G feedstocks originate from edible crops like oil palm, corn, sugarcane, sugar beets, rapeseed/canola, and wheat. Most of these food crops have a high water footprint and high nutrients demands, which are the main environmental impacts of choosing 1-G feedstocks (Doliente et al. 2020). The water consumption was 131–143 m^3/GJ to produce biojet fuel from rapeseed in North Dakota using the HEFA (Shi et al. 2017). Cox et al. (2014) observed a water footprint of 15.6–147 m^3/GJ to produce biojet fuel from sugarcane through direct sugar-to-hydrocarbons (DSHC) process. Staples et al. (2013) observed a consumption of water of 76.46–85.81 m^3/GJ and 63.65–106.79 m^3/GJ to biojet fuel produced from corn grain using the DSHC and soybean using the HEFA. Another problem of the 1-G feedstocks to biojet fuel is the use of arable land for the cultivation of crops, which can cause scarcity of food (Cox et al. 2014; Doliente et al. 2020; Wei et al. 2019). Moreover, to address the shortage of available land resources, the common approach has been to expand into forested areas. However, this strategy comes at the cost of deforestation and a reduction in biodiversity (Doliente et al. 2020). One consequence of using food-based products is the rise of food prices and food supply imbalances (Buchspies and Kaltschmitt 2018). Today, these feedstocks, especially the oleochemical/lipid, are the main raw material to biojet fuel in the market (IRENA 2021).

The subsequent generation of aviation biofuels (2-G) can be sourced from nonedible oil crops and waste biomass, and they harmonize the competition to food vs. fuel of 1-G feedstocks (Doliente et al. 2020; Wei et al. 2019). Notable examples include camelina, jatropha, and castor bean that can be transformed into biojet fuel through processes like esterification and isomerization via hydroprocessing (Gutierrez-Antonio et al. 2017; Wei et al. 2019). The demand for fertilizer of these energy crops is lower than the 1-G feedstock, allowing the cultivation in nonfertile and nonfood productive marginal lands (Doliente et al. 2020).

Moreover, certain by-products from industrial waste, such as crude tall oil from paper manufacturing and residual substances like soapstocks, oil sediments, and acid oils from edible oil refining, used cooking oils (UCOs), and waste animal fats (WAFs), can also serve as a viable feedstock for conversion into jet fuel using hydrogenation (Chiaramonti et al. 2014). UCO has been regarded as the most cost-effective and ecologically sound raw material for biojet fuel production (Doliente et al. 2020). Nevertheless, the inconsistency in its availability and the unpredictability of how much it truly aids in achieving greenhouse gas emission reduction goals might curtail its widespread use (Doliente et al. 2020).

Another avenue involves lignocellulose biomass encompassing wood and forestry leftovers, agricultural remains, halophytes, short-rotation woody crops, and municipal solid waste (Wei et al. 2019). Before being used, these materials need pretreatment with enzymes/microorganisms and/or thermochemical transformations for biojet fuel conversion, which are expensive and nonefficient (Doliente et al. 2020). This is the main issue of 2-G feedstock utilization. Gasification utilizing Fischer–Tropsch processes and other thermochemical/biochemical routes can convert lignocellulose feedstock into biojet fuel (Wei et al. 2019). Lignocellulose is a particularly promising feedstock to biojet fuel due to its relatively high abundance and low-use competition (Doliente et al. 2020; Wei et al. 2019).

Waste biomass presents several advantages compared to energy crops feedstock: lack of land requirements (being generated as by-products from agro-forestry, domestic, commercial, and industrial activities), minimal economic value, and reduced water footprints (Doliente et al. 2020). Stimulating the circular economies, waste management, and environmental protection are other positive aspects of using waste biomass as feedstock. On the other hand, the main disadvantages of waste biomass are logistical complexity and variable availability of waste biomass (Emmanouilidou et al. 2023; Staples et al. 2018).

The third-generation aviation biofuel relies on algal feedstock and is highly regarded as a prime candidate for biofuel production due to addressing CO_2 sequestration concerns and not competing with food production (Doliente et al. 2020; Wei et al. 2019). Another advantage of algal biomass is high biomass productivity and oil content, 10–200 times more oil than the energy of other terrestrial crops (soybean, palm oil) (Lim et al. 2021). Algae requires less land compared to many other biomass sources and simple nutrients (Doliente et al. 2020; Lim et al. 2021). Water demand by algae biomass is lower than 1-G feedstocks (Cox et al. 2014; Doliente et al. 2020; Wei et al. 2019). Cox et al. 2014 observed a water footprint of 6.40–13.9 m^3/GJ to produce biojet fuel from microalgae through the HEFA process.

Moreover, wastewater and/or water unsuitable for agriculture can be used as algal growing medium, decreasing the operational cost and improving the environmental conditions (Doliente et al. 2020). However, there are a few drawbacks to cultivation, harvesting, and oil extraction technologies, which are still inefficient and/or capital- and resource-intensive (Doliente et al. 2020; Lim et al. 2021; Wei et al. 2019).

The choice of feedstock for biojet fuel production is a complex decision influenced by economic, environmental, and technological considerations. Each feedstock presents unique advantages and challenges, and the optimal choice may vary depending on the regional conditions and priorities.

4.7 Final Remarks

Introducing alternative jet fuel stands as a crucial stride toward the aviation industry's decarbonization, enabling it to both reduce its carbon footprint and liberate itself from the constraints of finite fossil fuel resources. In this chapter, we described a range of feedstocks for biojet fuel production available. The choice of feedstock for biojet fuel production is pivotal in determining both environmental impact and economic feasibility. Microalgae holds potential for efficient resource utilization, though challenges in cost-effective cultivation and harvesting persist. Oilseed crops have established practices but face concerns over land and food production competition. Waste-based feedstocks mitigate environmental impact and offer waste management solutions. Municipal solid waste and used cooking oil conversion have the potential, contingent on waste management improvements. Nevertheless, the inconsistency in its availability and the collection can limit their application. Therefore, a holistic approach considering environmental sustainability and economic viability is essential in shaping the future of biojet fuel production.

References

Agusdinata DB, Zhao F, Ileleji K, DeLaurentis D (2011) Life cycle assessment of potential biojet fuel production in the United States. Environ Sci Technol 45(21):9133–9143

Ajala EO, Ighalo JO, Ajala MA, Adeniyi AG, Ayanshola AM (2021) Sugarcane bagasse: a biomass sufficiently applied for improving global energy, environment and economic sustainability. Bioresour Bioprocess 8(1)

Arockiasamy S, Kumpatla J, Hadole S, Yepuri V, Patil M, Shrivastava V et al (2021) Breeding and biotechnological efforts in Jatropha curcas L. for sustainable yields. Oil Crop Sci 6(4):180–191

Arunachalam V (2012) Oil palm. In: Genomics of cultivated palms. Elsevier, pp 29–48

Bashir MA, Lima S, Jahangiri H, Majewski AJ, Hofmann M, Hornung A, Ouadi M (2022) A step change towards sustainable aviation fuel from sewage sludge. J Anal Appl Pyrolysis 163

Bennion EP, Ginosar DM, Moses J, Agblevor F, Quinn JC (2015) Lifecycle assessment of microalgae to biofuel: comparison of thermochemical processing pathways. Appl Energy 154:1062–1071

Bhattacharjee S, Panja A, Kumar R, Ram H, Meena RK, Basak N (2023) Municipal solid waste compost: a comprehensive bibliometric data-driven review of 50 years of research and identification of future research themes. Environ Sci Pollut Res

Bjornsson L, Ericsson K (2022) Emerging technologies for the production of biojet fuels from wood-can greenhouse gas emission reductions meet policy requirements? Biomass Convers Biorefin

Buchspies B, Kaltschmitt M (2018) Sustainability aspects of biokerosene. In: Biokerosene. Springer Berlin Heidelberg, Berlin/Heidelberg, pp 325–373

Caputo AC, Palumbo M, Pelagagge PM, Scacchia F (2005) Economics of biomass energy utilization in combustion and gasification plants: effects of logistic variables. Biomass Bioenergy 28(1):35–51

Carvalheiro F, Duarte LC, Girio FM (2008) Hemicellulose biorefineries: a review on biomass pretreatments. J Sci Ind Res 67(11):849–864

Cervi WR, Lamparelli RAC, Gallo BC, Bordonal RD, Seabra JEA, Junginger M, van der Hilst F (2021) Mapping the environmental and techno-economic potential of biojet fuel production from biomass residues in Brazil. Biofuels Bioprod Biorefin 15(1):282–304

Cheng J, Li T, Huang R, Zhou JH, Cen KF (2014) Optimizing catalysis conditions to decrease aromatic hydrocarbons and increase alkanes for improving jet biofuel quality. Bioresour Technol 158:378–382

Chiaberge S, Siviero A, Passerini C, Pavoni S, Bianchi D, Haider MS, Castello D (2021) Co-processing of hydrothermal liquefaction sewage sludge biocrude with a fossil crude oil by Codistillation: a detailed characterization study by FTICR mass spectrometry. Energy Fuel 35(17):13830–13839

Chiaramonti D, Prussi M, Buffi M, Tacconi D (2014) Sustainable bio kerosene: process routes and industrial demonstration activities in aviation biofuels. Appl Energy 136:767–774

Chiaramonti D, Prussi M, Buffi M, Rizzo AM, Pari L (2017) Review and experimental study on pyrolysis and hydrothermal liquefaction of microalgae for biofuel production. Appl Energy 185:963–972. (7th International Conference on Applied Energy (ICAE))

Choi IH, Hwang KR, Han JS, Lee KH, Yun JS, Lee JS (2015) The direct production of jet-fuel from non-edible oil in a single-step process. Fuel 158:98–104

Chu PL, Vanderghem C, MacLean HL, Saville BA (2017) Process modeling of hydrodeoxygenation to produce renewable jet fuel and other hydrocarbon fuels. Fuel 196:298–305

Chuck CJ (2016) In: Chuck C (ed) Biofuels for aviation. Elsevier

Cisneros-Yupanqui M, Chalova VI, Kalaydzhiev HR, Mihaylova D, Krastanov AI, Lante A (2021) Preliminary characterisation of wastes generated from the rapeseed and sunflower protein isolation process and their valorisation in delaying oil oxidation. Food Bioprocess Technol 14(10):1962–1971

Cox K, Renouf M, Dargan A, Turner C, Klein-Marcuschamer D (2014) Environmental life cycle assessment (LCA) of aviation biofuel from microalgae, Pongamia pinnata, and sugarcane molasses. Biofuels Bioprod Biorefin 8(4):579–593

Cronin DJ, Subramaniam S, Brady C, Cooper A, Yang ZB, Heyne J et al (2022) Sustainable aviation fuel from hydrothermal liquefaction of wet wastes. Energies 15(4)

Cui Q, Hu YX, Yu LT (2022) Can the aviation industry achieve carbon emission reduction and revenue growth simultaneously under the CNG2020 strategy? An empirical study with 25 benchmarking airlines. Energy 245

da Silva RL, Olivi JB, da Silva AMP (2022) Corn harvest residues for co-generation assessment: availability at Mato Grosso do Sul state, and energy characteristics. Revista em Agronegócio e Meio Ambiente 15(1):1–14

De Jong J, Akselsson C, Egnell G, Löfgren S Olsson B (2018) Environmental impact of forest fuel extraction – a synthesis of current research based on Bränsleprogrammet hållbarhet 2011–2016 (In Swedish: Miljöpåverkan av skogsbränsleuttag – En syntes av forskningsläget baserat på Bränsleprogrammet hållbarhet 2011–2016). Energimyndigheten, Eskilstuna

de Oliveira JAR, da Conceicao AC, Martins LHD, Moreira DKT, Passos MF, Komesu A (2021) Evaluation of the technological potential of four wastes from Amazon fruit industry in glucose and ethanol production. J Food Process Eng 44(2)
de Souza LM, Mendes PAS, Aranda DAG (2020) Oleaginous feedstocks for hydro-processed esters and fatty acids (HEFA) biojet production in southeastern Brazil: a multi-criteria decision analysis. Renew Energy 149:1339–1351
Dias MOS, Junqueira TL, Cavalett O, Cunha MP, Jesus CDF, Rossell CEV et al (2012) Integrated versus stand-alone second generation ethanol production from sugarcane bagasse and trash. Bioresour Technol 103(1):152–161
Diaz MJ, Cara C, Ruiz E, Romero I, Moya M, Castro E (2010) Hydrothermal pre-treatment of rapeseed straw. Bioresour Technol 101(7):2428–2435
Doliente SS, Narayan A, Tapia JFD, Samsatli NJ, Zhao YR, Samsatli S (2020) Bio-aviation fuel: a comprehensive review and analysis of the supply chain components. Front Energy Res 8
Dornau A, Robson JF, Thomas GH, McQueen-Mason SJ (2020) Robust microorganisms for biofuel and chemical production from municipal solid waste. Microb Cell Factories 19(1)
dos Santos M, Monteiro FMB, Gomes CFS, de Lima JCS, Santos ACRD (2017) Aplicação do Método MACBETH na hierarquização de biocombustíveis para a aviação comercial brasileira : um estudo de viabilidade. In: XIV Simpósio de Excelência em Gestão – SEGeT 2017, Rio de Janeiro
Elkelawy M, Bastawissi HA-E, Radwan AM, Ismail MT, El-Sheekh M (2022) Biojet fuels production from algae: conversion technologies, characteristics, performance, and process simulation. In: Handbook of algal biofuels. Elsevier, pp 331–361
Eller Z, Varga Z, Hancsok J (2016) Advanced production process of jet fuel components from technical grade coconut oil with special hydrocracking. Fuel 182:713–720
Emmanouilidou E, Mitkidou S, Agapiou A, Kokkinos NC (2023) Solid waste biomass as a potential feedstock for producing sustainable aviation fuel: a systematic review. Renew Energy 206:897–907
Escalante ESR, Ramos LS, Coronado CJR, de Carvalho JA (2022) Evaluation of the potential feedstock for biojet fuel production: focus in the Brazilian context. Renew Sustain Energy Rev 153
Farooq D, Thompson I, Ng KS (2020) Exploring the feasibility of producing sustainable aviation fuel in the UK using hydrothermal liquefaction technology: a comprehensive techno-economic and environmental assessment, vol 1. Cleaner Eng Technol
Fortier MOP, Roberts GW, Stagg-Williams SM, Sturm BSM (2014) Life cycle assessment of biojet fuel from hydrothermal liquefaction of microalgae. Appl Energy 122:73–82
Galbe M, Zacchi G (2012) Pretreatment: the key to efficient utilization of lignocellulosic materials. Biomass Bioenergy 46:70–78. (1st International Conference on Lignocellulosic Ethanol)
Gutierrez-Antonio C, Gomez-Castro FI, de Lira-Flores JA, Hernandez S (2017) A review on the production processes of renewable jet fuel. Renew Sustain Energy Rev 79:709–729
Hao J, Xiao J, Song G, Zhang Q (2021) Energy and exergy analysis of bio-jet fuel production from lignocellulosic biomass via aqueous conversion. Case Stud Therm Eng 26:101006
Hari TK, Yaakob Z, Binitha NN (2015) Aviation biofuel from renewable resources: routes, opportunities and challenges. Renew Sustain Energy Rev 42:1234–1244
Harisankar S, Vinu R (2023) Comprehensive evaluation of municipal solid wastes and mixed feedstocks for commercial hydrothermal liquefaction in bio-refineries. Fuel 339:127236
Herr A, Braid A, Carter J, McIvor J, Murphy HT, O'Connell D et al (2016) Cut your grass and eat it too- is aviation biofuel production and grazing in the Australian tropics possible? Renew Sustain Energy Rev 53:1377–1388
ICAO (2013) ICAO environment report, 2013. https://www.icao.int/environmental-protection/Pages/EnvReport13.aspx. Accessed 7 Aug 2023
IRENA (2021) Reaching zero with renewables: biojet fuels. https://www.irena.org/-/media/Files/IRENA/Agency/Publication/2021/Jul/IRENA_Reaching_Zero_Biojet_Fuels_2021.pdf. International Renewable Energy Agency, Abu Dhabi

Kaza S, Yao LC, Bhada-Tata P, Van Woerden F (2018) What a waste 2.0: a global snapshot of solid waste management to 2050. http://hdl.handle.net/10986/30317

Kenar JA, Moser BR, List GR (2017) Naturally occurring fatty acids. In: Fatty acids. Elsevier, pp 23–82

Labeckas G, Slavinskas S (2015) Combustion phenomenon, performance and emissions of a diesel engine with aviation turbine JP-8 fuel and rapeseed biodiesel blends. Energy Convers Manag 105:216–229

Li X, Mupondwa E, Tabil L (2018) Technoeconomic analysis of biojet fuel production from camelina at commercial scale: case of Canadian prairies. Bioresour Technol 249:196–205

Lim JHK, Gan YY, Ong HC, Lau BF, Chen WH, Chong CT et al (2021) Utilization of microalgae for bio-jet fuel production in the aviation sector: challenges and perspective. Renew Sustain Energy Rev 149

Lin CH, Chen YK, Wang WC (2020) The production of bio-jet fuel from palm oil derived alkanes. Fuel 260

Liu SY, Zhu QQ, Guan QX, He LN, Li W (2015) Bio-aviation fuel production from hydroprocessing castor oil promoted by the nickel-based bifunctional catalysts. Bioresour Technol 183:93–100

Luo YH, Zhang R, He YT, Lou DF, Zhu R, Zhu C et al (2023) Preparation of bio-jet fuels by a controllable integration process: coupling of biomass fermentation and olefin polymerization. Bioresour Technol 382

Madlener R, Bachhiesl M (2007) Socio-economic drivers of large urban biomass cogeneration: sustainable energy supply for Austria's capital Vienna. Energy Policy 35(2):1075–1087

Morgan TJ, Youkhana A, Turn SQ, Ogoshi R, Garcia-Perez M (2019) Review of biomass resources and conversion technologies for alternative jet fuel production in Hawai'i and tropical regions. Energy Fuel 33(4):2699–2762

Ng KS, Farooq D, Yang AD (2021) Global biorenewable development strategies for sustainable aviation fuel production. Renew Sustain Energy Rev 150

Nilsson D, Hansson PA (2001) Influence of various machinery combinations, fuel proportions and storage capacities on costs for co-handling of straw and reed canary grass to district heating plants. Biomass Bioenergy 20(4):247–260

O'Neil GW, Knothe G, Reddy CM (2019) Jet biofuels from algae. In: Biofuels from algae. Elsevier, pp 359–395

Obnamia JA, MacLean HL, Saville BA (2020) Regional variations in life cycle greenhouse gas emissions of canola-derived jet fuel produced in western Canada. Glob Change Biol Bioenergy 12(10):818–833

Ogundari IO, Momodu AS, Famurewa AJ, Akarakiri JB, Siyanbola WO (2012) Analysis of sustainable cassava biofuel production in Nigeria. Energy Environ 23(4):599–618

Okolie JA, Awotoye D, Tabat ME, Okoye PU, Epelle EI, Ogbaga CC et al (2023) Multi-criteria decision analysis for the evaluation and screening of sustainable aviation fuel production pathways. Iscience 26(6)

Olusola Sanusi R, Lordbanjou D-T, Olalekan Ibrahim A, Babakatcha Abubakar M, Olalekan Oke O (2023) Cassava production enterprise in the tropics. In: Tropical plant species and technological interventions for improvement. IntechOpen

Ortiz R, Geleta M, Gustafsson C, Lager I, Hofvander P, Lofstedt C et al (2020) Oil crops for the future. Curr Opin Plant Biol 56:181–189

Panoutsou C, Germer S, Karka P, Papadokostantakis S, Kroyan Y, Wojcieszyk M et al (2021) Advanced biofuels to decarbonise European transport by 2030: markets, challenges, and policies that impact their successful market uptake. Energ Strat Rev 34

Perkis DF, Tyner WE (2018) Developing a cellulosic aviation biofuel industry in Indiana: a market and logistics analysis. Energy 142:793–802

Rahman WU, Patel M, Kurian V, Kumar A (2022) A comparative techno-economic assessment of fast pyrolysis, hydrothermal liquefaction, and intermediate pyrolysis of municipal solid waste for liquid transportation fuels production. Energy Convers Manag 267

Renouf MA, Wegener MK, Pagan RJ (2010) Life cycle assessment of Australian sugarcane production with a focus on sugarcane growing. Int J Life Cycle Assess 15(9):927–937

Richter S, Braun-Unkhoff M, Naumann C, Riedel U (2018) Paths to alternative fuels for aviation. CEAS Aeronaut J 9(3):389–403

Romero MJA, Duca D, Toscano G (2022) Advancements in the conversion of lipid-rich biowastes and Lignocellulosic residues into high-quality road and jet biofuels using nanomaterials as catalysts. Processes 10(2)

Ruan Z, Wang X, Liu Y, Liao W (2019) Corn. In: Integrated processing technologies for food and agricultural by-products. Elsevier, pp 59–72

SAFUG (2018) Our commitment to sustainable options. http://www.safug.org/safug-pledge/. Accessed 7 Aug 2023

Sajid M, Raheem A, Ullah N, Asim M, Rehman MSU, Ali N (2022) Gasification of municipal solid waste: progress, challenges, and prospects. Renew Sustain Energy Rev 168

Saynor B, Bauen A, Leach MA (2003) The potential for renewable energy sources in aviation. www.iccept.ic.ac.uk. Accessed 7 Aug 2023

Scaldaferri CA, Pasa VMD (2019) Production of jet fuel and green diesel range biohydrocarbons by hydroprocessing of soybean oil over niobium phosphate catalyst. Fuel 245:458–466

Sharno MA, Hiloidhari M (2022) Life cycle carbon, energy, water, and landuse footprints of agricultural residue-based biojet fuel in Bangladesh. Bioresour Technol Rep 20:101234

Shi R, Ukaew S, Archer DW, Lee JH, Pearlson MN, Lewis KC, Shonnard DR (2017) Life cycle water footprint analysis for rapeseed derived jet fuel in North Dakota. ACS Sustain Chem Eng 5(5):3845–3854

Shi R, Archer DW, Pokharel K, Pearlson MN, Lewis KC, Ukaew S, Shonnard DR (2019) Analysis of renewable jet from oilseed feedstocks replacing fallow in the US Northern Great Plains. ACS Sustain Chem Eng 7(23):18753–18764

Sobieralski JB (2023) Sustainable air transportation through the operational use of a social cost index. J Clean Prod 385

Speight JG (2019) Unconventional gas. In: Natural gas. Elsevier, pp 59–98

Staples MD, Olcay H, Malina R, Trivedi P, Pearlson MN, Strzepek K et al (2013) Water consumption footprint and land requirements of large-scale alternative diesel and jet fuel production. Environ Sci Technol 47(21):12557–12565

Staples MD, Malina R, Olcay H, Pearlson MN, Hileman JI, Boies A, Barrett SRH (2014) Lifecycle greenhouse gas footprint and minimum selling price of renewable diesel and jet fuel from fermentation and advanced fermentation production technologies. Energy Environ Sci 7(5):1545–1554

Staples MD, Malina R, Suresh P, Hileman JI, Barrett SRH (2018) Aviation CO2 emissions reductions from the use of alternative jet fuels. Energy Policy 114:342–354

Statens Offentliga Utredningar (2019) Biojet fuels for aviation (In Swedish: Biojet för flyget). SOU, Stockholm, p 11

Suresh P (2016) Environmental and economic assessment of transportation fuels from municipal solid waste. Massachusetts Institute of Technology

Su-ungkavatin P, Tiruta-Barna L, Hamelin L (2023) Biofuels, electrofuels, electric or hydrogen?: a review of current and emerging sustainable aviation systems. Prog Energy Combust Sci 96

Tan KT, Lee KT, Mohamed AR, Bhatia S (2009) Palm oil: addressing issues and towards sustainable development. Renew Sustain Energy Rev 13(2):420–427

Tanzil AH, Zhang X, Wolcott M, Brandt K, Stockle C, Murthy G, Garcia-Perez M (2021) Evaluation of dry corn ethanol bio-refinery concepts for the production of sustainable aviation fuel. Biomass Bioenergy 146

Tao L, Milbrandt A, Zhang YA, Wang WC (2017) Techno-economic and resource analysis of hydroprocessed renewable jet fuel. Biotechnol Biofuels 10

Tiago AV, Pedri ECM, Rossi FS, Santos LL, Lima JA, Cardoso ES et al (2020) Phenotypic characterization of cassava ethno-varieties in the state of Mato Grosso, Brazil. Genet Mol Res 19(2)

Tye YY, Lee KT, Abdullah WNW, Leh CP (2016) The world availability of non-wood lignocellulosic biomass for the production of cellulosic ethanol and potential pretreatments for the enhancement of enzymatic saccharification. Renew Sustain Energy Rev 60:155–172

Ukaew S, Shi R, Lee JH, Archer DW, Pearlson M, Lewis KC et al (2016) Full chain life cycle assessment of greenhouse gases and energy demand for canola-derived jet fuel in North Dakota, United States. ACS Sustain Chem Eng 4(5):2771–2779

Uludere Aragon NZ, Parker NC, VanLoocke A, Bagley J, Wang M, Georgescu M (2022) Sustainable land use and viability of biojet fuels. Nat Sustain 6(2):158–168

van Dyk S, Saddler J (2021) Progress in commercialization of biojet/sustainable aviation fuels (SAF): technologies, potential and challenges. IEA Bioenergy Task, 39. https://www.ieabioenergy.com/wp-content/uploads/2021/06/IEA-Bioenergy-Task-39-Progress-in-the-commercialisation-of-biojet-fuels-May-2021-1.pdf. Accessed 7 Aug 2023

Vasquez MC, Silva EE, Castillo EF (2017) Hydrotreatment of vegetable oils: a review of the technologies and its developments for jet biofuel production. Biomass Bioenergy 105:197–206

Wang WC (2016) Techno-economic analysis of a bio-refinery process for producing hydro-processed renewable jet fuel from Jatropha. Renew Energy 95:63–73

Wang Y, Zhang QY, Hao JW, Luo ZC, Xiao J (2023) Global warming potential of bio-jet fuel produced by biomass aqueous-phase conversion. Sustain Prod Consumption 37:284–293

Wei HJ, Liu WZ, Chen XY, Yang Q, Li JS, Chen HP (2019) Renewable bio-jet fuel production for aviation: a review. Fuel 254. (6th International Conference on Biomass Energy (ICBE))

Why ESK, Ong HC, Lee HV, Gan YY, Chen WH, Chong CT (2019) Renewable aviation fuel by advanced hydroprocessing of biomass: challenges and perspective. Energy Convers Manag 199

Why ESK, Ong HC, Lee HV, Chen WH, Asikin-Mijan N, Varman M, Loh WJ (2022) Single-step catalytic deoxygenation of palm feedstocks for the production of sustainable bio-jet fuel. Energy 239

Wood SM, Layzell DB (2003) A Canadian biomass inventory: feedstocks for a bio-based economy, Kingston

Wu XP, Jiang PW, Jin F, Liu JX, Zhang YH, Zhu LJ et al (2017) Production of jet fuel range biofuels by catalytic transformation of triglycerides based oils. Fuel 188:205–211

Yang J, Xin Z, He Q, Corscadden K, Niu HB (2019) An overview on performance characteristics of bio-jet fuels. Fuel 237:916–936

Yao GL, Staples MD, Malina R, Tyner WE (2017) Stochastic techno-economic analysis of alcohol-to-jet fuel production. Biotechnol Biofuels 10

Yao BZ, Xiao TC, Makgae OA, Jie XY, Gonzalez-Cortes S, Guan SL et al (2020) Transforming carbon dioxide into jet fuel using an organic combustion-synthesized Fe-Mn-K catalyst. Nat Commun 11(1)

Yoo E, Lee US, Wang MC (2022) Life-cycle greenhouse gas emissions of sustainable aviation fuel through a net-zero carbon biofuel plant design. ACS Sustain Chem Eng 10(27):8725–8732

Zhang L, Dang YR, Zhou XH, Gao P, van Bavel AP, Wang H et al (2021) Direct conversion of CO2 to a jet fuel over CoFe alloy catalysts. Innovation 2(4)

Zhao XH, Wei L, Julson J, Qiao QQ, Dubey A, Anderson G (2015) Catalytic cracking of non-edible sunflower oil over ZSM-5 for hydrocarbon bio-jet fuel. New Biotechnol 32(2):300–312. https://doi.org/10.1016/j.nbt.2015.01.004

Zhao XH, Wei L, Cheng SY, Julson J, Anderson G, Muthukumarappan K, Qiu CL (2016) Development of hydrocarbon biofuel from sunflower seed and sunflower meat oils over ZSM-5. J Renew Sustain Energy 8(1)

Chapter 5
Characteristics of Biojet Fuel

Priyancka Arora and Shubhankari Mishra

Abstract The aviation sector is the largest producer of greenhouse gases, contributing 2% of global emissions. The finite supply of fossil fuels also emphasizes the need for sustainable energy sources in the aviation sector, which are producing significantly lower emissions as well as renewable resources. This chapter discusses the important production routes of biomass-derived fuels, also called biojet fuels (BJFs), which must meet the ASTM International specifications and are clean and complete substitutes for present-day jet fuels. The production of these fuels uses a wide range of biomass; consequently, the fuels produced have very different compositions. The performance characteristics of the fuels based on the physiochemical properties of their constituents are discussed elaborately. It has been observed that there is a direct association between the chemical composition of the biofuels produced and their performance characteristics. Many researchers have suggested that the properties of bio-aviation fuels are appropriate as per the specifications provided by the ASTM standards. The concentration of aromatic carbons is pivotal in influencing the characteristics of fuels. The blends of biofuel with conventional fuels are also studied to improve fuel performance. For biojet fuels to become 100% drop-in fuels in commercial aviation usage, some drawbacks such as the price of production, feedstock availability, energy intensity of the process, and storage stability need to be addressed.

Keywords Biojet fuels · Performance characteristics · Greenhouse gas emissions · Aviation sector · ASTM specifications · Alternative jet fuels · Aromatic content

P. Arora (✉)
Department of Biotechnology, Amity University, Mohali, Punjab, India

S. Mishra
Department of Biotechnology, Faculty of Engineering and Technology, Rama University, Kanpur, Uttar Pradesh, India

A. Kuila (ed.), *Biojet Fuel: Current Technology and Future Prospect*, Clean Energy Production Technologies, https://doi.org/10.1007/978-981-99-8783-2_5

5.1 Introduction

The Air Transport Industry plays a significant part in the world economy by allowing global connectivity; the Air Transport Action Group (ATAG) reported that 87.7 million jobs were provided worldwide by this industry, producing $961.3 billion of GDP per year, about 4.3 times higher than other jobs. Aviation is expected to continuously expand and contribute $1.7 trillion to world GDP by 2038. In 2019, 4.5 billion passengers were served by the airline industry (ATAG 2020). This increase in air travelers requires a considerably substantial quantity of aviation fuels, but the extended utilization of fuels in the past few years has developed a noticeable decline in the petroleum supply (Pavlenko and Kharina 2018).

The huge utilization of jet fuel provides a considerable volume of greenhouse gases (GHG), around 2.1% of all CO_2 emissions that are generated by human activities and 12% of all aircraft emissions. The increasing demand for air transport results because there is aqua druple increase in the amount of emissions from 2015, which was 0.78 billion tons, and is expected to reach 3.1 billion tons of GHG emissions by 2050 (Doliente et al. 2020). Due to the effects of GHG on global warming, the airline industry is required to reduce 50% of CO_2 emissions by 2050 compared to CO_2 emissions in 2005 (ATAG 2020). The major challenge is to discover the most acceptable way to reduce GHG emissions to the determined target set at 50% less than the volume that was in 2005. The International Civil Aviation Organization (ICAO) and the International Air Transport Association (IATA) developed a few ways to accomplish the target: operational advancements, market-based measures, technological improvements, and sustainable jet fuel (SJF). Technological refinements are resulting in the reduction of GHG emissions. The improvements lower fuel utilization while traveling and give competency in mileage.

The majority of the reduction in GHG emissions can be achieved by substituting conventional jet fuel (CJF) with alternative jet fuel (AJF). The physiochemical properties of AJF must be similar to CJF like AJF should have 30,000 feet above elevation, lower carbon footprints than CJF, an adequate amount of energy density to fulfill the demands of long-haul flights, and temperature stability between −47 °C and 40 °C. Alternative jet fuel (AJF) like biofuel ensures immense reduction in GHG emissions (Doliente et al. 2020). Biomass-derived aviation fuels (biojet fuels) or BAF are used as an alternative to conventional jet fuels. The International Air Transport Association (IATA) recognized that biojet fuels are the guaranteed policy to bring down GHG release from the aviation sector. The aircraft that utilized BAF resulted in remarkably lower carbon emissions when weighed against CJF (Yang et al. 2019).

The American Society for Testing and Material (ASTM) D7566-10 is the international organization that decides the standard specifications for fuel quality. The fuel specification for the synthesized hydrocarbons as per this international body is that the fuel should contain up to 50% of any of the five types of synthesized paraffinic kerosene (SPK), which needs to be blended with CJF. In 2011, the ASTM approved one of the synthesized fuels called hydroprocessed esters and fatty acids

(HEFA) that can be blended up to 50% with CJF. In 2015, FT-SPK combined with Aromatics (FT-SPK/A) became part of ASTM D7566 standards. The aromatics content is deliberately elevated up to the highest 20% in FT-SPK/A. At the same time, synthesized iso-paraffins (SIP) were also approved and certified in ASTM D7566, but the quantity of blends approved was up to only 10% with CJF. Alcohol-to-jet (ATJ) produces ATJ-SPK by utilizing C_2-C_5 alcohols, where iso-butanol (C_4) and ethanol (C_2) were approved in 2016 and 2018, respectively, with up to 50% blending permitted (Yang et al. 2019). In recent years, global interest in BAF production has escalated, showing the necessity of lowering GHG release by the jet industry through AJF.

The central theme of this chapter is the processes of conversion technologies of biomass to biofuels and the characteristics of BJFs.

5.2 Properties of Fuels Used in Aircraft

The ASTM-D16522 is an international institution that defines the basic characteristics of aviation fuels as shown in Table 5.1. Jet fuels are composed of stringent characteristics compared to land transportation fuels. As per the specifications declared by the ASTM, the jet fuel should basically be comprised of a complex mixture of C_9–C_{16} range hydrocarbons. It should consist of a mixture of alkanes, which could be linear alkanes, slightly branched alkanes, cycloalkanes, and 20% arene hydrocarbons such as benzene and naphthalene. Moreover, other physiochemical properties, such as freeze point, energy density, flash point, viscosity, flammability limits, combustibility, sulfur content, density, and amount of hydrogen ions, are strictly adhered to for the purpose of operational certification. The carbon chain length and the amount of different kinds of alkanes should be maintained so as to match the guidelines of the jet fuels (Wang and Tao 2016).

Table 5.1 Summary of the characteristics of fuels as per the ASTM D1655 standards

Property	Specifications	Comment
Carbon content	C_9-C_{16}	80% alkanes (linear, iso, cyclic) 20% aromatic
Density	High-energy density	
Flash point	High	Minimum 38°C
Freezing point	Low	Maximum – 47°C
Sealing property	Good	The presence of aromatic compounds (~20%) enhances the swelling of elastomeric valves in the fuel system, thus improving the sealing property
Sulfur content	Low	Maximum 0.30% by mass
Heat of combustion	High	42.8 MJ/kg

Another significant attribute is the high flash points, which is the minimum temperature at which enough vapors of material are generated that can be ignited. The flash point of the jet fuel is required to be a minimum of 38°C as there is a low chance of fire hazards on board. The low freeze point is recommended such that at high altitudes it possesses good cold flow properties. The freezing point should be a maximum of −47°C, and it indicates the temperature at which wax that had been crystallized when the fuel was previously cooled completely melts when the fuel is rewarmed. The fuel is expected to have a high-energy density that aids in storage space. The proportion of aromatic hydrocarbons should be around 20% in the fuel. The aromatic hydrocarbons are shown to have negative effects on the combustion efficiency of fuel, but their presence in the fuel is unavoidable as they provide good sealing properties to the fuel that are required to avoid leakage (Kramer et al. 2022). The sulfur content of the fuel should also be controlled as it is involved in producing harmful emissions such as sulfur oxide. Most importantly, the exothermic release of energy, when the fuel is subjected to 100% combustion at constant pressure, is required to be at least 42.8 MJ/kg (ICAO 2018). This energy released is called net heat combustion, which should be kept high.

5.3 Different Production Technologies of Biojets

The idea behind designing the biojet fuel using more reliant renewable resources is to have the advantages of renewability, less dependence on petroleum, more sustainability, environment-friendly, and easy carbon dioxide recycling (Bozell et al. 2000). The waste biomass has garnered interest in its conversion to biojet fuels such as feedstocks having triglyceride-containing materials, lignocellulose-containing wastes, and sugar and starch wastes (Moreno-Gómez et al. 2020). The research on BJF production has been done using various raw materials, out of which jatropha, microalgae, and camelina have the most potential (Wei et al. 2019; Lim et al. 2021). The production methodology includes catalytic cracking, pyrolysis, trans-esterification, hydroprocessing, and fermentation. Different raw materials require different production processes and result in different final fuel properties. The production route also impacts the cost, its effect on the environment, and its ultimate composition (Shahid et al. 2021). Some of the methods of biojet manufacturing are discussed below.

5.3.1 Alcohol Oligomerization

This method is also called the alcohol-to-jet fuels (ATJ) route, which comprises three steps. For the purpose of biojet production, alcohol used is a short-chain fatty alcohol having C_2 or C_4 chain such as ethanol and butanol. Initially, the bio-alcohol is dehydrated to its olefin compound, for instance, ethanol yields ethylene upon dehydration, which is converted to its corresponding olefin derivative. Secondly, oligomerization of olefin is done to produce C_9-C_{16} large-chain olefins; in the case of ethanol and dimerization of olefins, it is required if the raw material is butanol. Finally, the oligomerized olefin is subjected to hydrogenation to produce the saturated hydrogenated product, which has properties similar to jet fuel (Wang and Tao 2016).

One of the most important components of the ATJ route of biojet conversion is the use of a catalyst for the oligomerization and dehydration step, which enhances the rate at which the conversion of alcohol to biojet fuels takes place (Sundararaj and Kushari 2019). Some of the most efficient catalysts for this purpose are zeolite, aluminum (III) oxide, and heteropolyacid (HPAs) (Sundararaj and Kushari 2019). Over the years, with further research, the ATJ conversion route has been practiced using other acidic catalysts, specifically for dehydration and oligomerization steps. The ATF process is perfected by ByogyRenewables by means of a catalytic process, which involves the production of heterogeneous long-chain hydrocarbons from ethanol. The mixture produced is dissociated into aviation fuel and gasoline using a selective distillation process (Han et al. 2019). The production route is illustrated using a flowchart in Fig. 5.1. The Byogy fuel has ASTM approval for commercial flights, with an increase of 50% blend ratio. The other manufacturer of biojet fuels using the ATJ route is Gevo, in which higher alcohol is used instead of ethanol. It makes use of propanol and butanol and leads to the production of aromatics (Díaz-Pérez and Serrano-Ruiz 2020). The alcohol-to-jet synthetic paraffinic kerosene (ATJ-SPK) conversion technology was used to convert isobutanol feedstock to biojet fuel, and this technology was standardized in 2018 (Geleynse et al. 2018).

5.3.2 Fermentation of Sugar and Platform Molecules

This process is called the fermentation to jet (FTJ) process and direct sugar-to-hydrocarbon (DSHC). The technology involves anaerobic fermentation for the synthesis of alkane-type fuels from sugars such as lignocellulosic sugar. Recently, attention has been paid to the use of simple sugars (sorghum, maize, sugarcane) or platform molecules (bio-derived molecules) as a feedstock for their ease of fermentation (Mawhood et al. 2015). The FTJ process is complex as the feedstock contains a variety of functional groups, is highly oxygenated, and contains a maximum of six carbon atoms. On the other hand, jet fuel comprises higher carbon atoms (C_9-C_{16}), is devoid of a variety of functional groups, and is less oxygenated (Díaz-Pérez and Serrano-Ruiz 2020). Therefore, there is a need for complex chemical reactions in the FTJ process such as dehydration, hydrogenation, and hydrogenolysis for the removal of oxygen, aldol condensation, ketonization, and oligomerization for C–C coupling reactions (Serrano-Ruiz et al. 2011).

Another procedure for obtaining biojet fuels through the fermentation of sugars and biomolecules is by using bioengineered microorganisms that are made compatible to feed on these sugars and produce biojet fuels. Producing such

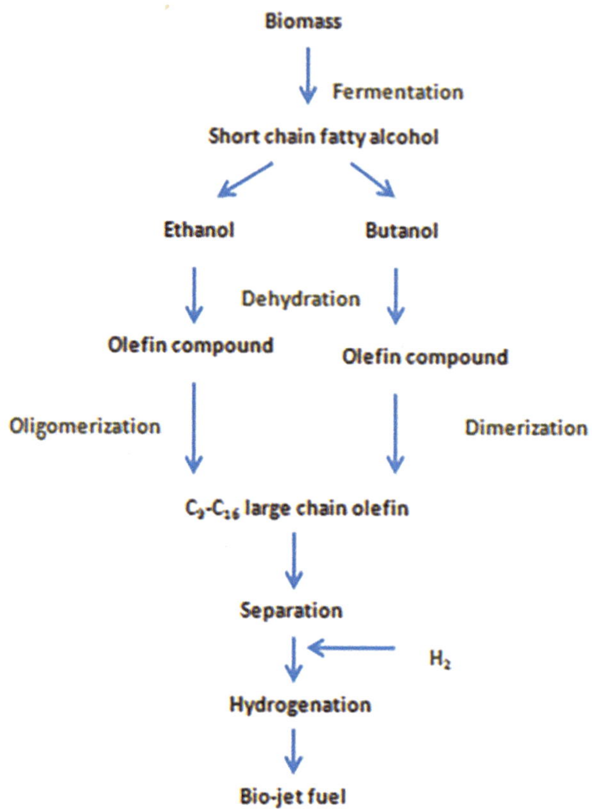

Fig. 5.1 ATJ production route at a glance

microorganisms through genetic engineering is difficult (Mawhood et al. 2015). Firstly, the hydrolytic catalysis of the preliminary treated biomass is required, followed by hydrosylate clarification. The engineering microorganisms are then introduced for the process of fermentation to occur. In the next step, the purification of the fermented products is carried out. Then, the hydrotreatment is given to the products before it is subjected to fractionation (Wang and Tao 2016). The steps involved in the synthesis of BJF using this method are explained in the flowchart in Fig. 5.2. The renewable fuel company, Virent, is involved in FTJ conversion for the production of sustainable fuels by BioForming (Díaz-Pérez and Serrano-Ruiz 2020).

5.3.3 Hydroprocessing

This method is involved in hydrocracking and hydrotreating of hydrogenated esters and fatty acids (HEFA) with the help of catalytic actions such as decarboxylation, hydrogenation, decarbonylation, cracking, and isomerization. This is a catalytic

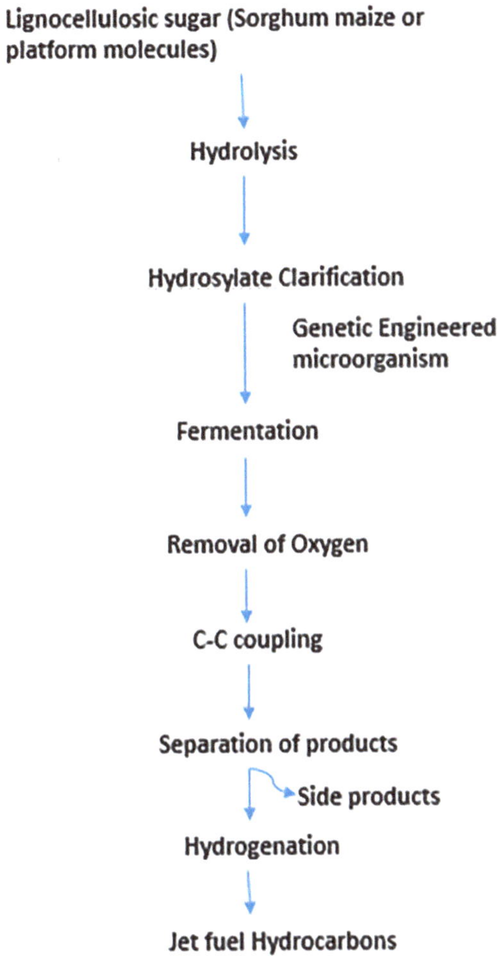

Fig. 5.2 Flowchart of the steps involved in direct sugar-to-hydrocarbon (DSHC) conversion to produce biojet fuel

process and the intermediate biofuels produced using this route are called hydroprocessed renewable jet (HRJ). This technique is also known as oils-to-jet fuels.

The methodology involved in the production of HRJ using hydroprocessing involves the oils from vegetables as a feedstock. These oils (e.g., soybean, palm, corn, jatropha, camelina, and canola) are enriched with triglycerides (TG), which help in the synthesis of straight-chain alkanes (Morgan et al. 2012). The *n*-alkanes serve as a biojet fuel component because they have a good combustible tendency and high-energy density (Lin et al. 2020). In the first step, hydrogenation of TG leads to the production of free fatty acids and propane. In the subsequent reactions, oxygen is removed from the product by the process of hydrodeoxygenation (HDO) and hydrodecarbonylation/hydrodecarboxylation (HDC) to produce alkanes (Ng et al. 2021). The resultant alkane in the case of HDO reaction is *n*-alkane, whereas HDO produces *n*-1 alkanes, thereby yielding low carbon yield in comparison to

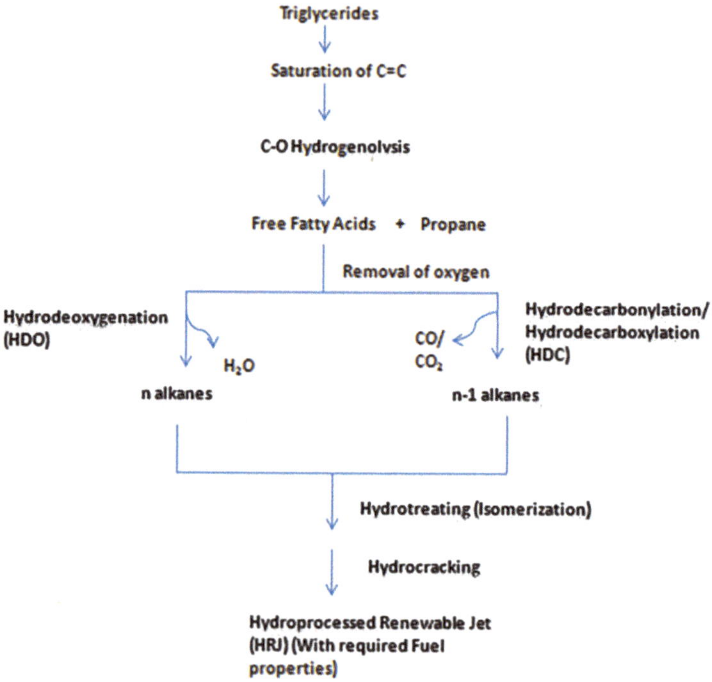

Fig. 5.3 Flowchart depicting the oils-to-jet fuel conversion method

HDO. In the second step, isomerization and hydrocracking are carried out to produce the hydrocarbons with the desired carbon chain length of C_9-C_{16} (Sundararaj and Kushari 2019). High temperature and hydrogen pressure are necessary for this HEFA process, which converts oils into biofuels. It is carried out along with heterogeneous catalysts such as transition metals or their bimetallic composites (Monteiro et al. 2022). The conversion of oil to jet fuel is shown in the form of a flowchart in Fig. 5.3. The most appropriate airplane fuel is HRJ biojet. Some of the characteristics that make HRJ biofuels most suitable for being a drop-in fuel are lesser aromatic carbons, more calorific content, zero sulfur content, and low emissions (Sundararaj et al. 2019). The companies that produce HRF fuels and meet the ASTM standards are Neste Oil and Honeywell Universal Oil Products (Tao et al. 2017).

5.3.4 *Hydrothermal Liquefaction*

An alternative method to develop biojets from vegetable oils is hydrothermal liquefaction (HTL). It is also known as catalytic hydrothermolysis (CH). It produces biofuels with a very low content of oxygen by the liquefaction reaction. This method

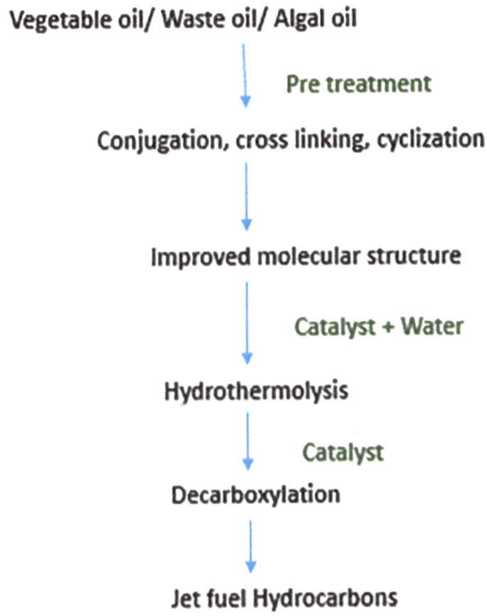

Fig. 5.4 Steps involved in the conversion of oils to hydrocarbon

is capable of producing 100% sustainable aviation fuel (SAF) in 2019, which was named 'ReadiJet." This production pathway of SAF or biojets is developed and patented by Applied Research Associate Inc., which contains algal oil or vegetable oils as biomass. This production pathway begins with pretreatment of the biomass, which helps in treating the triglycerides and unsaturated fatty acids using catalytic reactions such as conjugation, cross-linking, and cyclization that are essential to enhance its molecular structure. The conversion of oil to hydrocarbon is explained in the flowchart in Fig. 5.4. The advantage of this technique is that the feedstock is not required to be dewatered. The reaction conditions are kept such that the water stays in the fluid state and pressure is maintained around ~100–350 bars such that water is at a dense supercritical state in order to produce biofuel with high-energy efficiency (Grande et al. 2021). Water and catalysts are used to facilitate the catalytic hydrothermolysis process. In the succeeding steps, the amount of unsaturation and oxygenated content of the product is reduced by a catalytic decarboxylation reaction. The resultant fuel has a variety of alkanes ranging from C_6-C_{28} (Li et al. 2010).

5.3.5 *Hydrotreated Depolymerized Cellulosic Jet (HDCJ)*

It is a procedure that requires fast pyrolysis of biostock. Fast pyrolysis is the thermochemical treatment of the feedstock to convert it into liquid bio-oil, which is further processed to produce oils of biojet fuel standards. The procedure takes place in an oxygen-free environment at temperatures between 400 and 600 °C (Hu et al. 2020). The hydrotreatment steps of the process are carried out at mild conditions

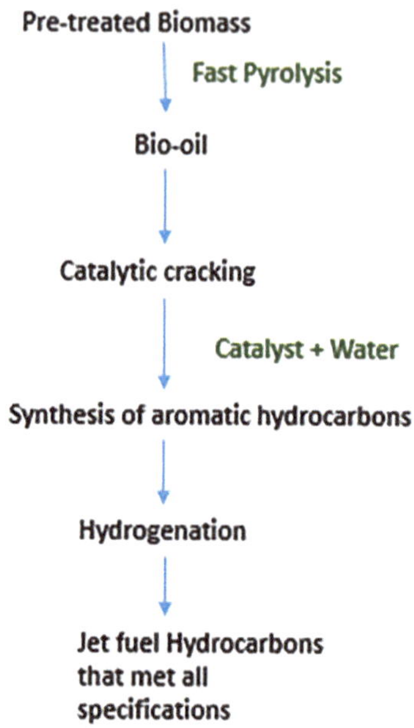

Fig. 5.5 Fast pyrolysis method to produce bio-oil from biomass

and with the presence of catalysts. Later, hydrogenation is performed under high temperatures. The treatment results in the synthesis of biofuels having less unacceptable properties as per the ASTM standards. The fuel production method is further improved by a three-step pathway, in which initially fast pyrolysis of biomass is carried out, which undergoes catalytic cracking; synthesis of aromatic hydrocarbons is the next step, followed by hydrogenation (Sundararaj and Kushari 2019). The bio-oil production from biomass using a fast pyrolysis method is depicted in Fig. 5.5. This technique is still in its initial stage, but various commercial groups, such as Ensyn, LLC, PNNL, UOP, and Tesoro, are dependent on this process for the production of biojet fuels (Abdullah and Battelle 2015).

5.3.6 Fischer–Tropsch (F-T) Synthesis

This method, also called gas to jet, is a catalytic process to transform biomass to jet fuel hydrocarbons with the intermediate step of gasification. The major advantage of producing bio-derived fuels using this method is that it can take any carbon-containing biomass as a feedstock, emits no net carbon dioxide upon combustion, and fits well with environmental regulations (Hu et al. 2012). The conversion process starts with the preliminary treatment of the biomass, which includes screening, drying, and reducing the particle size. It is essential for efficient heat transfer and depletion in the hydrogen content of the gas product. Some of the other pretreatment methods required for proper F-T synthesis are torrefaction, pyrolysis, and compression of the biomass to produce cylindrical pellets (Hu et al. 2012). Further, the gasification of the pretreated biomass is performed in the gasifiers and in the presence of gasification agents. The gasification method is dependent on biomass and the gasifier design. The syngas is produced at the end of gasification and carbon monoxide, hydrogen, carbon dioxide, nitrogen, and methane (Nwokolo et al. 2020). To obtain the required content of hydrogen and carbon monoxide, the optimization of gasification is needed. The remaining impurities are subjected to catalytic cracking and other reactions. Following this, the Fischer–Tropsch synthesis is carried out, which is a set of reactions that occur in the presence of a suitable catalyst to convert syngas to liquid hydrocarbons. The F-T process requires conditioned syngas so as to adjust its ratio of H_2 and CO, which happens using a water gas shift (WGS) reaction. The flowchart containing the synthesis of BJF by utilizing F-T synthesis is described in Fig. 5.6. Ruthenium is the most efficient catalyst, which is also responsible for increasing the cost of the reaction. In comparison, iron is a cheaper alternative to the catalyst that can be used, but it comes with certain disadvantages such as catalyst agglomeration and low product selectivity (Ma and Dalai 2021). This method of production is being utilized to create biojet, which is blended with traditional fuels (SWAFEA 2011). The process does have the advantage of using a variety of feedstock, but it is the most expensive method of all the others discussed (Roberts 2008).

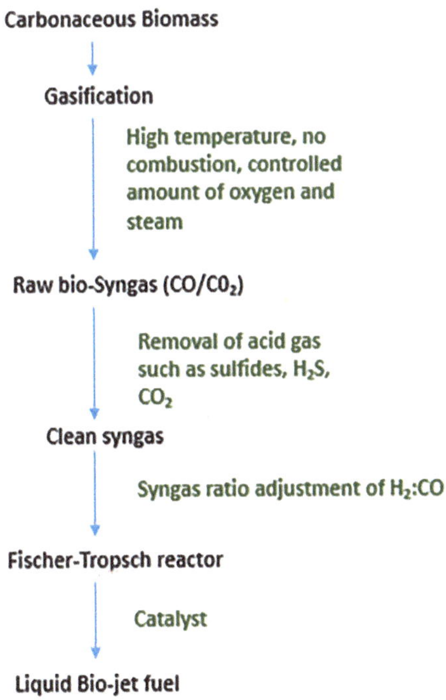

Fig. 5.6 Flowchart outlining the basic steps involved in the synthesis of biojet using Fischer–Tropsch synthesis

5.4 Performance Attributes of BJFs

The fuels are considered to be drop-in alternative jet fuels if they are produced from bio-hydrocarbon, function similarly to existing fuel, and are compatible with existing jets. The performance characteristics play a significant part in evaluating the viability of "drop-in" alternative jet fuels. The performance characteristics of biojet fuels need to be assessed in order to ensure fuel safety, dependability, compatibility with supporting aero-engines and airframe components, and conformity with the ASTM D7566-18 requirements. Here, with a solid grasp of the interplay between their physical and chemical properties, we address the performance attributes of BJFs. Although it becomes arduous to correctly estimate the fuel properties as biofuels are composed of different complex hydrocarbons (Wang et al. 2021). This chapter groups the BJFs' performance characteristics into several physiochemical qualities that need to be examined in accordance with the ASTM guidelines. These characteristics and comparison with traditional gasoline are covered in detail in the rest of the chapter.

5.4.1 Low-Temperature Fluidity

The major characteristic of drop-in fuel is that it should be able to maintain its fluidity even at high altitudes, where the temperature is very low, or at places with extreme climates. Failing to do so, the fuel flow to the engine will be poor or equal to zero. The freezing point and the kinematic viscosity of the fuel are the two parameters that control the low-temperature fluidity of biojet fuels. These two factors are reliant on intermolecular forces between the components of the fuels and, hence, on their molecular structure. To meet the need for proper fluidity of biojet fuels at very low ambient temperatures in high altitudes and low freezing points, kinematic viscosity of the fuel is required to make certain the flow of the fuel in the turbine engine is not affected (Benavides et al. 2021).

5.4.1.1 Freezing Point

The lowest temperature at which a certain fuel does not form hydrocarbon crystals and maintains enough fluidity to allow unobstructed fuel flow from the aviation system's tanks to the engine is referred to as the fuel's freezing point (Benavides et al. 2021). The fuel's freezing point is one factor that affects how biojet fuels behave at low temperatures. The ASTM D2386-19 standard test procedure for aviation fuel freezing point measurement is used to measure it. Differential scanning calorimetry (DSC) is used to calculate the crystallization onset temperature (Tco) (Benavides et al. 2021). The synthesized iso-paraffins (SIP) technique of producing biofuels has a maximum freezing point of −60 °C, while other biofuels created using FT-SPK, HEFA, FT-SPK/A, and ATJ-SPK have a maximum freezing point of −40 °C (Yang et al. 2019). The other two parameters that are used to determine biofuel fluidity are the pour point and the cloud point. The pour point is defined as the measure of the propensity of the fuel to gain viscosity and cease to flow when the temperature is low and the cloud point is explained as the temperature at which the paraffin in biojet fuels begins to separate and becomes cloudy at cold conditions (Demirbas 2009). The freezing point for biojet fuels is found to be lower than the ranges of the pour point (−35 to −15 °C) and cloud points (−15 to 5 °C) that are often used to assess the fluidity of diesel and biodiesel (Yang et al. 2019).

The composition of biojet fuels is majorly responsible for their freezing point. Components having higher viscosity have lower freezing points and, therefore, better fluidity at low temperatures (Pires et al. 2018). In addition to the length of the carbon chain of bio-paraffins, the amount of iso-paraffins and alkylated aromatics in the fuel also affects the freezing point. The appearance of a large amount of branched paraffin contributes to the very low freezing point, such as Sasol FT-SPK, which has a freezing point of about < −77 °C (Renninger et al. 2010). On the contrary, the presence of branched alkanes such as farnesane in SIP fuel contributes to an even lower freezing point of −90 °C (Renninger et al. 2010). The various production routes also result in different composition of biojets, which alters their characteristics. For instance, the freezing point of coconut HEFA-1 and HEFA-2 is higher than −40 °C, and in the event of isomerization, the freezing point of HEFA-2 (−18.5 °C) is lower than that of HEFA-1 (9.5 °C). Similarly, the freezing point of −80 °C is lower for biofuels containing branching cyclohexane that are made from furfural alcohols and aromatic oxygenates via alkylation and hydroxygenation (Han et al. 2017).

There are many other approaches for the production of biofuels, other than those mentioned in this chapter. In one such method, called the H_2SO_4 catalytic one-pot method, the liquid pretreatment and saccharification take place in one vessel. In this process, cyclic alcohols such as cyclohexanol and cyclopentanol, along with branched cycloalkanes like methylcyclohexane and methylcyclopentane, are utilized to produce branched decalins (also called decahydronaphthalene) at room temperature. Branched decalins are excellent components of jet fuel, with properties like high density, high thermal stability, and low freezing points, but their availability by fossil resources is finite (Nie et al. 2018). With a freezing point of less than −51 °C and a high heating value of ~42 MJ/kg, the decalin fuel is a potential jet fuel mixing (Nie et al. 2018). Furthermore, some constituents, like highly branched diamyl ether (DAE), have a freezing point as low as −92 °C. It has the capability of blending with fossil fuels like QAV-1 in various proportions, which results in an adequate freezing point. This DAE can be produced from the thermal cracking of iso-amyl alcohol and C_5 hydrocarbon using an insulated bioreactor with minimum heat transfer (Cataluna et al. 2018).

According to reports, alkylated aromatics have an effect on the freezing point of biojet fuels because propylbenzene reduces the freezing point of HEFA proportionate to the volume injected (Hong et al. 2013). The range of raw materials, including acidified oil, waste cooking oil, soyabean oil, and rubber seed oil, are utilized to create biojet fuels with a freezing point of −37 °C. A significant amount (60–77%) of linear C_8–C_{15} hydrocarbons was obtained by pyrolyzing the source material at 350–450 °C with 5% base catalyst weight. Because of a higher freezing point (−40 °C) than HEFA, HZSM-5 zeolites convert linear hydrocarbons into aromatics at 350 °C for 6 hours, and then aromatics are converted into cycloalkanes using PD/AC for 6 hours at 200 °C. The finished mixture has a freezing point of −47 °C (Li et al. 2018).

As previously indicated, in addition to their composition, the carbon chain's length in bio-paraffins significantly affects the freezing point of created BJFs. Fuels with short carbon chain lengths exhibit desirable low freezing points. The generated hydrocarbons must be hydrocracked in order to reduce the length of the carbon chain (Monteiro et al. 2022). As seen in an example, a biojet fuel substitute carrying short carbon chain limonene (C_{10}) has a freezing point of −97 °C, whereas farnesane (C_{15}) has a much higher freezing point of −40 °C (Yang et al. 2019). In a similar example, bio-kerosene produced as an end product of the catalytic distillation of triglyceride-based oils showed characteristics that were not up to the requirements, especially with regard to the freezing point. Research reported that the freezing point of coconut bio-kerosene is −10 °C and palm kernel bio-kerosene is −15 °C. The probable reason for high freezing points could be due to the carbon chain length without proper hydrocracking (Llamas et al. 2012). Upon blending 20% of palm kernel bio-kerosene with Jet A-1, the freezing point (−41.5 °C) higher than the ASTM D7566-18 specifications was obtained, which was not satisfactory (ElGalad et al. 2018).

The concentrations of iso-paraffins, alkylated aromatics, and the carbon chain length of bio-paraffins are all positively correlated with the freezing point of BJFs, according to a summary of the relationship between the composition of biofuels and their freezing points. Higher alkylated aromatics and iso-paraffin content led to a lower freezing point. Biojet fuels with a short carbon chain composition have a lower freezing point; hydrogenated algal oil had to be hydrocracked in order to lessen the carbon chain length.

5.4.1.2 Kinematic Viscosity at −20 °C

Kinematic viscosity at −20 °C is another criterion that typically characterizes the low-temperature fluidity of aviation gasoline. Kinematic viscosity (KV) is usually described as the internal resistance of the fuel under the effect of gravitational force. It is associated with chain length and degree of saturation of carbon chains (Gouveia et al. 2017). In spite of the fact that ASTM D7566 standards did not specify the limitations of kinematic viscosity of synthesized hydrocarbon fuels, the KV value of 8 mm^2/s at −20 °C is required to be maintained for a blended jet fuel to be considered as a drop-in fuel (Chuck and Donnelly 2014). The kinematic viscosity of fuel should not be very high because it causes various complications like poor atomization, pumping difficulties, incomplete combustion, and the blocking of fuel injectors. The viscosity of the blended jet fuel at a low blend level is suitable for aviation kerosene (Chuck and Donnelly 2014).

At −20 °C, the blended biofuels exhibited kinematic viscosities of less than 8 mm^2/s. However, certain bio-kerosenes, which are obtained through the catalytic distillation of triglyceride-based oils, had high viscosities. According to the research, at −20 °C, the viscosities of the castor HEFA and its even equal blended jet fuel were 5.3 mm^2/s and 3.3 mm^2/s, respectively (Liu et al. 2015). At −20 °C, the kinematic viscosity of the FT-SPK mix, even when blended equally with Jet A-1, is 4.65 mm^2/s (Lobo et al. 2011). Another research shows that ATJ-SPK fuel had a kinematic viscosity of 4.795 mm^2/s at −20 °C. However, SIP fuel shows more kinematic viscosity than FT-SPK, HEFA, and ATJ-SPK, which is 14.28 mm^2/s at −20 °C. The 50 volume % blends of SIP fuel with Jet A-1 had a viscosity of 8.37 mm^2/s and 20 volume % had a viscosity of 5.66 mm^2/s at −20 °C, respectively (Scheuermann et al. 2017).

There is insubstantial information that highlights the association of kinematic viscosity and chemical compositions of biojet fuels. In a study by Chuck and Donnelly (2014), the kinematic viscosities of a few biofuels, such as methyl linolenate, farnesane, *n*-butanol, butyl levulinate, limonene, butyl butyrate, *n*-hexanol, ethyl octanoate, and ethyl cyclohexane, were measured at temperatures between −30 °C and 40 °C. The researchers concluded that the viscosities of biofuels increased with decreasing temperature in a manner similar to an ideal fluid and that *n*-butanol and *n*-hexanol had high viscosities at −20 °C, 12.84 mm^2/s, and 36.21 mm^2/s, respectively, likely because of hydrogen bonding between alcohol groups. Butyl butyrate (C_8) and ethyl octanoate (C_{10}) show beneficial viscosities that are less than 8 mm^2/s at −20 °C, whereas methyl linolenate (C_{18}) had a viscosity of 20.68 mm^2/s and its blended fuel had viscosity of 12.77 mm^2/s at −20 °C (Chuck and Donnelly 2014). This illustrates that at the temperature of −20 °C, a reduced carbon chain length results in a lesser kinematic viscosity.

Likewise, SIP (UQJ-1) fuel had a higher kinematic viscosity of 7.714 mm^2/s at −20 °C when 90 volume % of farnesane (C_{15}) and 10 volume % of limonene (C_{10}) are blended with fuel. But when the SIP fuel contains 97.1 volume % of short-chain limonene, then kinematic viscosity is 3.818 mm^2/s at −20 °C. Rather than hydrocarbon classes, the molecular mass of chemical compounds determines the degree of viscosity of propellant. Due to the higher likelihood of high molecular weight molecules missing the viscosity test for biojet fuel, diaromatics predominated over monoaromatics (Scheuermann et al. 2017).

Most of the biojet fuels like FT-SPK, HEFA, and ATJ-SPK had acceptable kinetic viscosities except SIP; due to the presence of long-chain farnesane (C_{15}) content, the kinetic viscosity was relatively high.

5.4.2 *Stability During Thermal Oxidation*

In biojet fuels, thermal oxidation stability is categorized into two different aspects, thermal stability and oxidation stability. One crucial performance attribute needed in fuels is the biojet fuel's capacity to withstand thermal oxidation at the aircraft's operating temperature. Thermal oxidation stability should be high. The quartz crystal microbalance (QCM) is a suitable technique for measuring the thermal equilibrium of aircraft fuel (Corporan et al. 2011).

5.4.2.1 Thermal Stability

The capacity of biojet fuel to tolerate high temperatures under operating conditions without experiencing noticeable degradation is known as thermal stability, and it may be measured by the amount of deposits that accumulate in the engine fuel system (Lin and Tavlarides 2013). To estimate the thermal stability of biojet fuels, the jet fuel thermal oxidation stability test (JFTOT) is performed, which is standardized under the ASTM D3241 (Christison et al. 2019). Jet fuel deposit formation can be evaluated using two metrics provided by JFTOT: the surface deposit on the test tube and the pressure drop following fuel degradation (Jia et al. 2020). The ASTM D7566-18 standards provide information on these metrics to guarantee the thermal stability of BJFs. The JFTOT test requires that the pressure decrease after 2.5 hours is less than 25 mm Hg and that the surface deposit on the test tube is less than 3 at a temperature of 325 °C (Yang et al. 2019).

In general, biojet fuel has superior thermal stability than traditional jet fuels; nevertheless, there has not been much research done to estimate this thermal stability (Corporan et al. 2011). Fully synthesized jet fuel (FSJF) had very good thermal stability at more than the standard temperature, that is, 360 °C. HEFA had less than the standard value at tube deposit metrics, where almost no pressure drop was detected at 325 °C after 2.5 hours (Amara et al. 2016).

As we know, the thermal stability of biojet fuel is better compared to current in-use jet fuels; this notion is supported by the literature stating that biojet fuels are not much deteriorated under high temperatures than the JP-8, and also, the fully synthetic jet fuels are more resistant to deposit formation under high temperatures in comparison to conventional jets (Corporan et al. 2011). The existence of heteroatom-containing hydrocarbons accounts for contemporary jet fuel's reduced tolerance to high-temperature stress. Benzothiophenes (C_8H_6S) with cyclic sulfur structures may be the reason for the bad thermal stability of conventional jet fuel. As benzothiophene (C_8H_6S) was not present in FSJF, as a result, it shows better thermal stability than conventional jet fuel. Some researchers also concluded that the omission of heteroatom-containing compounds results in better thermal stability (Westhuizen et al. 2011).

Moreover, the presence of aromatic compounds also affects the thermal stability of the biojet fuels. Biojet fuels are almost devoid of aromatics as these are mainly composed of *n*-paraffins, iso-paraffins, and cyclo-paraffins, whereas conventional jet fuel contains about 10–20 weight % of aromatics. The constituents of biojet fuel such as paraffinic compounds were not shown to have the desirable capability of forming deposits at high temperatures, thereby improving their thermal stability. Amara et al. conducted experiments to evaluate the thermal stability of HEFA with the addition of several aromatic compounds, such as xylene, 1-methyl-naphthalene, and tetralin. The results showed that the addition of aromatic compounds in fuels had an effect on pressure drop, with the addition of 1-methyl-naphthalene resulting in a greater than threefold increase in deposit rate (Amara et al. 2016).

To conclude the findings on the comparison of the thermal stability of biojet fuel and conventional fuels, conventional jet fuels have poor thermal stability due to the presence of heteroatom-containing compounds and aromatic compounds. The biojet fuels show better performance in this regard.

5.4.2.2 Oxidative Stability

The term oxidative stability signifies the propensity of the fuel to react with oxygen at moderate temperature. In other words, it is the quantification of the resistance of a fuel to oxidize in the availability of oxygen at a temperature range between 100 and 160 °C (Jia et al. 2020). To measure the extent of fuel degradation by oxidation, the induction period (IP) of the fuel is calculated, which is the time when the fuel achieves the highest oxidation rate (Ben Amara et al. 2014). The thermal oxidative stability of propellant is determined by the physical conditions of the fuel as well as its chemical composition. It is dependent on ambient temperature, the amount of oxygen in the physical environment, the hydrocarbon molecular structure of its compositions, and the concentration of heteroatomic compounds (Odziemkowska et al. 2018). The oxidative stability of biodiesel at a temperature of 110 °C is more than 3 h IP according to the ASTM D6751 standards, whereas it is more than 8 h IP according to the EN 14214 (Moser and Vaughn 2010).

IP for HEFA was around 60 minutes at 140 °C and 7 bar of oxygen pressure. Because of the inclusion of aromatic chemicals, the IP for Jet A-1 was approximately 2.3 hours. To enhance the oxidative stability of HEFA, they blend it with Jet A-1 containing aromatic compounds. Before addition, they evaluate the effect of molecular structure on IP. Now blend of 25 volume % of Jet A-1 with HEFA gives an IP of about 3 hours. It is observed that diaromatic compounds like 1-methyl-naphthalene show higher IP values than monoaromatics compound and hydrocarbons show lower IP values than aromatic compounds (Amara et al. 2016). Further, it is noticed that the oxidative stability of HEFA was enhanced from 1 hour to 8 hours by blending 5 volume % of 1-MN. HEFA's oxidative stability shows average improvement by blending monoaromatics compounds, whereas cyclic alkane shows no improvement in HEFA's oxidative stability. Apart from this, FT-SPK and HEFA had better oxidative stability than conventional JP-8 due to the high oxygen consumption rate and lack of aromatics compounds in FT-SPK and HEFA. Besides, fossil jet fuels contain phenolic antioxidants that also lead to the low oxidative stability of BJFs (Tomar et al. 2023).

In conclusion, compared to commercial jet fuels, the oxidative resistance of biojet fuels was worse because they lacked antioxidant and aromatic components.

5.4.3 Combustion Characteristics

Fuel combustion characteristics are computed in order to examine the impact of biojet fuels, particularly with regard to their effect on climate change and rising greenhouse gas emissions (GHG). In the research led by Sundararaj et al., the biofuels containing camelina and jatropha have better emission characteristics, once tested against fossil fuel-based fuels. For the purpose of assessing how well biofuels burn, several gaseous emissions are taken into account, including carbon monoxide, soot, nitrogen oxides, and unburned hydrocarbons. The study involving the blends of biofuels suggested that the more the amount of camelina in the blend, the lesser the emission of these gases. The release of nitrogen oxides is also dependent on combustion temperature; therefore, there is an increase of nitrogen oxide emission with increasing camelina. Whereas jatropha-based biofuel blends do not follow the same trend and give mixed values (Sundararaj et al. 2019). In an aviation turbine engine, biojet fuel ignites and vaporizes with rapid hot air. Incomplete combustions are the outcome of particulates and unburned hydrocarbons. If the concentration of particulates is high, then it will be seen as smoke or soot. The metrics used to assess the BJF's combustion properties include the derived cetane number (DCN), smoke point, particle matter (PM), carbon dioxide (CO_2), and monoxide (CO) emissions (Yang et al. 2019).

5.4.3.1 Smoke Point

The temperature at which a particular fuel starts to produce smoke is known as its "smoke point." A fuel with a high smoke point is thought to have a low tendency to produce smoke. The fuel's smoke point is measured with the specific wick-fed test lamp, where the height of the highest flame produced (in millimeters) is checked, which is given off without soot breakthrough (Jiao et al. 2015), thereby assessing the combustion properties of the fuel. For instance, the smoke point of fossil jet fuels is 25 mm in height of flame without smoke production (Saffaripour et al. 2011).

The smoke produced is influenced by the amount of heavy hydrocarbon particles present in fuel. The lesser the concentration of aromatic hydrocarbons, the greater its smoke point, therefore, the better its burning quality. FT-SPK and HEFA show remarkable combustion performance with smoke points higher than 40 mm. The currently used jet fuel JP-8 had a smoke point of 25 mm and FT-SPK had a higher smoke point than JP-8, which is more than 50 mm. The difference between the smoke points of FT-SPK and JP-8 is due to the presence of aromatic contents. The soot-forming capacity is higher in aromatic compounds; thus, JP-8 shows a lower smoke point, whereas FT-SPK, which is free from aromatic compounds, shows a higher smoke point. Blending conventional aviation fuel with biojet fuel raised the smoke point of the fuel. By adding 20 volume % of bio-kerosene (palm kernel biodiesel) into Jet A-1, there is a minute increase in smoke point from 27.1 to 29.1 mm (Corporan et al. 2007). All the biojet fuel blends have high smoke points because of the least aromatic content, low density, and higher hydrogen concentration than conventional fuels (Sundararaj et al. 2019). If we consider the example of biofuel blend 3, which is made of 90% universal oil products – synthetic paraffinic kerosene (UOP-SPK) and 10% Van-Sol 53, its chemical composition is composed of the least aromatic content, lowest density, and highest hydrogen content of all of the blends possible, consequently having high smoke point (Sundararaj et al. 2019).

In addition, the threshold sooting index (TSI) assesses the soot-forming capacity of conventional jet fuels and biojet fuels, which is also used to test the combustion characteristics of fuel. The TSI is linearly associated with the density of the fuel and the smoke point. The TSI of biojet fuels is relatively lower; for example, if we consider the TSI of Shell FT-SPK (9.11), Sasol FT-SPK (17.28), camelina HEFA (11.99), and tallow HEFA (11.58), whereas conventional jet fuels JP-8 have TSI of 19.28. After this study, a new fuel oil substitute will be created using an advanced optimization methodology to measure composition that satisfies sooting capacity, physiochemical properties, and optimized mole fraction for decalin (0.1449), toluene (0.2591), iso-octane (0.0195), iso-cetane (0.2059), and *n*-dodecane (0.3706) (Yu et al. 2018).

5.4.3.2 Particulate Matter (PM) Emissions

Particulate matter emissions are caused by noncombustible fuel components that have the potential to produce smog, which is harmful to both human health and the environment (Tiwari et al. 2023). The PM emissions in alternative jet fuel depend on the amount of aromatics compounds present. As FT-SPK has extremely low aromatics content, there is 52% of PM number reduction. Hence, the reduction in PM number and PM mass is achieved by blending FT-SPK with jet fuels. When PM emissions from aircraft are assessed, it is found that blended fuels, such as a camelina HEFA blend with Jet A, lower mass and PM emissions (Moore et al. 2017). Biojet fuels producing PM have a particle size smaller than fossil jet fuels. Farnesane showed a low potential from soot intermediates in the kinetic modeling of its burning, whereas *p*-cymene produces comparatively more naphthalene (Oßwald et al. 2017). This implies that low PM emissions of biojet fuels are due to the absence of aromatic content.

5.4.3.3 Gaseous Emissions

During the different phases of flight, such as take-off, climb, and cruise, the emission from the burnt fuel contains different concentrations of gases. Some of the gases released from jet fuel are carbon monoxide, carbon dioxide, nitrogen oxides (NOx), and unburned hydrocarbons (UHC) (Gaspar and Sousa 2016). In the different studies conducted, it was seen that the gaseous emission of biojet fuel blends was less than fossil-derived jet fuels. In an experiment conducted by Timko et al., it was observed that FT-SPK when blended with conventional jet fuels in a ratio of 1:1 emits 5% gaseous emission, whereas pure jet fuel produces 10% of NOx, in particular. There is a slight reduction in CO emission also in comparison to standard jet propellant (Timko et al. 2011). A study conducted by Corporan et al. suggests that the emission of NOx and CO_2 is similar in both biojets (tests conducted in biofuels produced using FT-SK and HEFA routes) and JP-8 (with no blends with biofuel). Although a 10–25% reduction in CO and UHC emission index was seen, owing to the fact that lesser aromatic hydrocarbons were present in biofuels produced by this method (Corporan et al. 2011).

The data emphasizes the fact that there is a moderately lower emission of gases upon combustion of biojet fuels in contrast with jet fuels, which is not very significant. The explanation for this result could be the improper mixing of the fuel blends, the difference in viscosity and density of fuel blends, and the high fuel-to-air ratio (Sundararaj et al. 2019).

5.4.3.4 Derived Cetane Number (DCN)

DCN constitutes characteristics of ignited fuels considering the minimum standards that are set by various countries (Prak et al. 2021). With more combustion of fuel, there is an increase in DCN value, which also indicates that there is a decrease in ignition delay time. Therefore, higher DCN specifies better combustion performance, in addition to lower harmful emissions. The DCN number in the fuel is affected by the amount of aromatic hydrocarbons present in the fuel. The DCN value of Jet A fuel is calculated to be 49.35, which is comparatively lesser than FT-SPK and HEFA, which have a DCN value of 33.46 (Hui et al. 2012).

5.4.4 Consistency with the Current Aviation Fueling Infrastructure

Biojet fuels are functionally equal to or better than fossil-derived jet fuels as they reflect the excellent characteristics mentioned above. Nonetheless, it is crucial to take into account how well biojet fuels and elastomers work together. Also, 10–20% of conventional jet fuel contains aromatic compounds; biojet fuels do not contain aromatic compounds, which can cause fuel leakage because they cause O-ring seals to harden and shrink. A blend of biojet fuel and conventional jet fuel permits enough quantity of aromatic compounds to ensure the purity of engine seals (IRENA 2017). We go into great detail regarding the volume swell of sealant and lubricity in this section.

5.4.4.1 Volume Swell of Seal Materials

Alternative fuel enhancement is limited by the volume swell of seal material compatibility. It is necessary to evaluate whether BJF is suitable with engine seals prior to commercialization. The low density of biojet fuel due to the lack of aromatic compounds causes shrinkage in the seal. Because of this seal shrinking, we have seal failure, which further causes damage to the system. The two primary parameters in the aircraft system that determine the volume swell of sealant are the strength of the interaction between aircraft fuel and seal materials. Because of their large molecular weight, aromatic chemicals, such as naphthalene, have excellent interaction with seal polymers.

The three most commonly used seals in aircraft engines are fluorocarbon, fluorosilicone, and nitrile seals. Nitrile rubber is usually used as an O-ring seal in aircraft engines because it shows a greater response toward aromatic compounds than fluorosilicate and fluorocarbon seals (Moses 2008). Leakage in the hydraulic system and engine is prevented by the elastomers, like O-ring seals. The sealing function of O-ring elastomer is because of deformation when it is crushed between two parts of the engine (Qin et al. 2019).

In the O-ring, two effects that are usually seen are swelling and shrinking of the O-ring; an increase in seal volume is defined as swelling of the O-ring, here elastomer absorbs chemical components of fuel that result in swelling, whereas a decrease in seal volume is defined as shrinking of the O-ring, here the O-ring degrades when some components are released into the fuel and absorbed by the seal. The seal defiance with regard to fuel is indicated by swelling of the O-ring (Liu and Wilson 2012).

However, adding aromatic chemicals to biojet fuels improves their compatibility and may also lead to an increase in PM emissions. Both concentrations of aromatic compounds and types of aromatic compounds used are correlated to the PM emissions and volume swell. The concentration of aromatic compounds is directly proportional to the PM emissions and volume swell. An increase in the molecular weight of aromatic compounds causes an increase in PM emissions (DeWitt et al. 2008).

Less than 10% of aromatic compounds with greater molecular weight and more than 10% of arene compounds with a lower molecular weight must be added in order to produce biojet propellant that meets the required output standards for volume swell and PM emissions (Yang et al. 2019).

5.4.4.2 Lubricity

Lubricity is the capacity of the fuel to reduce wear or friction between two surfaces of engine components in relative motion. Good lubricity of fuel is important for the engine to run smoothly. The lubricity of a substance or fuel depends on the fuel composition, and it is not an intrinsic property. In ASTM D7566-18, the synthesized hydrocarbons do not have any specific lubricity limits (Elkelawy et al. 2022). The presence of polar compounds in biojet fuels is directly associated with the lubricity of the fuel (Hari et al. 2015).

The main disadvantage of biojet fuel production approaches of BJFs is that they comprise various steps of hydrotreatment processes due to which the compounds containing oxygen, nitrogen, and sulfur are removed from synthesized hydrocarbons, thus ensuing lubricity of below standard (Hari et al. 2015). This limitation of BJF is withdrawn by making blends of BJFs with suitable conventional jet fuels as it contains 700 ppm (parts per million) sulfur or adding additives like fatty acid methyl ester (FAME), which is used as an additive in HEFA fuel to enhance the lubricity. The amount of FAME in HEFA is limited due to its poor low-temperature fluidity as according to the ASTM D7566 standards it should be less than 5 ppm (McNutt 2016).

As per the ASTM D7566 standards, the compatibility and characteristics of BJF combustion are balanced by blending about 8 weight % of aromatic compounds in the final blend (Lahijani et al. 2022). This implies that biojet fuels show poor lubricity due to the absence of naturally occurring compounds like oxygen, nitrogen, and sulfur and the absence of compounds with polarity.

5.4.5 Volatility of Fuel

Fuel volatility is defined as the fuel's ability to evaporate quickly. Fuel volatility is caused by two key features, which are covered in the sections that follow: the distillation property and the flash point.

5.4.5.1 Distillation Property

The distillation property describes the percentage of fuel vaporized with the increase in temperature, that is, it tells us about the percentage of recovery fraction when fuel is burnt (del Coro Fernández-Feal et al. 2017). The distillation property is determined by the concentration of volatile substances present in the fuel and the amount of residue left after the combustion. This can be tested using a distillation test (ASTM D1160 2015). The temperature of the boiling point (BP) of the fuel has an impact on its vaporization and combustion (Kook and Pickett 2010). The BP is defined as initial BP, mid-BP, and final BP. The initial BP is the temperature at which the fuel starts to evaporate, mid-BP is the temperature at which half of the fuel has been vaporized, and final BP is the temperature at which 100% of the fuel sample is evaporated (Sundararaj et al. 2019). The fuel with the low BP has the advantage of vaporizing readily and thus complete combustion of the fuel (Maly et al. 2007). The complete combustion leads to low PM emissions. Although the complete combustion of fuels with low BP leads to low PM emission, there is a release of more nitrogen oxide (NOx). The reason for NOx emission is that the quick evaporation of the fuel causes more of the fuel to get mixed with the air before the actual combustion starts, which increases the volume of the flammable mixture and, thus, more heat emission (Kook and Pickett 2010). The higher final BP is linked with more smoke and PM release. This distillation property needs to be critically considered to understand energy penetration so that product optimization is done during fuel production (Acosta-Solórzano et al. 2016).

The standard distillation range selected is as temperature at 10% recovery (T10) should be less than 205 °C and final BP should be less than 300 °C. Upon investigation of the distillation property of alternative jet fuels, it was seen that FT-SPK and HEFA have a distillation range within the set standard range (Wierzbicki et al. 2014). The distillation range of some of the BJFs is mentioned in Table 5.2.

The fuel's constituents have an effect on the fuel's distillation range as well. For example, ethyl cyclohexane (C_8H_{16}) has a lower boiling point than Jet A-1 because its carbon chain is shorter. The HEFA fuel and its blends have distillation temperatures at all fractions because of their higher chain length (C_{17}) (Scheuermann et al. 2017).

5.4.5.2 Flash Point

The lowest temperature at which a liquid's vapors are concentrated enough to create an ignitable vapor in the presence of an ignition source is known as the flash point. It represents fuel volatility (Kong et al. 2003). Fuels can be classified as combustible, flammable, or gasoline based on their flash point. It is commonly used to evaluate the handling as well as hazards of flammable substances during storage and shipping (Hassan et al. 2023). Fuels are classified as combustible when their flash point exceeds 37.8 °C and flammable fuels when their flash point falls below 37.8 °C (Kong et al. 2003). According to the ASTM D7566-18 standards, FT-SPK, HEFA, FT-SPK/A, and ATJ-SPK should have a minimum flash point of 37.8 °C. But

Table 5.2 Biojet fuels' spectrum of distillation properties

Biojet fuel	T10	Final BP	References
FT-SPK	179 °C	225 °C	Wierzbicki et al. (2014)
HEFA	179 °C	255 °C	Wierzbicki et al. (2014)
50 vol.% FT-SPK/JP-8	–	268 °C	Corporan et al. (2007)
20 vol.% farnesene/jet A-1	205 °C	–	Chuck and Donnelly (2014)
50 vol.% farnesene/jet A-1	220 °C	–	Chuck and Donnelly (2014)
Jet A-1	167.2–175.3 °C	243.7–258.5 °C	Yang et al. (2019)

because farnesane (C_{15}) has a long carbon chain and a high flash point, synthetic iso-paraffins (SIP) fuel needs a minimum flash temperature of 100 °C. Pensky Martens Flash Point Tester is the equipment used to measure the flash point of BJFs (Hristova 2013). The flash points of various alternative fuels were studied, and it was concluded that biofuels with low-BP aliphatic components have low flash points, in contrast to the fuels with high-BP aromatic compounds, which have high flash points (Scheuermann et al. 2017). Thus, the fuel's flash points are likewise influenced by the chemical components' BJFs.

5.4.6 Fuel Metering and Aircraft Range

The fuels in its liquid state are not combustible. Correct air and fuel mixture are required for the proper and complete combustion of fuel. The fuel metering system is a device that allows the proper fuel flow while maintaining the air/fuel ratio required for the clean combustion of the fuel at existing engine operating conditions (Hideg 1982). The jet load and jet range are very well impacted by the density of the fuel. Since the fuel occupies the engine of the aircraft volumetrically, the density of the fuel is the major criterion in deciding the flow calculations, adjusting the fuel metering device, and calculations with respect to thermal expansions of the fuel (Vozka et al. 2019). The amount of heat energy produced upon the combustion of fuel is directly proportional to fuel density, fuel volume, and net heat of combustion. The fuel with higher energy extent allows more aviation range and higher payload. Besides, the reduced thermal energy generated by the full combustion of fuel leads to a significant increase in fuel consumption, which raises the expense of jet operations (Yang et al. 2019). The fuel density and composition of fuel that contribute to their densities, along with the net heat of combustion of BJFs, are elaborated below.

5.4.6.1 Density of Fuel at 15 °C

As per the standard density values for the BJFs to be a drop-in fuel, ASTM D7566-18 has decided the density range of 730–770 kg/m^3 at 15 °C (Green et al. 2020). Table 5.3 displays the densities of a few biojet fuels at 15 °C.

The densities of FT-SPK, HEFA, and ATJ-SPK are all within the ideal range for them to generate a considerable amount of heat energy when they burn. Whereas SIP does not have the optimum density because of the presence of a large content of long-chain farnesene. The blends of SIP also do not provide a satisfactory density range. The best-suited SIP fuel with optimum density at 15 °C is a blend of 90 volume % farnesene from SIP and 10 volume % limonene (Chuck and Donnelly 2014).

The biojet fuel with a relatively higher amount of aromatics provides even more high fuel density. Therefore, the blends of biojet fuels with current fuels are expected to have better densities in terms of enhancing their energy content. Scheuermann et al. in 2017 tested the fuel density of blends of ATJ-SPK/A with 15.8 volume % of

Table 5.3 Fuel density of various biojet fuels

Fuel	Density at 15 °C	References
FT-SPK	737 kg/m^3	Corporan et al. (2011)
HEFA	751 kg/m^3	Corporan et al. (2011)
ATJ-SPK	757.1 kg/m^3	Scheuermann et al. (2017)
SIP	765–780 kg/m^3	Chuck and Donnelly (2014)
Pure farnesene	795 kg/m^3	Chuck and Donnelly (2014)
20 vol.% farnesene/Jet A-1	790 kg/m^3	Chuck and Donnelly (2014)
50 vol.% farnesene/Jet A-1	785 kg/m^3	Chuck and Donnelly (2014)
SIP with 90 vol.% farnesene/10 vol.% of limonene	778 kg/m^3	Chuck and Donnelly (2014)
Jet A-1	803 kg/m^3	Corporan et al. (2011)

aromatics resulting in a higher density of 785.9 kg/m^3. These findings suggested that the concentration of aromatics in the biofuels is directly related to the density of the fuel and eventually to the heat energy production of fuel upon combustion.

5.4.6.2 Net Heat of Combustion

For both conventional and blended fuels, the ASTM D7566-18 specifies that the net heat combustion value must be greater than 42.8 MJ/kg. The biofuels are also known to have optimum net heat combustion. The net heat of combustion of various biofuels is SIP has 43.93 MJ/kg (Brennan et al. 2012), farnesene has 47 MJ/kg (Rude and Schirmer 2009), FT-SPK and HEFA have 44 MJ/kg (Hui et al. 2012), and 50 volume % FT-SPK/Jet A-1 blend fuel had lower net heat of combustion (43.7 MJ/kg), in comparison to pure FT-SPK (Timko et al. 2011). There is a slight dip in the net heat of combustion when biofuels are used as blends as the conventional jet fuels have availability of aromatics in their composition. The decrease in the ratio of H/C (hydrogen/carbon content) has reportedly shown lower net heat of combustion and aromatics having a lesser H/C ratio as it contains one or more double bonds (Lobo et al. 2011).

5.5 Challenges and Future Look

This chapter has thoroughly described the characteristics of biofuels and made the comparison of biofuels with fossil-based fuels. Consequently, it can be stated that biojet fuels are the better choice for the selection of fuel as they are technologically advanced and ecologically sustainable. There are a few areas in which biojet fuel needs improvement to overcome the small challenges it faces to replace fossil fuel-based aviation fuels completely. Using the techniques outlined in this chapter to generate BJFs for commercialization is expensive and currently unable to satisfy fuel demand. The cost of the production is affected by the feedstock used for the production of biojet fuels. The selection of the production route and raw material feedstock can be worked upon to reduce the cost involved. Furthermore, it has been tested that not all production routes and choice of feedstock are capable of reducing greenhouse gas emissions (GHG) (Cui et al. 2018).

Less information is available with regard to the correct amount of constituents present in the biojet fuels, and thereby, their relation with the performance characteristics of the fuel. More research and proper characterization are required to deeply understand the role of each component present in fuels (ElGalad et al. 2018). Not all of the characteristics of biojet fuels are covered in the already published work, such as the presence of gum, corroding properties, water separation trait, and electrical conductivity. These properties are not readily studied while selecting a fuel, but these also impact the performance characteristics of the fuel in a great way. For example, studying the gum-existent feature of the fuel gives information about the contamination of high-BP oils and particulate matter in the fuel. Moreover, the gum in the fuel makes it difficult to store (Yang et al. 2019).

Further research on the fuels' characteristics, such as soot generation paths, combustion species profiles, laminar flame speeds, and extinction limits, is necessary before considering biojet fuels. Research is required in this direction as the long-term combustion of biojets and blends is not documented much (Yang et al. 2019).

There is an insufficiency of effective government policy incentives to promote the switch from traditional fuels to biofuel. Moreover, there are strict guidelines to be followed, which pose difficulty in the production of BJFs. The field is also facing a lack of investments owing to the fact that the returns expected from biojet fuels are uncertain. There is also a negative perception associated with safety while using biojet (Lim et al. 2023). Lastly, some of the undesirable properties are witnessed with the synthesis of biojet fuels through the methods mentioned, such as constituents of fuel with long-chain carbon atoms or fuels with oxygen in distillate having properties that do not adhere to the guidelines. The evaluation of each constituent is therefore important to understand the performance characteristics of the fuel designed. Alternative approaches to producing BJFs are being adopted. One such approach is the catalytic synthesis of high-density BJFs using bio-derived furfurals as biomass. This process uses alkylation, aldol condensation, and hydrodeoxygenation (Han et al. 2017). These alternative methodologies are the main scope of biojet fuels, which needs to be characterized more.

5.6 Conclusion

With an increased demand for aircraft travelers, emission reduction has become the main area of research nowadays. The application of biojet fuels has become quintessential to dealing with deficient fossil fuel supply and environmental problems that come with fuels. Many production routes have been optimized for the production of biofuels such as FT-SPK, HEFA, FT-SPK/A, SIP, and ATJ-SPK as specified in the ASTM D7566-18. The evaluation of performance characteristics suggested that the chemical composition of the fuel is highly influencing its performance. The BJFs demonstrated acceptable low-temperature fluidity in fuels with more levels of iso-paraffins, short-chain paraffins, and alkylated aromatic content. Moreover, high kinematic viscosity is also observed in SIP fuels having high farnesane content, thereby increasing the low-temperature fluidity. Biofuels have relatively greater thermal stability, but the oxidative thermal stability is still questionable due to the presence of high paraffin in biofuels. Less particle emission, gaseous release, a high smoke point, and derived cetane number are among the combustion properties of the BJFs that also meet the required standards. However, while blending with the current aviation fueling system, BJFs' lubricity and compatibility are unsatisfactory. The amount of aromatics that is near zero is not compatible with the volume swell of seal materials and thus can lead to shrinkage and leakage. Due to the ideal chain length of the components that make up these fuels, the distillation property of the BJFs is adequately good. The flash point of these is also fitting within the range due to the presence of low-BP aliphatic components. The fuel metering and aircraft range are acceptable as per specified standards.

This information is helpful in understanding the practicality of biojet production. Further research such as the life cycle study of the fuel is necessary to comprehend the carbon footprints and efficiency of biojets. Efforts are required to improve the performance characteristics as well as ensure the storage stability of the fuels.

References

Abdullah Z, Battelle MI (2015) Upgrading of biomass Fast Pyrolysis Oil (Bio-oil)

Acosta-Solórzano AD, Guerrero-Farfán O, Ramírez-Márquez C, Gómez-Castro FI, Segovia-Hernández JG, Hernández S et al (2016) Controllability analysis of distillation sequences for the separation of Bio-Jet fuel and Green diesel fractions. Chem Eng Technol 39(12):2273–2283

Amara AB, Kaoubi S, Starck L (2016) Toward an optimal formulation of alternative jet fuels: enhanced oxidation and thermal stability by the addition of cyclic molecules. Fuel 173:98–105

ASTM D1160 (2015) Standard test method for distillation of petroleum products at reduced pressure. Annual Book of Standards

ATAG (2020). Air Transport Action Group- Facts & figures. https://atag.org/facts-figures.

Ben Amara A, Lecointe B, Jeuland N, Takahashi T, IIda Y, Hashimoto H, Bouilly J (2014) Experimental study of the impact of diesel/biodiesel blends oxidation on the fuel injection system. SAE Int J Fuels Lubr 7. https://doi.org/10.4271/2014-01-2767

Benavides A, Benjumea P, Cortés FB, Ruiz MA (2021) Chemical composition and low-temperature fluidity properties of jet fuels. PRO 9(7):1184

Bozell, J. J., Moens, L., Wang, Y., Neuenswander, G. G., Fitzpatrick, S. W., Bilski, R. J., & Jarnefeld, J. L. (2000). The use of renewable feedstocks for the production of chemicals and materials. National Renewable Energy Laboratory 1617 Cole Boulevard Golden, CO, 8040

Brennan TC, Turner CD, Krömer JO, Nielsen LK (2012) Alleviating monoterpene toxicity using a two-phase extractive fermentation for the bioproduction of jet fuel mixtures in Saccharomyces cerevisiae. Biotechnol Bioeng 109(10):2513–2522

Cataluna R, Shah Z, Venturi V, Caetano NR, da Silva BP, Azevedo CM et al (2018) Production process of di-amyl ether and its use as an additive in the formulation of aviation fuels. Fuel 228:226–233

Christison KM, Lorenz RM, Xue L, Sparkman OD (2019) Exploring the molecular origin of jet fuel thermal oxidative instability through statistical analysis of mass spectral data. Energy Fuel 33(2):830–836

Chuck CJ, Donnelly J (2014) The compatibility of potential bioderived fuels with jet A-1 aviation kerosene. Appl Energy 118:83–91

Corporan E, DeWitt MJ, Belovich V, Pawlik R, Lynch AC, Gord JR, Meyer TR (2007) Emissions characteristics of a turbine engine and research combustor burning a Fischer– Tropsch jet fuel. Energy Fuel 21(5):2615–2626

Corporan E, Edwards T, Shafer L, DeWitt MJ, Klingshirn C, Zabarnick S et al (2011) Chemical, thermal stability, seal swell, and emissions studies of alternative jet fuels. Energy Fuel 25(3):955–966

Cui Q, Li Y, Lin JL (2018) Pollution abatement costs change decomposition for airlines: an analysis from a dynamic perspective. Transp Res A Policy Pract 111:96–107

del Coro Fernández-Feal MM, Sánchez-Fernández LR, Sánchez-Fernández B (2017) Distillation: basic test in quality control of automotive fuels. In: Distillation innovative applications and modeling, p 77

Demirbas A (2009) Characterization of biodiesel fuels. Energy Sources Part A 31(11):889–896

DeWitt MJ, Corporan E, Graham J, Minus D (2008) Effects of aromatic type and concentration in Fischer– Tropsch fuel on emissions production and material compatibility. Energy Fuel 22(4):2411–2418

Díaz-Pérez MA, Serrano-Ruiz JC (2020) Catalytic production of jet fuels from biomass. Molecules 25(4):802

Doliente SS, Narayan A, Tapia JFD, Samsatli NJ, Zhao Y, Samsatli S (2020) Bio-aviation fuel: a comprehensive review and analysis of the supply chain components. Front Energy Res 8:110

ElGalad MI, El-Khatib KM, Abdelkader E, El-Araby R, ElDiwani G, Hawash SI (2018) Empirical equations and economical study for blending biofuel with petroleum jet fuel. J Adv Res 9:43–50

Elkelawy M, Bastawissi HAE, Radwan AM, Ismail MT, El-Sheekh M (2022) Biojet fuels production from algae: conversion technologies, characteristics, performance, and process simulation. In: Handbook of Algal biofuels. Elsevier, pp 331–361

Gaspar RMP, Sousa JMM (2016) Impact of alternative fuels on the operational and environmental performance of a small turbofan engine. Energy Convers Manag 130:81–90

Geleynse S, Brandt K, Garcia-Perez M, Wolcott M, Zhang X (2018) The alcohol-to-jet conversion pathway for drop-in biofuels: techno-economic evaluation. ChemSusChem 11(21):3728–3741

Gouveia L, Oliveira AC, Congestri R, Bruno L, Soares AT, Menezes RS, Tzovenis I (2017) Biodiesel from microalgae. In: Microalgae-based biofuels and bioproducts. Woodhead Publishing, pp 235–258

Grande L, Pedroarena I, Korili SA, Gil A (2021) Hydrothermal liquefaction of biomass as one of the most promising alternatives for the synthesis of advanced liquid biofuels: a review. Materials 14(18):5286

Green R, Cooper S, McManus M (2020) Fuel standards summary. University of Bath

Han P, Nie G, Xie J, Xiu-tian-feng E, Pan L, Zhang X, Zou JJ (2017) Synthesis of high-density biofuel with excellent low-temperature properties from lignocellulose-derived feedstock. Fuel Process Technol 163:45–50

Han GB, Jang JH, Ahn MH, Jung BH (2019) Recent application of bio-alcohol: Bio-jet fuel, vol 1. IntechOpen, pp 109–121

Hari TK, Yaakob Z, Binitha NN (2015) Aviation biofuel from renewable resources: routes, opportunities and challenges. Renew Sust Energ Rev 42:1234–1244

Hassan SH, Attia NK, El Diwani GI, Amin SK, Ettouney RS, El-Rifai MA (2023) Catalytic hydrocracking of jatropha oil over natural clay for bio-jet fuel production. Sci Rep 13(1):13419

Hideg L (1982) U.S. Patent no. 4,357,923. U.S. Patent and Trademark Office, Washington, DC

Hong TD, Soerawidjaja TH, Reksowardojo IK, Fujita O, Duniani Z, Pham MX (2013) A study on developing aviation biofuel for the tropics: production process—experimental and theoretical evaluation of their blends with fossil kerosene. Chem Eng Process Process Intensif 74:124–130

Hristova M (2013) Measurement and prediction of binary mixture flash point. Cent Eur J Chem 11:57–62

Hu J, Yu F, Lu Y (2012) Application of Fischer–Tropsch synthesis in biomass to liquid conversion. Catalysts 2(2):303–326

Hu Y, Bassi A, Xu CC (2020) Energy from biomass. In: Future Energy. Elsevier, pp 447–471

Hui X, Kumar K, Sung CJ, Edwards T, Gardner D (2012) Experimental studies on the combustion characteristics of alternative jet fuels. Fuel 98:176–182

ICAO Environment (2018) Sustainable aviation fuel guides. Version 2

IRENA (2017) Renewable Energy Statistics 2017. The International Renewable Energy Agency, Abu Dhabi

Jia T, Gong S, Pan L, Deng C, Zou JJ, Zhang X (2020) Impact of deep hydrogenation on jet fuel oxidation and deposition. Fuel 264:116843

Jiao Q, Anderson JE, Wallington TJ, Kurtz EM (2015) Smoke point measurements of diesel-range hydrocarbon–oxygenate blends using a novel approach for fuel blend selection. Energy Fuel 29(11):7641–7649

Kong D, am Ende DJ, Brenek SJ, Weston NP (2003) Determination of flash point in air and pure oxygen using an equilibrium closed bomb apparatus. J Hazard Mater 102(2–3):155–165

Kook S, Pickett LM (2010) Effect of fuel volatility and ignition quality on combustion and soot formation at fixed premixing conditions. SAE Int J Engines 2(2):11–23

Kramer S, Andac G, Heyne J, Ellsworth J, Herzig P, Lewis KC (2022) Perspectives on fully synthesized sustainable aviation fuels: direction and opportunities. Front Energy Res 9:782823

Lahijani P, Mohammadi M, Mohamed AR, Ismail F, Lee KT, Amini G (2022) Upgrading biomass-derived pyrolysis bio-oil to bio-jet fuel through catalytic cracking and hydrodeoxygenation: a review of recent progress. Energy Convers Manag 268:115956

Li L, Coppola E, Rine J, Miller JL, Walker D (2010) Catalytic hydrothermal conversion of triglycerides to non-ester biofuels. Energy Fuel 24(2):1305–1315

Li F, Jiang J, Liu P, Zhai Q, Wang F, Hse CY, Xu J (2018) Catalytic cracking of triglycerides with a base catalyst and modification of pyrolytic oils for production of aviation fuels. Sustainable Energy Fuels 2(6):1206–1215

Lim JHK, Gan YY, Ong HC, Lau BF, Chen WH, Chong CT et al (2021) Utilization of microalgae for bio-jet fuel production in the aviation sector: challenges and perspective. Renew Sust Energ Rev 149:111396

Lim M, Luckert MM, Qiu F (2023) Economic opportunities and challenges in biojet production: a literature review and analysis. Biomass Bioenergy 170:106727

Lin R, Tavlarides LL (2013) Thermal stability and decomposition of diesel fuel under subcritical and supercritical conditions. J Supercrit Fluids 75:101–111

Lin CH, Chen YK, Wang WC (2020) The production of bio-jet fuel from palm oil derived alkanes. Fuel 260:116345

Liu Y, Wilson CW (2012) Investigation into the impact of n-decane, decalin, and isoparaffinic solvent on elastomeric sealing materials. Advances in Mechanical Engineering 4:127430

Liu S, Zhu Q, Guan Q, He L, Li W (2015) Bio-aviation fuel production from hydroprocessing castor oil promoted by the nickel-based bifunctional catalysts. Bioresour Technol 183:93–100
Llamas A, García-Martínez MJ, Al-Lal AM, Canoira L, Lapuerta M (2012) Biokerosene from coconut and palm kernel oils: production and properties of their blends with fossil kerosene. Fuel 102:483–490
Lobo P, Hagen DE, Whitefield PD (2011) Comparison of PM emissions from a commercial jet engine burning conventional, biomass, and Fischer–Tropsch fuels. Environ Sci Technol 45(24):10744–10749
Ma W, Dalai AK (2021) Effects of structure and particle size of iron, cobalt and ruthenium catalysts on Fischer–Tropsch Synthesis. Reactions 2(1):62–77
Maly RR, Schaefer V, Hass H, Cahill GF, Rouveirolles P, Röj A et al (2007) Optimum diesel fuel for future clean diesel engines. SAE Trans:28–47, Vol116
Mawhood RK, Gazis E, Hoefnagels R, De Jong S, Slade R (2015) Technological and commercial maturity of aviation biofuels: emerging options to produce jet from lignocellulosic biomass
McNutt J (2016) Development of biolubricants from vegetable oils via chemical modification. J Ind Eng Chem 36:1–12
Monteiro RR, dos Santos IA, Arcanjo MR, Cavalcante CL Jr, de Luna FM, Fernandez-Lafuente R, Vieira RS (2022) Production of jet biofuels by catalytic hydroprocessing of esters and fatty acids: a review. Catalysts 12(2):237
Moore RH, Thornhill KL, Weinzierl B, Sauer D, D'Ascoli E, Beaton B et al (2017) Biofuel blending reduces aircraft engine particle emissions at cruise conditions. Nature 543(7645):411
Moreno-Gómez AL, Gutiérrez-Antonio C, Gómez-Castro FI, Hernández S (2020) Production of Biojet fuel from waste raw materials: a review. In: Process systems engineering for biofuels development, pp 149–171
Morgan T, Santillan-Jimenez E, Harman-Ware AE, Ji Y, Grubb D, Crocker M (2012) Catalytic deoxygenation of triglycerides to hydrocarbons over supported nickel catalysts. Chem Eng J 189:346–355
Moses CA (2008) Comparative evaluation of semi-synthetic jet fuels. Contract 33415(02-D):2299
Moser BR, Vaughn SF (2010) Coriander seed oil methyl esters as biodiesel fuel: unique fatty acid composition and excellent oxidative stability. Biomass Bioenergy 34(4):550–558
Ng KS, Farooq D, Yang A (2021) Global biorenewable development strategies for sustainable aviation fuel production. Renew Sust Energ Rev 150:111502
Nie G, Zhang X, Pan L, Wang M, Zou JJ (2018) One-pot production of branched decalins as high-density jet fuel from monocyclic alkanes and alcohols. Chem Eng Sci 180:64–69
Nwokolo N, Mukumba P, Obileke K (2020) Gasification of eucalyptus wood chips in a downdraft gasifier for syngas production in South Africa. Int J Renewable Energy Res (IJRER) 10(2):663–668
Odziemkowska M, Czarnocka J, Wawryniuk K (2018) Study of stability changes of model fuel blends. In: Improvement trends for internal combustion engines, pp 1–17
Oßwald P, Whitside R, Schäffer J, Köhler M (2017) An experimental flow reactor study of the combustion kinetics of terpenoid jet fuel compounds: Farnesane, p-menthane and p-cymene. Fuel 187:43–50
Pavlenko N, Kharina A (2018) Policy and environmental implications of using HEFA+ for aviation. IntCounc Clean Transp 9:1–9
Pires AP, Han Y, Kramlich J, Garcia-Perez M (2018) Chemical composition and fuel properties of alternative jet fuels. Bioresources 13(2):2632–2657
Prak DL, Cooke J, Dickerson T, McDaniel A, Cowart J (2021) Cetane number, derived cetane number, and cetane index: when correlations fail to predict combustibility. Fuel 289:119963
Qin K, Zhou Q, Zhang K, Feng Y, Zhang T, Zheng G et al (2019) Non-uniform abrasive particle size effects on friction characteristics of FKM O-ring seals under three-body abrasion. Tribol Int 136:216–223
Renninger NS, Newman JD, Reiling KK (2010) U.S. Patent no. 7,854,774. U.S. Patent and Trademark Office, Washington, DC

Roberts WL (2008) Bio jet fuels. ile 4(4.00):3–75
Rude MA, Schirmer A (2009) New microbial fuels: a biotech perspective. Curr Opin Microbiol 12:274–281
Saffaripour M, Zabeti P, Kholghy M, Thomson MJ (2011) An experimental comparison of the sootingbehavior of synthetic jet fuels. Energy Fuel 25(12):5584–5593
Scheuermann SS, Forster S, Eibl S (2017) In-depth interpretation of mid-infrared spectra of various synthetic fuels for the chemometric prediction of aviation fuel blend properties. Energy Fuel 31(3):2934–2943
Serrano-Ruiz JC, Luque R, Sepúlveda-Escribano A (2011) Transformations of biomass-derived platform molecules: from high added-value chemicals to fuels via aqueous-phase processing. Chem Soc Rev 40(11):5266–5281
Shahid MK, Batool A, Kashif A, Nawaz MH, Aslam M, Iqbal N, Choi Y (2021) Biofuels and biorefineries: development, application and future perspectives emphasizing the environmental and economic aspects. J Environ Manag 297:113268
Sundararaj RH, Kushari A (2019) A review of Bio-jet production pathways. Recent Adv Petrochem Sci 7(1):555702
Sundararaj RH, Kumar RD, Raut AK, Sekar TC, Pandey V, Kushari A, Puri SK (2019) Combustion and emission characteristics from biojet fuel blends in a gas turbine combustor. Energy 182:689–705
Sustainable Way for Alternative Fuels and Energy in Aviation (SWAFEA) (2011) formal report, European Commission
Tao L, Milbrandt A, Zhang Y, Wang WC (2017) Techno-economic and resource analysis of hydroprocessed renewable jet fuel. Biotechnol Biofuels 10:1–16
Timko MT, Herndon SC, De La Rosa Blanco E, Wood EC, Yu Z, Miake-Lye RC et al (2011) Combustion products of petroleum jet fuel, a Fischer–Tropsch synthetic fuel, and a biomass fatty acid methyl ester fuel for a gas turbine engine. Combust Sci Technol 183(10):1039–1068
Tiwari R, Mishra R, Choubey A, Kumar S, Atabani AE, Badruddin IA, Khan TY (2023) Environmental and economic issues for renewable production of bio-jet fuel: a global prospective. Fuel 332:125978
Tomar M, Abraham A, Kim K, Mayhew E, Lee T, Brezinsky K, Lynch P (2023) Bio-derived sustainable aviation fuels—on the verge of powering our future. In: Combustion chemistry and the carbon neutral future. Elsevier, pp 521–598
van der Westhuizen R, Ajam M, De Coning P, Beens J, de Villiers A, Sandra P (2011) Comprehensive two-dimensional gas chromatography for the analysis of synthetic and crude-derived jet fuels. J Chromatogr A 1218(28):4478–4486
Vozka P, Moderegger BA, Park AC, Zhang WTJ, Trice RW, Kenttämaa HI, Kilaz G (2019) Jet fuel density via GC× GC-FID. Fuel 235:1052–1060
Wang WC, Tao L (2016) Bio-jet fuel conversion technologies. Renew Sust Energ Rev 53:801–822
Wang X, Jia T, Pan L, Liu Q, Fang Y, Zou JJ, Zhang X (2021) Review on the relationship between liquid aerospace fuel composition and their physicochemical properties. Trans Tianjin Univ 27:87–109
Wei H, Liu W, Chen X, Yang Q, Li J, Chen H (2019) Renewable bio-jet fuel production for aviation: a review. Fuel 254:115599
Wierzbicki TA, Lee IC, Gupta AK (2014) Performance of synthetic jet fuels in a meso-scale heat recirculating combustor. Appl Energy 118:41–47
Yang J, Xin Z, Corscadden K, Niu H (2019) An overview on performance characteristics of bio-jet fuels. Fuel 237:916–936
Yu W, Tay K, Zhao F, Yang W, Li H, Xu H (2018) Development of a new jet fuel surrogate and its associated reaction mechanism coupled with a multistep soot model for diesel engine combustion. Appl Energy 228:42–56

Chapter 6
Upgrading Biomass-Derived Pyrolysis Bio-Oil to BioJet Fuel Through Catalytic Cracking and Hydrodeoxygenation

Moumita Bishai

Abstract The changeover from fossil bio-oil to biojet fuel is an imperative footstep in the direction of plummeting aviation sectors' global warming. Biomass-to-liquid thermochemical methods will be the major biological choice for creating long-lasting hydrocarbon fuels, including biojet fuel, in the near future. Fast pyrolysis of waste from biomasses is a potentially renewable and sustainable energy resource creation of bio-oil. Pyrolytic bio-oil has a deprived heating value in the direction of the occurrence for many aerated molecules and a greater aqua composition, which causes it to be chemically unbalanced, viscid, and corrosive. The application of bio-oil directly is not possible to drop-in fuel owing to its poor quality, and, hence, extensive improvement is required before its utilization as mixed oil. The most effective catalytic post-treatment strategies for improving bio-oil and purifying it to a final product have been demonstrated to be catalytic cracking of fast pyrolysis vapor, along with hydrodeoxygenation. The current review emphasizes both the catalytic cracking and hydrodeoxygenation of bio-oil from biomass into jet fuel-range hydrocarbons. It also delivers a painstaking summary of the trials and utmost novice development processes in forming biojet fuel from pyrolytic bio-oil using the methodologies, with introspection on both the reconstruction processes. As a result of the complicated configuration of crude bio-oil, there has been very little study on enhancing the molecular components of raw bio-oil, with the bulk of the studies concentrating on specific model compounds. As a result, research opportunities for long-term studies are highly desired, which will drive and boost to intensify the economy of a country in the direction of the aviation sector.

Keywords Bio-oil · Biojet oil · Biomass · Catalytic cracking · Hydrodeoxygenation

---

M. Bishai (✉)
Department of Botany, Gurudas College, Kolkata, India

A. Kuila (ed.), *Biojet Fuel: Current Technology and Future Prospect*, Clean Energy Production Technologies, https://doi.org/10.1007/978-981-99-8783-2_6

6.1 Introduction

The worldwide population is growing at an exponential rate, which is directly boosting global energy consumption. The current energy is having a substantial impact on both the universal economy and the environment. Natural gas, coal, and petroleum are among the most important energy sources. Currently, fossil fuels provide 80% of the ecosphere's energy (https://www.eesi.org/topics/fossil-fuels/description). This energy problem has ignited the search for an alternative fuel source that can replace fossil fuels, particularly for powering the transportation sector. Renewable energy sources such as hydro, wind, solar, and thermal have provided consolation and comfort to the energy industry (https://www.eesi.org/topics/fossil-fuels/description).

Renewable fuels are gaining popularity as a means of replacing orthodox fuel and filling the energy discrepancy. Biofuel, being the potential renewable energy solution, is acquiring fame throughout the world since it creates fuel through comparable functionality to crude oil. Biofuels are created by converting biomass through thermochemical or biochemical processes. Microorganisms break down organic waste or biomass to produce liquid and gaseous fuel. Biofuels like biogas and ethanol are the consequences of microbial digestion and fermenting, respectively, whereas biogas is produced by the anaerobic degradation of organic waste (Das 2022).

Travel by air is another important sector of biofuel for the growing socio-global treaties and economic activity. As per the data from the International Air Transport Association (IATA), additionally 3.8 billion people and 54.9 million metric tons of goods valued around $5.5 trillion traveled by air, accounting for nearly 35% of global trade in terms of value (https://www.iata.org/en/programs/cargo/sustainability/benefits/). Forecasts also suggested that over the next 20 years, air passenger numbers will double (Wei et al. 2019), meaning that airline consumption of fuel will increase in lockstep. Jet fuel use in the world exceeded 12.48 quadrillion btu, accounting for 12% of overall mobility power use; while jet fuel use is predicted to rise by 10 quadrillion btu between 2010 and 2040 (Statistical Review of World Energy 2021).

Fuel prices continue to be a key issue for the global airline industry, accounting for around 27% of overall airline operating expenditures and being heavily impacted by oil prices (https://www.mckinsey.com/industries/travel-logistics-and-infrastructure/our-insights/why-rising-fuel-prices-might-not-be-as-bad-for-the-airline-sector-as-it-seems). As a result, for the purpose of maintaining sustainable growth, it is necessary to find a substitute for renewable fuel to meet the expanding demand while reducing the dependency on fossil fuels. As a result, production of renewable propellant for aircraft from biomass is being widely explored in order to minimize carbon emissions and accomplish the long-term development of the aviation sector.

Biomass is the only renewable source that contains carbon that is capable of absorbing CO_2 directly from the environment to produce biological material. Since it is free of carbon, biofuel not only decreases dependency on petroleum and coal, but also reduces emissions throughout the life cycle. Sustainable aviation fuels have the potential to cut greenhouse gas emissions by as much as 80% during its life cycle. Additionally, professional jet fuels must meet certain physical as well as chemical specifications. Biojet fuels include minimal sulfur material, low emissions from the tailpipe, excellent thermal capacity, and superb cold flow properties (Doliente et al. 2020). Furthermore, biojet fuels have a benefit over other alternatives to petroleum (such as ethanol) in regard to interoperability with current motors and energy frameworks. Biojet fuels may be used without altering the engine and do not pose any worries about fuel quality (Doliente et al. 2020). Because of its poor fuel properties, ethanol has been ruled unsuitable for use as an aviation fuel. Furthermore, the most important issue is incompatibility with present technological systems upgrading. All aviation engines if run on a new type of fuel, in this regard, would cost billions of dollars. Hence, biojet biofuels have been evaluated satisfactorily (Wei et al. 2019).

Jet fuel is a kind of aviation fuel designed primarily for commercial and military airplanes. Processing crude oil from 205 to 260 °C yields conventional jet fuel. Jet fuel is made up of C8–C16 hydrocarbons, which include alkanes, iso-alkanes, naphthenic derivatives, and aromatic chemicals. The quantity of every element is closely connected to the properties of jet fuel. Alkanes to high hydrogen–carbon ratio may ensure the fuel's power density. Naphthene helps reduce the point of freezing, which is essential in high-altitude flights (Bjornsson and Ericsson 2022). The volatile nature offers fluidity to enhance material compatibility and minimize leaks in certain aircraft, but excessive volatility has a negative impact on fuel cleanliness, thus the amount of volatile compounds should be kept within a permissible limit (Yang et al. 2019). Professional jet fuels must additionally fulfill severe rules for sulfur content, weight, temperature of ignition, uniformity, smoke point, naphthalene, and conductivity, in addition to the aforementioned features (Neves et al. 2020).

The two basic technical strategies for biofuel production are the physiological route and the thermochemical method. The enzymes, along with other microbes, are often used in the biochemical process to produce biofuel. The thermochemical approach is capable of converting synthesis gas produced by decomposition or gasification to bioenergy. Examples include oil-to-jet, Fischer–Tropsch process, alcohol-to-jet, and sugar-to-jet (Ambaye et al. 2021). Several scientists are currently focusing on the manufacture of jet fuel-range hydrocarbons using lignocellulose-derived chemical platforms (Wang et al. 2022).

There have lately been a flurry of reviews focusing on the production of biojet fuels. Kumal et al. 2020 investigated biojet fuel production from three angles: routes, opportunities, and challenges. Environmental challenges such as food scarcity, soil quality decrease, and water shortages were included as part of their research. There was, however, little quantitative analysis of environmental concerns. Meanwhile, as public awareness of environmental issues grows, environmental consequences such as greenhouse gas (GHG) emissions, utilization of water, and land usage should be given more attention, which has been lacking in previous publications (Kumal et al. 2020).

6.2 Bio-Oil Composition and Characterization

Other names for pyrolysis include pyrolysis oil, bio-oil, and others. Michailos and Bridgwater (2020) describe the crude bio-oil as black in appearance and essentially identical to the original biomass (Michailos and Bridgwater (2020). Low molecular weight ketones and aldehydes are responsible for the scent of the bio-oil liquid. The oil has a comparatively high density, around 1.2 kg/L, as opposed to 0.85 kg/L for lightweight fuel oil. This has serious implications for the layout and requirements of machinery such as compressors. The consistency of new bio-oil at 40 °C or higher may range from 25 to 1000 cst, dependent on a raw material, amount of water, collecting technique, and additional variables. Once dampened and compressed, pyrolytic vapors are unable to completely distilled by boiling. It swiftly polymerizes at temperatures over 100 °C, producing a solid remnant containing around half of the initial fluid in addition to a solution called distillate containing organic compounds that are volatile and liquid. Low-temperature distillation under vacuum or liquid separation is utilized to fractionate the oil (Michailos and Bridgwater (2020).

Traditionally, chemical characterization requires fractionating bio-oil into a number of chemical characteristics, then performing GC/MS analysis of the fractions. Later, simpler characterization methods, including splitting into soluble and insoluble fractions. The insoluble portion is composed of lignin-derived material with a high molecular mass, while the water-soluble fraction is composed of water, aromatic acids, alcohols, and diethyl ether (Dane and Volmer 2023).

Bio-oils are acidic, viscous, and thermally unstable. They are a complex mixture of over 200 chemical elements, comprising hydroxy aldehydes, hydroxy ketones, sugars, carboxylic acids, and phenolics in addition to water. Because of condensation and polymerization reactions catalyzed by acids and traces of inorganic elements, the existence of up to 45–50 wt% carbon and other reactive polar molecules in charcoal causes a rise in stickiness during storage (https://www.iata.org/en/programs/cargo/sustainability/benefits/). Phounglamcheik et al. (2017) carried out extensive research to verify the impact of charcoal in triggering the course of ageing. The experiment is conducted by topically adding carbon to the oil while contrasting the method of aging with the process of purified bio-oil. In comparison to the viscous of filtered fresh bio-oil, the inclusion of char increased the pace of viscosity growth. The char particles are supposed to be generating "aging" because of perceived tar aggregates of particles (Phounglamcheik et al. 2017).

The temperature and duration of the pyrolysis process, as well as the presence of catalysts, have a significant impact on the chemical makeup of bio-oils (Dane and Volmer 2023). The quantity of variation in molecular weight in the resulting oils is reduced as cracking severity rises, culminating in more gases. Dehydrogenation activities are accelerated at very high temperatures, resulting in the formation of larger polynuclear aromatic hydrocarbons, which leads to increased carbonization. The relationship between the various types of particles in the final lead and the degree of heat suggested that as warmth grew alkyl groups detached from aromatic compounds, which led to the formation of hydrocarbons that are polycyclic aromatic, also known as PAHs, at higher temperatures (Patel et al. 2020).

Despite being called "bio-oil," the pyrolysis fluid does not mix with hydrocarbon liquids due to its inherent high level of polarities and aqueous nature. In contemporary industrial-scale bio-oil combustion tests, it is shown to be technically suitable for replacing heavy petroleum products in applications involving heating. This form of replacement, on the other hand, requires more suitable metallurgy, especially for parts that encounter bio-oil. In general, the solids, water, and nitrogen content of bio-oil have a big impact on the emissions. As a result, in order to build highly efficient, trustworthy bio-oil combustion systems, bio-oil grades must be standardized. Similar restrictions are also applied to other bio-oil applications (Chan et al. 2020).

6.3 Biosynthetic Pathway of Bio-Oil

Bio-oils are generally created utilizing sunshine, CO_2, and water from various materials, which include algal cells, agro and other kinds of waste, etc. Such bio-oils are created during the course of pyrolysis and must then be improved for commercial usage like biojet oil (Fig. 6.1).

6.4 Methods for Bio-Oil Production

Pyrolysis occurs in an O_2-free environment established by emptying the inside of the reactor with an inert gas such as N or Ar. Pyrolysis induces heat breaking of the biopolymers that make up lignocellulosic biomasses, resulting in a variety of smaller-molecule products that recondense to form bio-oil and biochar, as well as molecules that are not condensable like carbon monoxide, carbon monoxide, as well as hydrogen. Bio-oil offers the potential to be employed to create artificial fuels for transportation as well as biochemical substances. Biochar, on the other hand, is a carbon-rich solid product that is potentially used in farming, absorbing carbon, catalytic support, absorbent, graphite synthesis, gas storage, tailored carbon-based products, and therapeutics (Das et al. 2021). The generating gas contains combustible ingredients such as carbon monoxide, hydrogen, and methane.

There are two types of thermochemical transformation. The primary process is gasification of feedstock and hydrocarbon conversion. The following method is used to rapidly liquefy biomass utilizing high-temperature pyrolysis or high-pressure liquefaction. These techniques turn recyclable materials into high-energy products that are profitable. The nature and amount of feedstock from biomass, the expected kind of energy, that is, final application specifications, laws governing the environment, finances, and site-specific considerations all impact converting choice of processes. Thermal liquefaction and pyrolytic techniques are the two main thermochemical methods that have been actively exploited for the generation of

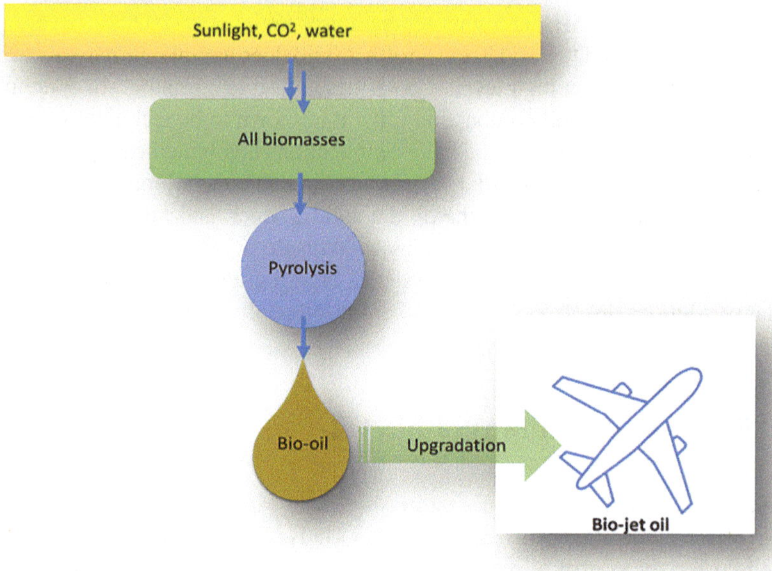

Fig. 6.1 Biosynthetic pathway of bio-oil

powerful bio-oil from garbage (Zhang et al. 2019). Ignition generates heat and electricity, whereas gasification generates gaseous by-products and hydrothermal carbonization aids in the production of materials. The process type and feedstock variables both influence the level of quality of bio-oils generated by hydrothermal liquefaction (HTL) and pyrolysis. Pyrolysis oils are less sticky, have a smoky scent, and possess an inferior thermal value. Also, pyrolysis bio-oil has a higher oxygen content than HTL bio-oil (Zhang et al. 2019).

With the combination of the level of heat, thermal rate, and gaseous retention time, pyrolysis is classified as slow, fast, or flash. Slow pyrolysis is employed to optimize biochar development due to its moderate warming rate, process temperatures, and longer vapors residence durations (Al-Rumaihi et al. 2022). Fast pyrolysis, on the contrary, yields more bio-oil owing to its faster warming rates, greater process temperatures, and shorter vapor residence times. Finally, flash pyrolysis operates at higher temperatures, with quicker heating rates and a shorter vapor residence duration. Because flash and fast pyrolysis processes have shorter vapor residence times, hydrocarbon smoke and volatile species are quickly quenched and condensed, resulting in higher bio-oil yields (Al-Rumaihi et al. 2022).

All the organic molecules improve the oxygen-to-carbon ratio despite reducing the thermal value in pyrolytic bio-oil. Catalytic pyrolysis can assist in reducing the quantity of oxygen-rich organic molecules in bio-oils while increasing the hydrogen-to-carbon ratio. Catalytic pyrolysis employs a diverse spectrum of homogeneous and heterogeneous catalysts (e.g., zeolites, aluminosilicate, MgO, Na_2CO_3, and other catalysts) (Nanda et al. 2021). It improves organic thermal cracking, boosts aromatic content, and raises bio-oil heating parameters synchronously. Apart from powdered zeolites, numerous metal-supported zeolites (Y-zeolites with Fe, Co, Ga, Mo, and Ru) have been established to be good catalytic materials for improving bio-oil quality (Nanda et al. 2021).

Fast pyrolysis procedures produce 60–75 wt% fluid bio-oil, 15–25 wt% solid char, and 10–20 wt% non-condensable gases, depending on the substrate (Park et al. 2019). A fast pyrolysis procedure has four main characteristics: very high warming and heat transfer rates, which necessitate a coarsely ground biomass feedstock, meticulously controlled pyrolytic temperature, short vapor residence times and quick quenching, along with refreshing of the pyrolysis vapors and aerosols to produce bio-oil (Uddin 2018). Such a technique is appealing because the biomass gets swiftly transformed into liquid products. These liquids have benefits in terms of transportation, storage, combustion, retrofitting, and the manufacture and distribution of mobility (Demirbas 2009).

6.5 Production of Bio-Oil from Biomass

Bio-oil is a sort of liquid fuel made from biomass resources such as algal bodies, crops cultivated for agriculture, municipal garbage, and agricultural, as well as forestry by-products (Demirbas 2009).

Wood, medicinal plants, crops, human and animal waste, and scrap from factories are all examples of biomass. The broad categorization of biomass resources is shown in Table 6.1. The use of biomass is determined by its physicochemical qualities and lignocellulosic composition. Organic configuration and assembly, as well as Vigor needs, are significant aspects of biomass resources that influence the conversion course or somewhat precise subsequent dispensation problem (Inayat et al. 2022). The cellulose, lignin, and hemicellulose content of lignocellulosic biomass varies. Such biomass may also be differentiated by fundamental and final inspection, which assists in the selection of the process and final product from converting biomass [40]. In general, proximate and ultimate analysis assists in assessing a material's energy content by measuring the ratio of flammable to noncombustible stuff (Inayat et al. 2022).

6.6 Impact of Operating Parameters on Bio-Oil Production

6.6.1 Impact of Cellulose, Hemicelluloses, and Lignin Content

Each biomass material has varying quantities of hemicellulose, cellulose, and lignin. The number of hemicelluloses and cellulose in a biomass determines its production and chemical makeup. These three key constituents of biomass degrade at a

Table 6.1 List of different types of biomasses used to produce bio-oil

Types of biomasses	Type of material	Condition	Remark	References
Woody biomass	Mallee wood	Temperature 350–600 °C, with biomass particle size 100–600 μm.	The upper layer of softwood bark oil comprised 16 weight percent of the total bio-oil, with more than 50 weight percent of extractive-derived chemicals, whereas the upper layer of hardwood bark represented just 1.3 weight percent of the bio-oil.	Modupe (2019)
	Pinewood	Catalytic pyrolysis at 450 °C in a fluidized bed reactor with acidic zeolite.	The structures had no effect on the yield of the pyrolysis product phases, but the chemical composition of the bio-oil was affected by the structure of acidic zeolite catalysts.	Nisar et al. (2022)
Agriculture wastes	Rice straw and bamboo sawdust	In a bubbling fluidized-bed reactor, the temperature range for rice straw is 415–540 °C and for bamboo sawdust is 350–510 °C.	Bamboo sawdust has the highest bio-oil output (70 wt%). The principal components of bio-oil, according to compositional analysis, were phenolics, furfural, acetic acid, levoglucosan, guaiacol, and alkyl guaiacol.	Landrat et al. (2022)
	Fruit bunches	Fluidized bed reactor with a residence time of 0.79–1.32 s and a temperature range of 400–600 °C.	Maximum bio-oil yield was found at 450 °C and gas yield increased as temperature climbed. Furthermore, the ash concentration and particle size both have an influence on product yield.	Thu et al. (2020)
	Jute sticks, sugarcane, and wood and agricultural residues	Bio-oil generation using flash pyrolysis of biomass and biopolymer waste. Cost–benefit analysis and Monte Carlo simulations were used to report on 1:1 w/w ratio blends of willow and several biopolymer waste streams.	When compared to pure willow flash pyrolysis, the economics of flash co-pyrolysis of biomass with biopolymer waste improved. Polyhydroxybutyrate (PHB) was believed to be the most promising biopolymer under development, followed by Eastar, Biopearls, potato starch, polylactic acid (PLA), maize starch, and Solanyl in decreasing order of profitability.	Sarkar and Wang (2020)

(continued)

Table 6.1 (continued)

Types of biomasses	Type of material	Condition	Remark	References
Algae waste	*Chlorella protothecoides* and *Microcystis aeruginosa*	Pyrolyzed in a fluidized bed reactor at 500 °C and a heating rate of 600 °C.	The saturated and polar fractions accounted for 1.14 and 31.17% of the microalgae bio-oils, respectively, which were greater than those of wood bio-oil.	Devi et al. (2022)
	Chlorella algae	Pyrolytic conversion to liquid fuels using Na_2CO_3 as a catalyst. TGA combined with MS was used to conduct thermal breakdown investigations on algal samples.	By lowering the decomposition temperature, pretreatment of *Chlorella* with Na_2CO_3 impacts the primary conversion. The bio-oil produced via catalytic runs had a better heating value and a lower acidity. Increased aromatics paired with increased heating value showed potential for up to 40% output.	Tirapanampai et al. (2019)
Municipal solid waste	Potato skin, a food industry waste	Pyrolysis is carried out in three distinct atmospheres: Static, nitrogen, and steam.	At 550 °C, the bio-oil output was 24.77 wt.% in stable environment and 27.11 wt.% in an atmosphere of nitrogen. The usage of steam increased bio-oil output to 41.09 wt.%. The viability of the bio-oil production method was proven by TG-DTA, FT-IR, and NMR analyses.	Yildiz (2022)
	Sludge collected from pulp and paper-making industries	Pyrolysis is carried out to perform thermal analysis.	According to the TGA analysis report, the losing weight procedure for decontamination sludge was a non-pyrolytic product.	He et al. (2021)
Plants and shell cake of different oil seeds	*Jatropha curcas* L. nutshell	Continuous bench-scale pyrolyzer at a feeding rate of 2.27 kg/h at 480 °C and atmospheric pressure.	Bio-oil: 50 wt.%.	Romuli et al. (2018)
	Rungam oil cake	Pyrolysis.	High yield of bio-oil.	Chhabria et al. (2022)

variety of temperatures and heating rates. The pyrolysis characteristics of three key components of biomass were investigated using a thermal analyzer. During the thermal research, hemicellulose pyrolysis was seen at 220–315 °C while cellulose pyrolysis was observed between 315 and 400 °C. Lignin, on the other hand, was more difficult to degrade since its weight loss occurred across a temperature range of 160–900 °C and 40% solid residue was generated (Ansari et al. 2019).

6.6.2 Impact of Product on Biomass Breakdown and Dynamics Research

Temperature affects biomass degradation during pyrolysis. The percentages of lignin, hemicelluloses, and cellulose in all biomass materials vary. The higher the percentage of cellulose and hemicellulose, the more bio-oil is generated and the higher the concentration of lignin, the more charcoal is created. As a result, biomass thermal degradation progresses as follows: hemicelluloses outnumber cellulose and lignin in abundance. All biomass materials have a high carbon and hydrogen content. When studying feedstock thermal breakdowns and momentum rate equations, it is critical to include the effects of heat and biomass mechanics. The ambient humidity, heating rate, and duration of the reactor's residency all have an influence on biomass heating breakdown (Zhang et al. 2019).

6.6.3 Impact of Particle Dimensions

Mass and heat transmission rates as well as surface rates of chemical reactions are all affected by particle dimension. It also has an effect on the reactor's pressure drop. The flow characteristics of biomass materials are also important in determining the best type of reactor. It was discovered that particle size influenced bio-oil output and was an important factor in boosting bio-oil production. Small particle size promotes the transfer of heat among particles of biomass during the pyrolysis process because of low thermal conductivity. Particle size and heating should be small in order to obtain a greater amount of liquid fuels from biomass (Qureshi et al. 2021).

6.6.4 Impact of Moisture Content

Because of the substantial moisture level, biomass is unsuitable for ignition and hydrolysis. It has an impact on thermal stability, which has an impact on bio-oil yield and gas production component (CO, CO_2, CH_4, and other gas mixes), as well as heating value. A low moisture content is favorable because the steam initiates a steam gasification process, resulting in higher bio-oil and gas quality. It is well known that the moisture content of biomass affects the quality of thermal processing products. While pyrolyzing a pinewood sample in a reactor with a fluidized bed observed that the moisture content of the biomass had a substantial impact on the dispersion of flammable volatiles through the fluidized bed's sectional dimension (Fonseca et al. 2019).

6.6.5 Impact of Fixed Carbon Content

The static carbon concentration in the feedstock samples determines the degree of oxidation and lessening processes as well as the residence duration in the reactor. The composition of the gas is also affected by this value (Velez et al. 2018).

6.6.6 Impact of Volatile Matter

Volatile biomass resources have a significant role in the synthesis of bio-oil. More volatile materials imply more bio-oil production. Araújo et al. in 2018 extracted 45% bio-oil from sunflower oil cake containing 73.8% volatile materials. Chen et al. (2020) produced 46% bio-oil from canola with 86.04% volatile compounds. However, some biomass, such as barley, bagasse from sugarcane, and the seeds of rap and grapevine bagasse, do not follow the predictable pattern. This might be related to the original moisture level of the feed material. Singh in 2020 identified a 43% bio-oil output using barley, which includes 98.7% volatile constituents. The primary oil with a boiling point of 36 MJ/kg was obtained at 300 °C and 12 MPa N_2 pressure. The dense viscosity and nitrogen, oxygen, and sulfur contents of the primary oil were reported to be removed by hydro treatment of the bio-oil with a $NiMo/Al_2O_3$ catalyst at a starting H_2 pressure of 10 MPa. The bio-oil had a yield of 43% at 350 °C.

6.6.7 Impact of Ash Content

The abundance of alkaloid metals in plant waste has an effect on the formation of bio-oil. The lesser the bio-oil yield, the greater the ash content. The formation of charcoal is caused by greater amounts of alkali and other metals such as Na, Ca, Mg, Zn, and Cd. Alkali metals have a higher ash quantity and are accountable for a reduction in volatile substance, which has caused sludge formation during biomass pyrolysis (Tomczyk et al. 2020).

6.6.8 Impact of Temperature

During pyrolysis, temperature is an important parameter. Temperatures for slow and rapid pyrolysis differ as well. At 350–550 °C temperature, 10–12 °C/min heating rate, 200–500 cm^3/min sweep gas flow rate, and particle size 0.425–0.850 mm, the greatest bio-oil yield (59%) was obtained. Singh in 2020 obtained 46.1% bio-oil by pyrolyzing rapeseed (0.85–1.80 mm) at 500 °C temperatures. Rapeseed (0.224 mm) thicknesses provided 42.9% bio-oil at the same temperature. The influence of reaction circumstances on the characteristics of bio-oil has been reported. It was observed that the optimal pyrolysis temperature for creating bio-oil was between 400 and 450 °C. With increased flow and feeding rates, its manufacture was more efficient. Applying the gas as the fluidizing solution resulted in the highest bio-oil output. With the possible exception of temperature, no single operational factor had a substantial influence on the physicochemical properties of the bio-oil (Singh 2020).

6.6.9 Impact of Heating Rate

The pace of heating during the pyrolysis process is an important factor in creating a larger volume of bio-oil. Biomass feed decomposed effectively in a briefer residence time at a quicker heating rate, resulting in enhanced bio-oil yield. In several investigations, a rate of combustion of (5–40) °C/min was used in the slow pyrolysis technique. A higher heating rate may degrade lignin, hemicellulose, and cellulose more quickly, boosting bio-oil yield. Higher heating rates and lower dimensions of particles aided bio-oil synthesis in general (Yogalakshmi et al. 2022).

6.7 Weakness of Bio-Oil for Use as Bio-Jet Fuel

Moisture, heteroatoms, and heavy metals are abundant in bio-oil produced by pyrolysis. Furthermore, throughout pyrolysis and liquefying, numerous oxygenated compounds are generated, increasing the degree of tartness of the bio-oil. The presence of carboxylic acids triggers evaporation and polymerization steps, which increase the acidity and viscosity of the oil. While the acidity of bio-oil causes corrosion of crushers and storage vessels, the increased viscosity poses issues with transport. When contrasted with fossil fuels, these oxygenation molecules diminish the thermal efficiency of bio-oil by nearly 50%. Unprocessed bio-oil might have 15–30% moisture, reducing heating properties and causing delayed ignition. Bio-oil includes distinct types of oxygenated compounds, along with carboxylic acids as well as ethanol (Dalai et al. 2021). The existence of such highly reactive organic compounds in the environment makes the bio-oil thermochemically volatile. The high ash level has the potential to cause serious damage to the reactor and pipelines.

Crude oil's poor lubricity may also lead to the deterioration of metallic parts in engines, chemical plants, fuel pumps, metering, and pipelines (Dalai et al. 2021). However, by upgrading and blending bio-oil with normal gasoline oils, the fluidity may be improved to satisfy fuel specifications. Unprocessed bio-oil contains a greater concentration of nitrogen and sulfur-containing compounds, which may result in the production of sulfur oxides (SOx), nitrogen oxides (NOx), and particulate matter (PM) during combustion (Dalai et al. 2021).

6.8 SWOT Analysis of Bio-Oil and Its Upgradation

Several thermochemical techniques for producing biocrude oil, then transformed into fluid transportation fuels, have been developed. Nonetheless, improving biofuel efficiency, output, and profitability is vital for market survival. When compared to typical fossil fuels, thermochemical biofuel manufacturing technologies (such as combustion, gasification, pyrolysis, and liquefaction) strive for environmental friendliness and financial effectiveness (Shahbaz et al. 2021). Combining pyrolysis and liquefaction with certain bio-oil upgrading technologies might be a promising method for generating biofuels efficiently and cheaply (Zhang et al. 2019). An integrated strategy might save capital and operating expenses while also simplifying processing and using the majority of by-products. Additional studies and development are required to determine an effective integrated conversion method.

Co-refining, co-processing, or blending drop-in biocrude with synthetic distillates is an alternate method for producing diesel, petrol, and jet fuel (Dyk et al. 2019; Lindfors et al. 2023). Co-processing hydrotreated biocrude with traditional crude oil might result in the production of hydrocarbon fractions such as petrol and diesel (Lindfors et al. 2023). Although this technology has been proven to be economically favorable for commercialization, it still requires biocrude pretreatment to remove oxygenated compounds prior to merging or coprocessing. Thus, improving the existing process by concurrent monitoring and product characterization may result in co-refining technologies that are both economical and sustainable (Zhang et al. 2019). Despite the benefits, the disadvantages of improving bio-oil must be considered. Catalytic hydrogenation, cracking, and esterification are all important steps in the upgrading process. However, because of the high process temperatures and costly catalysts utilized in these catalyst-assisted upgrading procedures, various problems might occur. Some common difficulties with catalytic upgrading procedures include catalyst deactivation, catalyst poisoning and sintering, and catalytic point obstruction. Tar accumulation decreases petrol performance and purity (Zhang et al. 2019). As a result, the creation of coal and paraffin in catalytic boosting processes raises the cost of process cleanup and equipment maintenance. Rehabilitation and regrowth might boost both the time spent and input of energy to catalytic update procedures while maintaining catalyst recyclability (Inayat et al. 2022). Contaminants in bio-oil may hinder the improvements process, which leads to low-carbon conversion and increased processing costs. By eliminating oxygenated components and lowering viscosity, noncatalytic bio-oil upgrading techniques, notably supercritical fluids, significantly increase the fuel effectiveness of improved bio-oil. However, the costs of liquids and apparatus for pilot-scale implementation are too expensive. Additionally, supercritical water at elevated pressures and temperatures in the presence of certain catalysts may be harmful to stainless steel vessels owing to salt formation and precipitation (Inayat et al. 2022).

There is an evident need to develop biocrude upgrading methods if it is to become a viable solution to the problems associated with manufacturing liquid transportation fuels. Pyrolysis bio-oil must be improved to meet the standards of conventional jet fuels. When it comes to removing oxygen, this is crucial since the inclusion of oxygenated molecules impacts the rate of oxidation permanency and thermal efficiency of the produced biofuel. According to the ASTM D1655 and D7566, the concluding fuel composition must include 99.5% C and H_2 and the presence of any kind of oxygenated molecule disqualifies it as a viable biojet fuel (Wang et al. 2022). As a result, eliminating oxygen is critical for boosting the H/C mole ratio (1.9–2.2), the stability of biojet fuel and its distribution with typical jet fuel (Wang et al. 2022).

Hence, some specific catalytic methods, such as hydrogenation, catalytic cracking, esterification and transesterification, and noncatalytic methods, such as emulsifier solvent addition, supercritical fluids, electrochemical stabilization, and other organic reactions, are used to improve the fuel characteristics of bio-oil (Zhang et al. 2019). Given the negative impact of oxygenated molecules on bio-oil characteristics, hydrodeoxygenation is critical in removing oxygen by generating H_2O, CO_2, and CO. The depletion of carbon as a result of CO_2 and CO, as well as the undesirable creation of coke, which compromises catalyst selectivity and causes catalyst poisoning, is a significant drawback of this upgrading process. Catalytic cracking with zeolites, which eliminates excess oxygen during the cracking operations, is another approach to enhancing crude bio-oil (Dane and Volmer 2023). The approach is limited by the negative effects of catalyst poisoning produced by tar deposition during condensed procedures.

6.9 Catalytic Cracking of Bio-Oil

The combination of fast pyrolysis with catalytic improvement of pyrolysis vapor to generate a more stable bio-oil with better properties is a watershed moment in bio-oil development. Pyrolysis oil is enhanced by using a catalyst that acts in the state of vapor before condensing into a liquid (Al-Rumaihi et al. 2022). By inducing particular reactions that minimize activation energy, restrict product dispersion, decrease nitrogen- and oxygen-containing chemicals, and decrease acidity, the catalyst may decrease the pyrolytic temperature, boost feedstock transformation, and enhance bio-oil properties (Rangel et al. 2023). It also provides a bio-oil intermediate with substantially less oxygen. However, catalytic pyrolysis provides lesser bio-oil than heat quick pyrolysis since the usual catalysts used in catalytic fast pyrolysis, zeolites, possess a high breakdown action, limiting bio-oil production (Al-Rumaihi et al. 2022). Oxygen restriction cracking, aromatization, aldol condensation, ketonization, restructuring, and hydroprocessing are the key chemical reactions that occur during catalytic pyrolysis of material. Catalytic deoxygenation might efficiently lower the oxygen component in bio-oil. Deoxygenation is frequently caused by activities that include decarboxylation, which removes oxygen in the type of CO_2, decarbonylation, which removes oxygen in the form of CO and dehydration, which removes oxygen in the form of H_2O. Catalytic cracking can convert large molecules into tiny molecules by breaking C–C bonds, isomerization, proton movement, deoxygenation, and aromatic side-chain splitting. Aromatization is an intricate procedure that uses the hydrocarbon reservoir principle to transform compounds with small molecules into aromatic hydrocarbons. Aldol condensation may transform carbonyl and carboxyl components into longer-chain petroleum products, CO_2, and H_2O. Reforming and hydroprocessing, which yield H_2 and H_2O, are two more efficient processes for offline upgrading of bio-oil. For further information on the linked reactions, the literature has been carefully explored (Rangel et al. 2023; Al-Rumaihi et al. 2022). To enhance product yields, certain reactions can be increased using a suitable catalyst. Various catalysts are being created and utilized in this regard to selectively encourage particular reactions in catalytic fast pyrolysis, like conventional zeolites, metal-modified zeolites, metal oxides, and metal-supported catalysts, as extensively reviewed in other works on the subject. Zeolites are the most often used family for catalytic fast pyrolysis, with their acid sites encouraging cracking and dehydration processes. Furthermore, for catalytic fast pyrolysis, mesoporous catalysts and metal oxides, transition metal oxides and metal complex oxides, and carbon-based catalysts and biochar have been used (Lahijani et al. 2022).

According to the interaction mechanism of the pyrolytic vapor with the catalyst, catalytic rapid pyrolysis of plant matter can be performed in situ or ex situ. Table 6.2 provides a comparative overview of both systems' strengths and faults (Muneer et al. 2019).

It is especially true when the perfect pyrolysis and effective cracking temperatures are not identical. As an outcome, each stage may be managed under optimal operating conditions, increasing bio-oil extraction and making this technology more versatile and attractive. However, the greater overall capital cost of constructing two distinct ex situ power plants, as well as the cost of heating separate reactors, must be considered. To reduce total expenses, small-scale ex situ catalytic pyrolysis reactors usually support both the pyrolysis and catalysis processes concurrently, although in physically distinct zones (Ambaye et al. 2021; Al-Rumaihi et al. 2022). Fast pyrolysis combined with catalytic improvement, whether with a catalyst that is installed in the pyrolysis unit or through a downstream converter to improve the pyrolysis vapor after it condenses into a liquid, can considerably increase bio-oil quality. However, the laborious process of turning biomass into high-quality fuels cannot be accomplished without more modifications and purifying this substantially improved bio-oil. Nonetheless, several investigations have demonstrated that jet fuel spectrum hydrocarbons may be generated by catalytic rapid pyrolysis of biomass, either alone

Table 6.2 Comparative account of in situ and ex situ catalytic pyrolysis

In situ catalytic pyrolysis	Ex situ catalytic pyrolysis
The catalyst is contained within pyrolysis reactor and product production and upgrading occur in a single phase It is commonly carried out in fluidized-bed reactors, where the bed material acts as both a catalyst and a heat transmission medium It allows for immediate interaction between catalyst and pyrolytic fragments while minimizing repolymerization of the initial pyrolytic products, which leads to a better yield of the desired product The removal of the catalyst after the reaction, which has become entangled with the char, is a major difficulty	Pyrolytic products are formed initially, followed by catalytic reforming of the pyrolysis vapor in a second stage immediately following the reactor It is chosen for catalyst renewal and the recovery of char as a valuable solid by-product from pyrolytic reactors It also prevents catalyst particles from coming into touch with biomass minerals, reducing the minerals' contribution to catalyst deactivation The ex situ technique prevents catalyst deactivation due to coke or char development on the catalyst surface Another benefit is that pyrolysis of biomass and catalytic conversion are performed in distinct reactors with autonomous operating conditions

or in combination with other substrates. Ex situ catalytic rapid pyrolysis of lignin in the vicinity of a maize stover-based activated carbon substrate to create jet fuel-range aromatics (Ambaye et al. 2021; Al-Rumaihi et al. 2022).

Because lignin had a H/C ratio just below 0.3, which was unfavorable for excellent bio-oil and could contribute to the deposition of coke on the catalyst, the material was co-pyrolyzed with soapstock to provide hydrogen. Meanwhile, for additional deoxygenation, an acidic catalyst produced from maize stover and activated chemically using a phosphoric acid (H_3PO_4) solution was employed. Ex situ catalytic fast pyrolysis of the identical soapstock feedstock over a comparable catalyst was utilized in another investigation. The product oil's major components were C8–C16 aromatic compounds and alkanes, which might be regarded a predecessor to jet fuel and petroleum range HCs. At a pyrolysis temperature of 500 °C and a soapstock/catalyst ratio of 1:1.5, the highest response of jet fuel using petroleum spectrum hydrocarbons was 98.78 and 91.03%, respectively. Zhang et al. (2019) also reported ex situ catalytic combustion from pine sawdust employing a bimetallic Pt-Ni/-Al_2O_3 catalyst to produce jet fuel spectrum hydrocarbon-rich bio-oil. Pyrolysis-gas chromatography/mass spectrometry (Py-GC/MS) was used for the catalytic pyrolysis experiments. The most common aromatic hydrocarbons were found to include benzene and its analogs, with naphthalene and anthracene being relatively rare. Duan et al. in 2019 looked into the potential of synthesizing jet fuel profile hydrocarbons using corncob via co-catalysis rapid pyrolysis. They suggest a two-step method of conversion that consists of (1) biomass pyrolysis generating volatile compounds and (2) a catalytic reaction in the presence of a downstream catalyst (Duan et al. 2019).

Sun et al. (2018) investigated the synthesis of biojet fuel utilizing a dual-functional activated carbon catalyst with co-pyrolysis of DF and LDPE. To improve the pyrolysis vapor, an H_3PO_4-activated carbon catalyst was modified using Fe. With this approach, they may be able to produce a superior bio-oil, which includes aromatics, alkanes, and phenols. According to the findings, raising the pyrolysis temperature and the catalyst-to-feedstock ratio boosted aromatic sensitivity while decreasing phenol and alkane selective. Applying the stated technique, the acidic catalyst proved successful in generating aromatics.

Considering the fact that catalytic degradation of bio-oil is a cheaper way of enhancing bio-oil, it usually ends in low-carbon hydrocarbons, which do not fulfill jet fuel criteria. Catalytic cracking is usually performed at heats ranging from 350 to 550 °C to ensure that oxygenated particles disintegrate quickly. This situation produces fairly substantial gaseous output and coke productivity, whereas liquid hydrocarbon products concentrate in petrol-range chemicals (Nanda et al. 2021).

Consequently, extra operations such as hydroprocessing are required to change the number of carbons distribution in hydrocarbons in order to meet jet fuel specifications. Hydroprocessing, also known as hydrotreating, is a popular method for turning liquid oily materials such as vegetable oil, plant oil, fatty acids, waste cooking oil, and bio-oil into hydrocarbons by eliminating oxygen-rich substances at high temperatures and pressures with the help of a catalyst (Doliente et al. 2020). Hydroprocessing has long been used by the petrochemical industry to remove heteroatoms that are from fossil crudes such as sulfur (hydrodesulfurization), nitrogen (hydrodenitrogenation), oxygen (hydrodeoxygenation), and metals (hydrodemetallization) (Nanda et al. 2021). Bio-oil, on the other hand, is thermally unstable because it includes a lot more oxygen plus heteroatoms compared to crude oil. Therefore, the hydroprocessing catalysts-associated reaction conditions for bio-oil are significantly distinct from those for hydroprocessing a fossil crude oil. Furthermore, deoxygenation is significantly more crucial than sulfur and nitrogen reduction. Hydroprocessing operations are frequently divided into four categories: (1) hydrogenation, (2) hydrogenolysis, (3) hydrodeoxygenation, and (4) others (non-H reactions). Why et al. (2019) demonstrated that it is the simplest and most practical method for manufacturing biojet fuels from biomass oil feedstock while the volume generated is large, the carbon number is kept, and the only by-product is water. Michailos and Bridgwater examined three bio-oil upgrading methods, including hydroprocessing, zeolite cracking, and gasification, followed by Fischer–Tropsch synthesis, to undertake a techno-economic study of biojet fuel generation. According to modeling, hydroprocessing had the highest total energy efficiency and jet fuel energy efficiency (Michailos and Bridgwater 2020).

Hydrogenation was suggested as a method of creating jet fuel-range hydrocarbons using pyrolytic bio-oil. High-density aromatics, which are common in catalytic pyrolysis-derived bio-oil, are sensitive to hydrogenating into cycloalkanes, like minor aromatics. Hydrogenation is a hydrogen-saving method by which proton is changed into cycloalkanes instead of water by hydrogenation-saturated double bonds or aromatic functional groups. First-row transition metals like nickel and cobalt have demonstrated significant potential for hydrogenating bio-oil, with inherent hydrogenation ability substantially greater than cobalt (Nanda et al. 2021; Lahijani et al. 2022; Al-Rumaihi et al. 2022).

A group of researchers did extensive research on the application of Raney-type nickel in the manufacturing of biojet fuel, an adaptable catalyst for the negative reduction of organic molecules in hydrogenation operations. They investigated the creation of jet fuel family alkanes by co-feeding biomass (DF) with plastic (LDPE) to a catalytic microwave-induced pyrolysis system (Shumeiko et al. 2019). The bio-oil was separated with *n*-heptane after microwave-assisted catalytic pyrolysis and then hydrogenated. The hydrogenation test was performed in a Parr reactor, where a mixture of liquid organics and *n*-heptane was in contact with a Raney Ni catalyst (20 wt%) for 2 hours at 500 psi pressure and 200 °C temperature. Catalytic microwave co-pyrolysis yielded liquid organics primarily made up of C8–C16 aromatics, which were regarded to be viable intermediates for jet fuel production. After hydrogenation, this bio-oil yielded a carbon yield of 38.4% for hydrogenated organics, with a selection toward jet fuel range alkanes of around 90%, and a high level of cycloalkanes contributing to 75% of the specificity. During the thermal degradation of lignocellulosic biomass, furan was largely formed from cellulose and hemicellulose, while lignin was broken down producing phenolic compounds. In contrast, plastic degradation created free radicals and long-chain carbons, probably by stochastic and chain-end splitting processes. At the same time, hydrogen transfer mechanisms may transform radical pieces into straight-chain hydrocarbons. While wax generated by the heat breakdown of plastic may be chemically split over a zeolite catalyst to generate light olefins, which may then mix with furans via the Diels–Alder process and dehydrate to form aromatic hydrocarbons. A mixture of aromatic and aliphatic hydrocarbons was generated as a consequence of catalytic pyrolysis in microwaves of co-fed biomass and plastic. Then, the Raney Ni catalyst was employed for hydrogenating unsaturated aliphatic olefins. In the event of the presence of this catalyst, aromatic hydrocarbons were transformed into cycloalkanes or hydroaromatic hydrocarbons. Meanwhile, hydro-isomerization among dimethylcyclohexanes may occur and hydrocracking events may produce a trace number of tiny hydrocarbons. Researchers studied the generation of jet fuel-range cycloalkanes from lignocellulosic biomass using the same microwave-assisted pyrolysis process as well as bio-oil hydrotreating with a similar catalyst. Under 500 °C temperatures for pyrolysis and a catalyst/biomass proportion of 0.25, aromatic hydrocarbons formed during the pyrolysis process in the *n*-heptane environment were completely hydrogenated to petrol spectrum cycloalkanes. In hydrogenation, the selectivity to high-density cycloalkanes is enhanced through raising the catalyst load (10 or 20 wt%) and reaction temperature (150, 200, and 250 °C). Among the biomass feedstocks examined, hybrid poplar showed the best cycloalkane selective of 95.2% at a hydrogenation temperature of 250 °C and in the presence of a 20-weight % Ni catalyst (Shumeiko et al. 2019).

In another significant paper, Wang et al. (2022) catalytically upgraded bio-oil generated by rapid pyrolysis of wheat stalk to jet and gasoline range HCs via three reaction steps. Initially bio-oil breakdown and deoxygenation of oxygenated substances were done in the context of a HZSM-5 zeolite catalyst to yield carbon-free aromatics (C6–C8) and light oils (C2–C4). The 1-butyl-3-methylimidazolium chloroaluminate ionic liquid was used to alkylate low-carbon hydrocarbons with light olefins, yielding C8–C15 aromatics with an 88.4% sensitivity. Following the hydrogenation procedure, C8–C15 cyclic alkanes (72.5 wt%) with a typical composition of C10.5H20.9 were produced. The quality of the produced gasoline met the majority of jet fuel criteria. As a consequence, the chosen approach successfully created C8 C15 aromatic hydrocarbons including cyclic alkanes from biomass-derived bio-oil under mild reaction conditions. The same three-step technique was utilized in another research, but the initial step was in situ catalytic pyrolysis, where the catalyst (HZSM-5) and biomass were mixed at an ordinary catalyst/sawdust mass ratio of 2 and pyrolyzed. They had been capable of obtaining the required C8–C15 aromatics with 92.4% purity. The final biofuel encompassed 80.4 wt% cycloalkanes (Wang et al. 2022).

6.10 Hydrodeoxygenation

Despite the fact that it takes a substantial amount of H_2 and operates under harsh conditions, hydrodeoxygenation is another adaptive approach for producing superior hydrocarbons from bio-oil. Hydrodeoxygenation enhances the enormous prospective of bio-oil as a biofuel and constitutes one of the greatest effective approaches for its carbon economy. It is fortunate that the necessary facilities for hydrodeoxygenation of bio-oil refining may be easily accessed in ordinary petrochemical plants. Notwithstanding this, the intense H_2 feed and difficult working conditions highlight the significance of stringent security protocols. Throughout bio-oil hydrodeoxygenation, which includes simultaneous hydrogenation and deoxygenation processes, the brittle bonds that are unsaturated are hydrogenated and the oxygen content is reduced by water generation. Pyrolytic bio-oil again polymerizes at 175–250 °C in the absence of a catalyst or hydrogen, resulting in char generation within a few minutes. However, in the condition of a catalyst and H, bio-oil will change into firm molecules under identical circumstances and hydrodeoxygenation happens when the degree of heat exceeds 250 °C. High pressures and cool temperatures are often used for hydrodeoxygenation and an H_2 source is necessary. Catalysts may catalyze hydrogenation and dehydration procedure at very low temperatures via their metal and moderately acidic sites, respectively. As a result, by removing oxygen from oxygenated compounds and hydrogenating aromatic rings, high-pressure hydrodeoxygenation can significantly improve bio-oil quality [92]. To preserve the aromatic quality of the fuel, hydrodeoxygenation ought to be prioritized while preventing saturation on the aromatic rings in hydroprocessing. The bio-oil hydrodeoxygenation reaction pathways are highly complex. The most commonly accepted method for bio-oil hydrodeoxygenation is divided into two steps. In the first step, known as stabilization, reactive functional groups such as carboxyl and carbonyl are converted into alcohols; these processes occur at temperatures ranging from 100 to 300 °C. In the subsequent phase, typically happens at degrees ranging from 350 to 400 °C, cracking and hydrodeoxygenation occur (Why et al. 2019; Lahijani et al. 2022; Inayat et al. 2022).

Deoxygenation in a hydrogen mixture occurs by numerous separate mechanisms, comprising decarbonylation, decarboxylation, and hydrodeoxygenation, generating biojet fuel hydrocarbons. During decarboxylation, the carboxyl group is eliminated, causing carbon dioxide and paraffinic HC. Carbonylation releases the carbonyl group by producing olefins and emitting carbon monoxide and water. Under enormous pressures of excessive H gas, hydrodeoxygenation is employed to break C–O bonds. The dissolution of H_2 throughout the catalyst active place, resulting in extremely reactive H_2 radicals; the association resulting from H_2 radicals alongside bonds of C–O in bio-oil, which produces hydroxyl groups, water and alkanes, are two of the biggest and most significant occurrences that take place during hydrodeoxygenation. Furthermore, H_2 can degrade the C–O bonds in bio-oil, resulting in water or alkanes. The amount of energy of bio-oil improves as the percentage of deoxygenation increases. In general, fuels with less oxygen, more carbon, and less unsaturation have a greater thermal value. Because hydrodeoxygenation procedures are very exothermic, full deoxygenation may result in the formation of hot spots and an electrical runaway; this is particularly problematic when applying a batch or fixed-bed reactor. Due to the inherent contradiction among deoxygenation level and bio-oil results, complete deoxygenation of natural oil is not a final objective; rather, as an interim goal, deoxygenation ought to be done to the extent that maintains the bio-oil. For aircraft biofuel use, an oxygen content of fewer than 5% by weight is most likely suitable (Dabros et al. 2018; Cordero-Lanzac et al. 2021).

The catalyst is essential for the successful conversion of hydrodeoxygenation into HC fuel. Catalyst action, choice, equilibrium, and expense are all important factors to consider when developing a catalyst for budget-effective biofuel synthesis. Many catalysts with various active stages and promoters have been used in the hydrodeoxygenation of bio-oil. The majority of the enzymes utilized for hydrodeoxygenation for bio-oil are now heavy metal (such as Pd, Pt, Rh, and Ru) or transition metals (such as Ni, Co, Mo, and W)-based catalysts along with bimetallic catalysts (such as Pt–Pd, Pt–Sn, Rh–Pd, NiMo, NiW, CoMo, and NiCo) endorsed on Al_2O_3, ZrO_2, TiO_2, MgO, activated carbon, and mesoporous zeolite. Hydrodeoxygenation catalyst effectiveness was enhanced by using supported bimetallic catalysts. By successfully adjusting electrical and geometric factors, a combination of precious metals and transition metals, in especially, can increase catalyst constancy and efficacy (Yeboah et al. 2020; Lahijani et al. 2022).

Comprehensive investigations of hydrodeoxygenation catalysts used for bio-oil upgrade to transportation fuel were reported by Dabros et al. in 2018. In the circumstance of multifunctional catalysts, the hydrodeoxygenation action route is connected to a synergistic relationship among the metallic site and the backing to produce hydrogenolysis goods hydrocarbons and water. Acidic sites are essential for deoxygenation here, whereas metal sites serve as vital hydrogenation centers. During hydrodeoxygenation, bifunctional catalysts with both metal and acid sites can catalyze a wide range of activities, having the metallic site activating H and the other oxophilic metallic point or acid support activating oxygen (Dabros et al. 2018).

Catalyst inactivation through hydrodeoxygenation is a frequent issue that can be caused by (Why et al. 2019)

(i) Bio-oil's high H_2O content
(ii) Metal, N, and S toxicity of the catalyst sites that are active
(iii) Metallic site rearranging and melting
(iv) Deposition of coke

Employing a thermally resistant catalyst and a low reaction temperature are done to avoid metallic particle melting and water degradation. Coke buildup on the catalyst surface is a major problem; it is the principal pathway for catalytic deactivation. Coke manufacture deactivates catalysts by masking their active regions and restricting the catalytic pores. As an outcome, the catalytic activity and efficiency of the catalyst are reduced (Wei et al. 2019).

Continuous hydrodeoxygenation occurs during coke synthesis or the deactivation of catalysts and can continue up to 100–300 hours. Many undesirable thermal by-products compete with hydrodeoxygenation. A number of these procedures have been connected to the generation of coke during the hydroprocessing of bio-oil. The chemical composition of the oxygenated atoms in bio-oil, the sort of hydrodeoxygenation catalyst used, and the seriousness of the reaction's operating conditions all influence the quantity of coke generated. The presence of numerous oxygenates, along with the highly reactive nature of bio-oil, encourages coke generation during hydrodeoxygenation. Carbonaceous deposits form as a result of the rapid re-polymerization of unstable and exceptionally reactive oxygenates in bio-oil, such as phenols, catechol, furfural, and guaiacols. The strength of the reaction circumstances influences coke formation as well. While high temperatures are required for the successful transformation of bio-oil organic elements, they also result in coke generation. By flooding the polymerization precursors across catalyst metal sites, high hydrogen pressure inhibits coke deposition. Longer residence periods required to achieve the desired degree of hydrodeoxygenation improve carbon deposition. High-acidity catalysts, especially those with Brnsted acidity, have a high probability of hydrogen protons dissociating and generating carbon ions, which are precursors to coke synthesis. Lower acidity catalysts, on the other hand, are less suited to trans-alkylation, hydrogenolysis, and hydrocracking. Activated carbon catalysts having moderate and low acidity encouraged the synthesis of aviation fuel-ranged alkanes and aromatic compounds, respectively, according to Zhang et al. in 2019. As a consequence, slightly alkaline catalysts with an acceptable mix of medium and low acidic locations, as well as proper pore sizes that allow coke precursors to be swept into the reaction fluid, can partially prevent deposition of coke. The tendency of various catalysts for coke production in hydrodeoxygenation, according to Dabros et al. in 2018, varies in an order of alumina > sulfided transition metal oxides > mono-metallic noble metal catalysts > bimetallic catalysts.

Hu et al. (2020) conducted a thorough review of the methods used to reduce the formation of coke to reduce blockages in bio-oil-enhancing processes, including multistage hydrodeoxygenation processes, various reactors designs, and the development of various catalysts. The content of the bio-oil, as well as the parameters of the catalyst and support, affects the performance of the catalysts in the hydrodeoxygenation operation. Furthermore, temperature, pressure, catalyst load, gas hourly space velocity (GHSV), and batch length all have an effect on the catalyst's activity and selectivity. Hydrogen pressure is a key factor determining the distribution of products in the hydrodeoxygenation method. Larger H_2 pressures enhance hydrogen solubility in bio-oil and consequently an increased supply of H_2 around the catalyst accelerates the reaction activity and reduces coking. Thermodynamically, full hydrodeoxygenation is attainable at temperatures that range from 250 to 400 °C. Under a low temperature, H quickly decreases reactive compounds such as aldehydes, ketones, and olefins, hence stabilizing the bio-oil. Higher temperatures promote deoxygenation for refractory phenolic compounds. Due to the intricate nature of raw bio-oil's composition, comparatively few studies on biomechanical elements involved in pure bio-oil hydrodeoxygenation have been done due to a plethora of associated reactions, side reactions, and reactant interactions. The vast bulk of research on the method and fundamentals of hydrodeoxygenation has focused on individual model molecules. Because of their high concentration in bio-oil and because they are primarily inert component of hydrodeoxygenation processes, lignin-derived elements have been used as model compounds in the bulk of such studies. Although these results cannot be utilized to reach broad inferences regarding the hydrodeoxygenation of actual bio-oil, they offer little knowledge into the reaction routes and efficacy of the catalysts used under specific conditions for reaction, which is particularly significant given the lack of research on the hydrodeoxygenation of legitimate bio-oil to jet fuel-range HCs (Dabros et al. 2018; Hu et al. 2020; Lahijani et al. 2022).

Bashir et al. (2022) recently presented a thermo-catalytic reforming system consisting of pyrolytic temperature at 450 °C and post-reforming temperature of 700 °C, resulting in bio-oil enhancing via successive two-step hydroprocessing, featuring hydrodeoxygenation and hydrocracking. Aside from hydrodesulfurization and hydro denitrification, the hydrodeoxygenation process was used to deoxygenate organic substances to saturating alkanes while the procedure of hydrocracking was employed for the catalytic transformation to alkanes of smaller chain (C8–C16), which correspond to jet fuel-range HCs, as well as isomerization, splitting, and hydrodealkylation reactions. For hydrodeoxygenation, a NiMo catalyst based on alumina was used, while for hydrocracking, a NiW catalyst based on silica–alumina was used. Both catalysts were warmed up for 4 hours using dimethyl disulfide at 20 bar, 350 °C, and hydrogen before each hydroprocessing test. The only variable in the procedure was hydrogen pressure, which was held constant throughout. The two-step hydroprocessing experiments were conducted for 4 hours in a bench-scale autoclave reactor at 350 °C with 30 or 60 bar hydrogen, with a catalyst loading of 1 g/10 g bio-oil. The results revealed a significant decrease in O and N levels and an increase in C and H values. Hydrodeoxygenation and hydrocracking accomplished considerable deoxygenation and denitrogenation at 60 bar H_2. The hydroprocessing at 60 bar generated better fuel properties, whereas atmospheric distillation recovered a 25 wt% jet fuel fraction comprised of *n*, iso, and cycloparaffins, as well as C8 C16 aromatics; green naphtha and diesel were the main by-products. The majority of jet fuel fraction metrics, including heating value, viscosity, weight, and freezing point, satisfied the ASTM D7566 standards; nevertheless, flash point, smoke point, and total acid number were predicted to improve with additional treatments. The crude bio-oil generated via the TCR system exhibited a low O_2 content, which made the subsequent deoxygenation procedure easier.

In another study, Chen et al. in 2020 employed fluidized bed fast pyrolysis, hydroprocessing, and hydrocracking to produce jet fuel from rice husk. The pyrolytic bio-oil generated by fast pyrolysis of biomass was directed toward the hydroprocessing and hydrocracking processes, which were catalyzed by Pd/AC and NiAg/SAPO-11 catalysts. The hydroprocessing unit was operated at 300 °C using a liquid hourly space velocity (LHSV) of 1 hour and a pressure of 60 bar, with an H_2-to-oil ratio of 1000. The reaction parameters during hydrocracking were 320–380 °C, 40 bar pressure, with the identical LHSV and H_2-to-feed proportion as for hydroprocessing. The results revealed that when the temperature rose, the concentration of aromatics decreased. At 340 °C, the most cycloalkanes were created; nevertheless, once the ambient temperature was raised to 380 °C, the number of cycloalkanes decreased while more traditional alkanes were formed. These studies demonstrated that using the NiAg/SAPO-11 catalyst at greater temperatures improved ring-opening reactions. The ring-opening processes generated alkenes, which resulted in 360 °C. Aromatics, *n*-alkanes, and cycloalkanes were chosen as hydroprocessed oil components. Despite this, the fraction of aromatics was dramatically reduced after hydrocracking, whereas the proportion of *n*-alkanes, cycloalkanes, and isoalkanes rose. The much higher aromatic percentage in hydroprocessed oils compared to straight-chain and cycloalkanes might be attributed to the unique appearance of ring-opening occurrences during hydroprocessing. However, because the catalyst was acidic, the quantity of straight-chain alkanes that consist of pentadecane, hexadecane, heptadecane, and octadecane increased after hydrocracking. Simultaneously, the catalyst's metal sites encouraged olefin oligomerization by carbonation and hydrogenation. These processes led to a 63% decrease in aromatics content, bringing the final fuel's composition closer to that of aviation fuel. The generated fuel was identical to Jet A-1 in terms of its molecular weight, H/C ratio, viscosity, and density. While the aromatic content and vapor pressure were much higher than in Jet A-1, the heat of ignition was somewhat lower. The finished fuel dissolved adequately in JP-5 fuel with a decreased aromatic content. In pyrolysis vapor or condensed bio-oil fraction, tiny oxygenates such as ketones, alcohols, acids, and anhydro-sugars are common (Why et al. 2019). Direct hydrodeoxygenation of such substances yields light olefins, which reduces the carbon length in desired hydrocarbon-based fuel range products and enhances biofuel costs. Light oxygenates, which make up around 20% of bio-oil, undergo dehydration directly on the hydrodeoxygenation catalyst's acid sites before finishing up in the light gas stream. This can be handled by catalyzing activities and acylation to create carbon-coupled extremely oxygenated products with a C–C coupling catalyst. These carbon-coupling reactions are particularly important in the conversion of biomass-derived compounds into petroleum-based range HC. Because of the containment of acids, carbonyls, and alcohols, light oxygenates are extremely reactive. These functionalities can undergo carbon chain development and oxygen removal activities in the context of bifunctional carbon-coupling catalysts. As a consequence, combining a coupling catalyst with a hydrodeoxygenation catalyst might be a potential technique for enhancing the carbon recovery from pyrolysis vapor. The long-chain oxygenates that result from the carbon-coupling mechanism are subsequently hydrodeoxygenated to form fuel spectrum HC. Nonetheless, merging multiple procedures in an individual unit is challenging (Bashir et al. 2022). One method for efficiently merging the various activities of the used catalyst is to use a dual-bed catalytic biovessel to carry out the cascading events in a single reactor. In this case, Yeboah et al. in 2020 employed a tandem dual-bed catalytic technology for transforming virtualized bio-oil to jet fuel spectrum HCs. The upstream catalysts were 0.2 weight percent X-TiO_2 (X: Au, Pd, Ru on TiO_2 pellet), whereas the downstream catalysts comprised Ru-MoFeP/Al_2O_3. A recreated bio-oil containing a characteristic makeup of wood-derived bio-oil gathered through fast pyrolysis, which includes distilled water, vinegar, acetol, and furfural as typical by-products produced through the breakdown of hemicellulose as well as phenol, guaiacol, and eugenol as lignin monomers. All of the tested dual-bed catalysts reduced light gas production by around 40% while improving HC output in the jet fuel range. Alkylated aromatics seemed the most common chemical while every kind of catalyst was employed. It was proposed that the connected light oxygenates be alkylated using phenolic compounds and then hydrodeoxygenated to produce higher hydrocarbons. The most efficient carbon-coupling catalyst for carbon-chain development was Au/TiO_2, which produced 71.8% C7 hydrocarbons. This was most likely owing to the fact that Au has a low hydrogenation characteristic, which accelerated an aldol condensation process.

The hydrodeoxygenation of phenolic compounds necessitates a high reaction temperature due to the supporting benzene ring and phenolic OH groups. Cross-coupling reactions between phenols are conceivable, which might result in carbonation, and is highly bad for catalyst performance. In the hydrodeoxygenation of phenols, high aromatic yields may only be obtained by selective breaking of the strong caryl-O link without hydrogenation of the aromatic ring (Shu et al. 2019).

Phenol-to-jet fuel hydrodeoxygenation occurs through many chemical routes (Cordero-Lanzac et al. 2021):

(a) Hydrogenation of the ring of phenol to cyclohexanone
(b) Direct deoxygenation of phenol via breaking of the Csp2-O link
(c) Phenol tautomerization

The method used for hydrodeoxygenation of cyclohexanone showed that the transformation of benzene happens completed through a pair of routes:

(a) Dehydrogenation of cyclohexanone followed by hydrodeoxygenation
(b) Cyclohexanone hydrodeoxygenation, which again is followed by dehydrogenation pathway

Other jet fuel HC are created in four stages:

Hydrogenation of tetralone to tetralol ➔ Dehydration of tetralol to dialin intermediate ➔ Hydrogenation of dialin's C–C bonds to tetralin ➔ Aromatic hydrogenation of tetralin with excess hydrogen to obtain isomers of *cis* and *trans* decalin.

6.11 Future Prospects

Biojet fuels are great prospects for addressing numerous challenges, with biojet fuels capable of reducing emission of carbon molecule by 80% during the entire life cycle of the fuel. Any replacement to fossil jet fuel should be inexpensive. Because the current rate for biojet fuel is roughly three times the cost of standard jet fuel, affordability continues the most significant impediment. Nevertheless, it is envisaged that administrations take the initiative for the growth of renewable fuels in the aeronautical segment, as well as encourage the application of biojet fuel in market-level airplanes, which might change the economic value of biojet fuel, also reduce the tax for carbon dioxide emission.

Fast pyrolysis is an established method. However, the crude bio-oil is of low quality. As a result, such bio-oil needs to be deoxygenated and improved to enhance its physicochemical qualities to a standard near to application for fuel. Discussion of the underlying chemical behavior of the responses tangled in pyrolytic bio stock, as well as the effect of process parameters, is critical for high bio-oil yield. As a result, a thorough examination of the pyrolysis condition and the aforementioned factors is critical. The catalytic cracking includes complicated chemical pathways. In this case, the design and deployment of an adequate catalyst are vital to the process's success. Furthermore, choosing a proper reactor architecture, either ex situ or in situ, is critical for achieving a high gradation for improvement.

All these factors should be considered when designing a catalytic pyrolysis procedure for a given research. Acidic supports are known to enhance catalytic deoxygenation, while active sites of metals stimulate hydrogenation. As a result, catalysts with multivariant property help catalyze various processes throughout the hydrodeoxygenation process. Yet, catalysts belonging to solid acid are susceptible to inactivation, thus increasing coke formation. As a result, finding a catalyst with the required strength is critical for achieving an equilibrium between catalytic activity and deactivation. Another conundrum is that the lignin-generated bio-oil contains a low level of oxygen atom. Besides, the movement and fussiness of the catalyst in the direction of the chosen artifact are affected by the hydrodeoxygenation temperature and hydrogen pressure, along with a dosage of the catalytic product.

Though the hydrodeoxygenation routes have been widely explored, their utilization is still novice to produce genuine bio-oil. As a result, a substantial breach existing at the research level for crude bio-oil broth is utilization as a substrate intended for jet fuel synthesis and in what way the intricate bio-oil concoction and the collaborative interaction between the compounds distresses the worldwide reaction remains unknown. A significant amount of research and development is required in this respect for manufacturing effective catalysts for hydrodeoxygenation for the formation of biojet fuel from bio-oil.

It is advantageous to employ preexisting units for bio-oil hydrodeoxygenation needed for downstream processing. With this incorporation, it would significantly cut expenses and make the product's introduction into the market easier. In terms of energy use, it also provides a more sustainable scenario. A potential and practical approach toward maximizing the value of feedstock as well as fluidic unit of catalysis would be the progressive adaption of eccentric waste products. However, this is no simple task. As a result, research opportunities for comprehensive investigations on the advancement of bio-oil to biojet fuel remain open in researched sites until the commercialization of such projects is implemented.

References

Al-Rumaihi A et al (2022) A review of pyrolysis technologies and feedstock: a blending approach for plastic and biomass towards optimum biochar yield, renew. Sust Energ Rev 167:112715. https://doi.org/10.1016/j.rser.2022.112715. ISSN 1364-0321

Ambaye TG et al (2021) Emerging technologies for biofuel production: a critical review on recent progress, challenges and perspectives. J Environ Manag 290:112627. https://doi.org/10.1016/J.JENVMAN.2021.112627

Ansari K et al (2019) Fast pyrolysis of cellulose, hemicellulose, and lignin: effect of operating temperature on bio-oil yield and composition and insights into the intrinsic pyrolysis chemistry. Ind Eng Chem Res 58(35):15838–15852

Araújo AMM et al (2018) Fast pyrolysis of sunflower oil in the presence of microporous and mesoporous materials for production of bio-oil. Catalogue 8(7):261. https://doi.org/10.3390/catal8070261

Bashir MA et al (2022) A step change towards sustainable aviation fuel from sewage sludge. J Anal Appl Pyrolysis 163:105498

Björnsson, L. and Ericsson, K., (2022) Emerging technologies for the production of biojet fuels from wood—can greenhouse gas emission reductions meet policy requirements? Biomass Conv Bioref. https://doi.org/10.1007/s13399-022-02916-0

Chan YH et al (2020) Fractionation and extraction of bio-oil for production of greener fuel and value-added chemicals: recent advances and future prospects Chem Eng J 397:125406. https://doi.org/10.1016/j.cej.2020.125406. ISSN 1385-8947

Chen Z et al (2020) Effect of volatile reactions on oil production and composition in thermal and catalytic pyrolysis of polyethylene. Fuel 271:117308. https://doi.org/10.1016/j.fuel.2020.117308. ISSN 0016-2361

Chhabria S et al (2022) Valorization of hazardous materials along with biomass for green energy generation and environmental sustainability through pyrolysis. J Chem:1–9. https://doi.org/10.1155/2022/2215883

Cordero-Lanzac T et al (2021) Advances and challenges in the valorization of bio-oil: hydrodeoxygenation using carbon-supported catalysts. Energy Fuel 35:17008–17031. https://doi.org/10.1021/acs.energyfuels.1c01700

Dabros TMH et al (2018) Transportation fuels from biomass fast pyrolysis, catalytic hydrodeoxygenation, and catalytic fast hydropyrolysis. Prog Energy Combust Sci 68:268–309. https://doi.org/10.1016/J.PECS.2018.05.002

Dalai AK, et al (2021) Catalytic and noncatalytic upgrading of oils, ACS Symposium Series, American Chemical Society, Washington, DC

Dane RL, Volmer DA (2023) Mass spectrometry-based methods for the advanced characterization and structural analysis of lignin: a review. Mass Spectrom Rev 42:144–188

Das P (2022) Pyrolytic bio-oil—production and applications. In: Tuli D, Kasture S, Kuila A (eds) Advanced biofuel technologies present status, challenges and future prospects. Elsevier, USA

Das C et al (2021) Incorporation of biochar to improve mechanical, thermal and electrical properties of polymer composites. Polymers 13(16):2663. https://doi.org/10.3390/polym13162663

Demirbas A (2009) Progress and recent trends in biodiesel fuels. Energy conversion and management. Energy Convers Manag 50:14–34. https://doi.org/10.1016/j.enconman.2008.09.001

Devi A et al (2022) Production of valuable platform chemicals through microalgal routes utilizing waste streams. Bioresour Technol Rep 18:101071

Doliente SS et al (2020) Bio-aviation fuel: a comprehensive review and analysis of the supply chain components. Front Energy Res 8:110. https://doi.org/10.3389/FENRG.2020.00110/BIBTEX

Duan D, Zhang Y, Lei H, Villota E, Ruan R (2019) Renewable jet-fuel range hydrocarbons production from co-pyrolysis of lignin and soapstock with the activated carbon catalyst. Waste Manag 88:1–9. https://doi.org/10.1016/j.wasman.2019.03.030

Dyk VS et al (2019) Potential yields and emission reductions of biojet fuels produced via hydrotreatment of biocrudes produced through direct thermochemical liquefaction. Biotechnol Biofuels 12:1–12. https://doi.org/10.1186/S13068-019-1625-2/TABLES/6

Fonseca FG et al (2019) Moisture content as a design and operational parameter for fast pyrolysis. J Anal Appl Pyrolysis 139:73–86. https://doi.org/10.1016/j.jaap.2019.01.012. ISSN 0165-2370

Fossil Fuel. (2021) Environmental and Energy Study Institute. https://www.eesi.org/topics/fossil-fuels/description

He S et al (2021) Catalytic upcycling paper sludge for the recovery of minerals and production of renewable high-grade biofuels and bio-based chemicals. Chem Eng J 420:129714. https://doi.org/10.1016/j.cej.2021.129714

Hu X, Zhang Z, Gholizadeh M, Zhang S, Lam JC-H, Xiong Z, Wang Y (2020) Coke Formation during Thermal Treatment of Bio-oil. Energy & Fuels XXXX. https://doi.org/10.1021/acs.energyfuels.0c01323

Inayat A et al (2022) Techno-economical evaluation of bio-oil production via biomass fast pyrolysis process: a mini review. Front Energy Res 9:770355. https://doi.org/10.3389/fenrg.2021.770355

Kumal RR et al (2020) Impact of biofuel blends on black carbon emissions from a gas turbine engine. Energy & fuels. J Am Chem Soc 34(4):4958–4966. https://doi.org/10.1021/acs.energyfuels.0c00094

Lahijani P et al (2022) Upgrading biomass-derived pyrolysis bio-oil to bio-jet fuel through catalytic cracking and hydrodeoxygenation: a review of recent progress. Energy Convers Manag 268:115956. https://doi.org/10.1016/j.enconman.2022.115956. ISSN 0196-8904

Landrat M et al (2022) Bio-oil derived from teff husk via slow pyrolysis process in fixed bed reactor and its characterization. Energies 15:9605. https://doi.org/10.3390/en15249605

Lindfors C et al (2023) Co-processing of biocrudes in oil refineries. Energy Fuel 37(2):799–804. https://doi.org/10.1021/acs.energyfuels.2c04238

Michailos S, Bridgwater AA (2020) Comparative techno-economic assessment of three bio-oil upgrading routes for aviation biofuel production. Int J Energy Res 43:7206–7228. https://doi.org/10.1002/ER.4745

Modupe A (2019) Polymer-modified bio-asphalt: a sustainable panacea to greenhouse gas emissions. In: Zhang H (ed) Asphalt and asphalt mixtures. Intechopen. https://doi.org/10.5772/intechopen.86832

Muneer B et al (2019) Influence of in-situ and ex- situ HZSM-5 catalyst on co-pyrolysis of corn stalk and polystyrene with a focus on liquid yield and quality. J Clean Prod 237:117762. https://doi.org/10.1016/J.JCLEPRO.2019.117762

Nanda S, et al (2021) Catalytic and noncatalytic upgrading of bio-oil to synthetic fuels: an introductory review. In: Dalai AK, et al (ed) Catalytic and noncatalytic upgrading of oils. ACS Publication, pp 1–28. Chapter-1

Neves RC et al (2020) A vision on biomass-to-liquids (BTL) thermochemical routes in integrated sugarcane biorefineries for biojet fuel production. Renew Sust Energ Rev 119:109607

Nisar J et al (2022) Production of bio-oil from de-oiled karanja (pongamia pinnata l.) seed press cake via pyrolysis: kinetics and evaluation of anthill as the catalyst. Sust Chem 3(3):345–357. https://doi.org/10.3390/suschem3030022

Park JY et al (2019) Production of bio-oil from fast pyrolysis of biomass using a pilot-scale circulating fluidized bed reactor and its characterization. J Environ Manag 234:138–144. https://doi.org/10.1016/j.jenvman.2018.12.104. ISSN 0301-4797

Patel AB et al (2020) Polycyclic aromatic hydrocarbons: sources, toxicity, and remediation approaches. Front Microbiol 11:562813. https://doi.org/10.3389/fmicb.2020.562813

Phounglamcheik A et al (2017) Pyrolysis of wood in a rotary kiln pyrolyzer: modeling and pilot plant trials. Energy Procedia 105:908–913. https://doi.org/10.1016/j.egypro.2017.03.413

Qureshi KM et al (2021) Effect of temperature and feed rate on pyrolysis oil produced pyrolysis oil produced via helical screw fluidized bed reactor. Korean J Chem Eng 38:1797–1809. https://doi.org/10.1007/s11814-021-0842-0

Rangel, M.d C et al (2023) Selecting catalysts for pyrolysis of lignocellulosic biomass. Biomass 3(1):31–63. https://doi.org/10.3390/biomass3010003

Romuli S et al (2018) Combustion performance of Jatropha curcas L. residues in a biomass burner with continuous feeding system. Biofuels 12:1–6. https://doi.org/10.1080/17597269.2018.1464876

Sarkar JK, Wang Q (2020) Characterization of pyrolysis products and kinetic analysis of waste jute stick biomass. PRO 8(7):837. https://doi.org/10.3390/pr8070837

Shahbaz M et al (2021) A comprehensive review of biomass based thermochemical conversion technologies integrated with CO2 capture and utilisation within BECCS networks. Resour Conserv Recycl 173:105734. https://doi.org/10.1016/j.resconrec.2021.105734

Shu R et al (2019) Efficient catalytic hydrodeoxygenation of phenolic compounds and bio-oil over highly dispersed Ru/TiO2. Fuel Process Technol 184:12–18. https://doi.org/10.1016/J.FUPROC.2018.11.004

Shumeiko B et al (2019) Efficient one-stage bio-oil upgrading over sulfided catalysts. ACS Sustain Chem Eng 8:15149–15167. https://doi.org/10.1021/ACSSUSCHEMENG.0C03896

Singh SK (2020) Utilization of plant biomass for the production of renewable and sustainable biofuels with zero carbon emission. In: Recent technologies for enhancing performance and reducing emissions in diesel engines. IGI Global, pp 26–43. https://doi.org/10.4018/978-1-7998-2539-5.ch002

Statistical Review of World Energy (2021) https://www.bp.com/content/dam/bp/business-sites/en/global/corporate/pdfs/energy-economics/statistical-review/bp-stats-review-2021-full-report.pdf

Sun K et al (2018) Catalytic pyrolysis of waste polyethylene into aromatics by h3po4-activated carbon. Energ Fuels 32:9772–9781

Thu K et al (2020) Simulation of reaction kinetics and heat transfer effects on product yields from fast pyrolysis of oil palm empty fruit bunch biomass in fluidized bed reactor. Bioenergy Res 13. https://doi.org/10.1007/s12155-020-10148-1

Tirapanampai C et al (2019) Effect of temperature and the content of Na_2CO_3 as a catalyst on the characteristics of bio-oil obtained from the pyrolysis of microalgae. J Anal Appl Pyrolysis 142:104644. https://doi.org/10.1016/j.jaap.2019.104644. ISSN 0165-2370

Tomczyk A et al (2020) Biochar physicochemical properties: pyrolysis temperature and feedstock kind effects. Rev Environ Sci Biotechnol 19:191–215. https://doi.org/10.1007/s11157-020-09523-3

Uddin KT (2018) An overview of recent developments in biomass pyrolysis technologies. Energies 11(11):3115. https://doi.org/10.3390/en11113115

Velez DCP et al (2018) Carbon fiber from fast pyrolysis bio-oil. Sci Technol Mater 30(1):16–22. https://doi.org/10.1016/j.stmat.2018.10.001. ISSN 2603-6363

Wang WC et al (2022) Techno-economic analysis of renewable jet fuel production: the comparison between Fischer-Tropsch synthesis and pyrolysis. Energy 239:121970. https://doi.org/10.1016/J.ENERGY.2021.121970

Wei H, et al (2019) Renewable bio-jet fuel production for aviation: a review. Fuel 254:115599. https://doi.org/10.1016/j.fuel.2019.06.007

Why rising fuel prices might not be as bad for the airline sector as it seems (2022) https://www.mckinsey.com/industries/travel-logistics-and-infrastructure/our-insights/why-rising-fuel-prices-might-not-be-as-bad-for-the-airline-sector-as-it-seems

Why ESK et al (2019) Renewable aviation fuel by advanced hydroprocessing of biomass: challenges and perspective. Energy Convers Manag 199:112015. https://doi.org/10.1016/J.ENCONMAN.2019.112015

Yang J et al (2019) An overview on performance characteristics of bio-jet fuels. Fuel 237:916–936. https://doi.org/10.1016/J.FUEL.2018.10.079

Yeboah I et al (2020) Jet fuel range hydrocarbon production from propanal: mechanistic insights into active site requirement of a dual-bed catalyst. ACS Sustain Chem Eng 8(25):9434–9446. https://doi.org/10.1021/acssuschemeng.0c02200

Yildiz D (2022) Paulownia Elongata odununun pirolizinde sürükleyici gaz hızı ve parçacık boyutunun ürün verimlerine etkisi ve katran karakterizasyonu. Gazi Üniversitesi Mühendislik-Mimarlık Fakültesi Dergisi 38. https://doi.org/10.17341/gazimmfd.1050524

Yogalakshmi KN, et al (2022) Lignocellulosic biomass-based pyrolysis: a comprehensive review, chemosphere 286(2):131824. https://doi.org/10.1016/j.chemosphere.2021.131824. ISSN 0045-6535

Zhang S et al (2019) Liquefaction of biomass and upgrading of bio-oil: a review. Molecules 24(12):2250. https://doi.org/10.3390/molecules24122250

Chapter 7
Bio-Aviation Fuel via Catalytic Hydrocracking of Waste Cooking Oil

Shally Pandit, Anand Prakash, and Arindam Kuila

Abstract The aviation industry faces critical challenges regarding environmental sustainability due to the drawbacks associated with conventional jet fuels. These fuels, primarily derived from fossil sources, contribute significantly to greenhouse gas emissions and global warming. Biomass-based alternatives have emerged as promising solutions, but they present their own set of challenges, including feedstock limitations and complex conversion processes. Waste cooking oil, readily available and abundant, emerges as a sustainable savior, offering an environmentally friendly feedstock option. Various methods for producing bio-aviation fuel like transesterification and Fischer–Tropsch from diverse sources like lignocellulosic biomass, and edible and nonedible plants seed-based oil have been investigated, with hydrocracking standing out as the preferred choice. It offers high conversion efficiency, superior product quality, and versatility in meeting stringent jet fuel specification protocols. The jet fuels produced from waste cooking oils through hydrocracking demonstrate exceptional compatibility with industry standards, ensuring seamless integration into existing aviation infrastructure. This chapter covers the promising aspect of waste cooking oil as a valuable feedstock for bio-aviation fuel production, highlighting the merits of catalytic hydrocracking in achieving sustainability objectives within the aviation sector.

Keywords Aviation fuel · Waste cooking oil · Catalytic hydrocracking · Transesterification · Fisher–Tropsch

S. Pandit · A. Prakash · A. Kuila (✉)
Department of Bioscience and Biotechnology, Banasthali Vidyapith, Newai, Rajasthan, India

A. Kuila (ed.), *Biojet Fuel: Current Technology and Future Prospect*, Clean Energy Production Technologies, https://doi.org/10.1007/978-981-99-8783-2_7

7.1 Introduction

The global population is continuously increasing, as is the demand for energy. However, there are growing concerns about the scantiness of resources, including food and water, as well as the depletion of fossil fuel reserves. A report indicates that air travel is expected to see a 5% annual increase until 2026, leading to a 3% growth in jet fuel demand. The aviation sector is a significant consumer of energy, with commercial aviation fuel consumption reaching 13.16 exajoules (EJ) in 2012, accounting for 12% of worldwide transportation energy consumption. In 2016, commercial operators consumed 278 billion liters of jet fuel, resulting in 781 million tons of CO_2 emissions. This consumption is projected to rise by another 10.5 EJ by 2040. Commercial and military jet fuels typically consist of cycloparaffins, alkanes, and aromatics, with carbon atoms ranging from C8 to C16, while road transportation fuels include diesel alkanes, aromatics, cycloparaffins, and a few oxygenates, with carbon atoms greater than C16. The carbon atom range for commercial jet fuel can be derived from the conversion of waste triglycerides (Asiedu et al. 2019; Goh et al. 2020).

The negative consequences of burning fossil fuels include the volatility in oil cost and the release of greenhouse gases. Given the anticipated growth in air travel passengers and the associated surge in fuel consumption, there is a clear need for clean alternative fuels within the aviation industry (Goh et al. 2020).

Producing alternative sources like biojet fuel presents a sustainable and viable option. Developing biojet fuel from renewable sources offers several benefits, including reducing reliance on fossil fuels and potentially achieving a significant reduction in CO_2 emissions, surpassing the aviation industry's 50% reduction target for 2050 by up to 68.1% (Khodadadi et al. 2020). Biofuels can be derived from various sources (Why et al. 2019), including both nonrenewable ones like conventional vegetable oil and renewable ones like industrial waste, biomass, animal fats, microalgae, and nonedible plant seed-based oils. Hence, the use of renewable resources holds promise as an alternative to address future challenges. The reutilization of biomass by techniques analogous to those used in petrochemicals permits the manufacture of valuable materials, chemicals, and fuels. However, the use of biomass creates concerns about competitiveness between food and nonfood item applications, along with its impact on land usage and water availability. Also, biomass naturally contains a significant weight percentage of oxygen, and its structural components encompass a wide spectrum of molecular weights, ranging from low to high. ASTM and EU regulations require bio-aviation fuels to be oxygen-free hydrocarbons with particular specifications for the length of the carbon chain and physicochemical characteristics. Achieving these specifications proves challenging with current biomass conversion methods. These stringent requirements, coupled with the inherent limitations of biomass, restrict the feasibility of producing drop-in hydrocarbon fuels to bio-kerosene exclusively (Wang et al. 2019). Using feedstock derived from industrial waste, such as waste cooking oil (WCO), can be preferred because it avoids competition between sectors and allows for a more sustainable and environmentally friendly second cycle of utilization (Asiedu et al. 2019; Khodadadi et al. 2020).

WCO is produced as a waste from the cooking and frying processes in households or restaurants and often includes oils like canola, soybean, sunflower, and many others that have been exposed to high temperatures and food particles during frying (Awogbemi et al. 2021).

Vegetable oils primarily consist of triacylglycerols, making up a significant portion (88–98%) of their composition. Over the years, global vegetable oil usage has seen a notable increase, rising from 150 million metric tons (MMT) in 2013/14 to 200 MMT in 2020/21. In 2019, it was valued at $5.50 billion and is estimated to reach approximately $8.48 billion by 2027. This growth likely reflects increased awareness of recycling and repurposing WCO, driven by environmental and economic considerations (Awogbemi et al. 2021).

Canada generates between 120,000 and 135,000 tons of WCO annually. In the United States, 0.6 million tons of yellow grease were produced in 2011. The UK and European Union countries produce approximately 700,000–1,000,000 tons of WCO per year. South Africa collects 60,000 tons of WCO annually, but an estimated 200,000 tons remain uncollected each year. Japan, China, and Malaysia generate 6000, 45,000, and 60,000 tons of WCO, respectively, on an annual basis. Despite these figures, it is estimated that more than 60% of globally generated WCO is improperly disposed of. During the frying process, vegetable oils undergo repeated exposure to high temperatures, typically ranging between 150 °C and 200 °C. This, along with moisture infiltration and contamination, leads to physical changes in properties such as color, viscosity, and density, as well as chemical modifications, including alterations in acid value and fatty acid composition. Additionally, severe thermal degradation results in the formation of total polar compounds within the oil structure. These thermal degradation processes induce changes in several oil properties, including saponification value, kinematic viscosity, moisture content, iodine value, density, specific heat, peroxide value, flash point, the number of single/double bonds, and the percentage of mono/polyunsaturated components as a consequence of usage (Awogbemi et al. 2021). Consuming food repeatedly prepared with reprocessed waste cooking oil (WCO) can lead to adverse health consequences, including conditions such as hypertension, diabetes, vascular inflammation, and other health issues. According to available statistics, in 2015, WCO contributed 17% of feedstock used in the production of 11.92 million tons of biodiesel in the European Union and 9% of the feedstock used for the production of 26.62 million tons of biodiesel globally. It is crucial to remember, however, that not all WCO gathered is used to produce biodiesel or other fuels. There are valid fears that fraudulent individuals may be filtering and repackaging discarded vegetable oil for resale to naïve customers. This raises concerns about the quality and safety of food prepared with such recycled oils and underscores the need for proper regulation and oversight in this industry to protect public health (Awogbemi et al. 2019).

Unfortunately, in many urban areas, WCO is improperly disposed of into water resources like rivers, polluting natural water bodies and decanting waste oil into sewers, drains, and open spaces like forests. This leads to several adverse consequences, including the generation of offensive odors, blockage of drainage systems, damage to concrete structures, and contamination of both terrestrial and aquatic habitats. Improperly disposed WCO can also induce froth formation, raise the level of organic matter in water sources, disrupt wastewater treatment operations, diminish dissolved oxygen levels in aquatic habitats, and upset ecological balance. However, there is potential for a more responsible approach. When catalyzed with certain agricultural waste materials, WCO can be used to generate biodiesel, which can be used to serve various applications, including the production of bio-aviation fuel. This conversion process can significantly hamper the harmful impacts of WCO on the environment. This would not only mitigate environmental problems but also contribute to more sustainable practices in handling waste cooking oil (Awogbemi et al. 2021). It is important to highlight that disposing of waste cooking oil (WCO) in natural environments can have a detrimental impact on plant and animal life, primarily because of its low water solubility. However, WCO can be utilized effectively without the need for extensive treatment in various industrial processes. It can serve as a primary raw material for activities like bio-lubricant and fuel production, as well as additives for applications like asphalt modification and animal feed.

Moreover, WCO has the potential for transformation through chemical or biochemical processes to yield valuable products, including biofuels, bio-plasticizers, synthetic gas (syngas), and sorbents for capturing volatile organic compounds. These versatile applications demonstrate the potential for WCO to be repurposed and contribute to sustainable and environmentally friendly solutions across multiple industries (Khodadadi et al. 2020).

7.2 Waste Cooking Oil (WCO): Feedstock for Bio-Aviation Fuel

Bio-aviation fuel is often referred to as renewable jet fuel, biojet fuel. Biofuel designed for the aviation industry is acknowledged as a strategy to help reduce greenhouse gas emissions within the aviation sector as a whole (Doliente et al. 2020). WCO is generated through a continuous oxidation process of virgin cooking oil, typically occurring during open-air frying. This process primarily operates through a free radical mechanism, resulting in the production of hydroperoxides as the main oxidation product. These hydroperoxides can further oxidize into toxic substances like 4-hydroxy-2-alkenals, making WCO a hazardous waste. Additionally, WCO is considered nonedible due to its potential to cause various adverse effects such as indigestion, diarrhea, stomach discomfort, and even gastric cancer.

There are several compelling reasons to consider using WCO as a feedstock for jet and diesel fuel: (a) abundance: WCO is readily available in humongous quantities, with annual generation of 29 million tons across the globe. (b) Environmental impact: discharging just 1 l of WCO into water bodies can contaminate approximately 500,000 l of water, adversely affecting aquatic ecosystems by obstructing sunlight and hindering the exchange of oxygen. (c) Cost-effectiveness: WCO is 3×

than virgin vegetable oil, making it an economically attractive option. WCO costs approximately $224 per ton compared with $771 per ton for soybean oil. (d) Environmental benefits: reusing WCO not only benefits the environment but also reduces the cost of wastewater treatment. (e) Chemical composition: WCO has a fatty acid composition with approximately 14–22 carbon atoms, and carboxylic acid is the primary functional group in waste triglycerides. These properties make it relatively straightforward to upgrade triglycerides into hydrocarbon fuels. Considering these factors, WCO has already been successfully used as a feedstock for the commercial production of biodiesel, with established processing facilities in place. This suggests that there is a feasible feedstock logistics system and potential to adapt existing infrastructure for the production of waste cooking oil-based biofuels, specifically for the aviation industry (Asiedu et al. 2019; Awogbemi et al. 2021). Table 7.1 indicates the values of various chemical and physical parameters of waste cooking oil and neat oil.

Table 7.1 Differences between various chemical and physical properties of waste cooking and vegetable oil

Property	Waste cooking oil	Neat oil	References
pH	5.34	7.38	Awogbemi et al. (2021)
Viscosity at 40 °C (mm^2/s)	31.381	28.744	Awogbemi et al. (2019)
Density (kg/m^3)	921.41 at 15 °C	919.21 at 20 °C	El-Sawy et al. (2020), Awogbemi et al. (2019)
Flash point (°C)	213	161–164	El-Sawy et al. (2020)
Pour point (°C)	7		El-Sawy et al. (2020)
Water content, wt.%	0.25	0.2%	El-Sawy et al. (2020), Negash et al. (2019)
Kinematic viscosity (mm^2/s)	48.8 at 37.8 °C	–	El-Sawy et al. (2020)
Elemental composition			
Sulfur content (wppm)	38	0.9	Bezergianni et al. (2009)
Nitrogen content (wppm)	47.42	0.69	Bezergianni et al. (2009)
C (wt %)	76.74	76.36	Bezergianni et al. (2009)
H (wt %)	11.62	11.62	Bezergianni et al. (2009)
O (wt %	11.6	12.02	Bezergianni et al. (2009)
Fatty acid composition (wt%)			
Oleic acid	0.8	–	Awogbemi et al. (2019)
Palmitic acid	0.36	32.21	Awogbemi et al. (2019)
Linoleic acid	0.10	21.98	Awogbemi et al. (2019)
Erucid acid	0.26	–	Awogbemi et al. (2019)
Caprylic acid	0.20	0.22	Awogbemi et al. (2019)
Stearic acid	1.14	9.27	Awogbemi et al. (2019)
Myristic acid	17.04	12.36	Awogbemi et al. (2019)

7.3 Hydrocracking Process

Various process technologies are available for converting materials based on biomass into substitutes for aviation fuel. These technologies range from commercial scale to those still in the research and development phase. Their applicability very much depends on the type of feedstock being used. Oil-based feedstocks are typically processed into biojet fuels using hydroprocessing technique. These include hydrotreating, deoxygenation, and isomerization/hydrocracking. Catalytic hydrothermolysis (CH) is another method developed for treating triglyceride-based oils. Solid feedstocks are processed into biomass-derived intermediates via processes like gasification. They can also be transformed into alcohols using biochemical and/or thermochemical methods, in sugars by biochemical processing method, and in bio-oils via pyrolysis. These intermediate products (alcohols, syngas, bio-oils, sugars) can then be further enhanced into bio-aviation fuel through various synthesis, fermentative, or catalytic processes. Bio-aviation fuels produced through Fischer–Tropsch synthesis and oil hydroprocessing technologies have received approval from the ASTM International (ASTM) method D7566 for blending into aviation fuel at levels up to 50%. Hydroprocessing technologies that use vegetable and waste oils are currently the most mature and ready for large-scale production. The cost of production is a critical factor in the commercial feasibility of bio-aviation fuel. The aviation industry consumes a vast amount of jet fuel, and even small price increases can result in significant additional costs. The cost of petroleum-derived aviation fuel is closely tied to crude oil prices, making long-term budgeting challenging. However, it is predicted that advancements in conversion technology could reduce the production cost of bio-aviation fuel to as low as $2.54 per gallon by 2030, potentially replacing a substantial portion of annual airline fuel consumption. The production cost of bio-aviation fuel is influenced by several factors, including the feedstock cost and composition, process design, conversion efficiency, valorization of co-products, and energy conservation. Reducing production costs requires improvements in feedstock productivity, extraction yield of oil or sugar from crops, energy-efficient processes, and maximizing the value of co-products. The development of biojet fuels involves various technologies and considerations, with a focus on reducing production costs to make these fuels economically competitive with traditional petroleum-based jet fuels. As advancements continue, bio-aviation fuels have the potential to play a significant role in reducing the aviation industry's carbon footprint (Wang and Tao 2016).

Catalytic hydroprocessing is a technology for bio-aviation fuels production that leverages the preexisting infrastructure of petroleum refineries. It is a well-established and widely adopted industrial process with numerous applications. This technology involves the use of catalysts and hydrogen to upgrade feedstocks, such as vegetable oils or triglycerides, into biofuels like biodiesel. The advantage of catalytic hydroprocessing is that it can be integrated into existing refinery operations, making it a practical and cost-effective approach to produce biofuels at scale. This helps in maximizing the utilization of resources and reducing the environmental impact by replacing or blending with traditional petroleum-based fuels. The process flow for hydrotreating and hydrocracking shares striking similarities, utilizing high-pressure hydrogen and catalysts to purge impurities from various petroleum fractions. Both processes involve some degree of conversion and make use of similar equipment and hardware. The historical evolution of these technologies can be traced back to the mid-twentieth century. In the 1950s, hydrotreaters were pioneered to primarily target the removal of sulfur from feedstocks, particularly before they entered catalytic reformers, thereby improving feedstock quality and reducing sulfur content in the final products. Subsequently, in the 1960s, the advent of hydrocrackers revolutionized the conversion of gas oil into naphtha, presenting a more intensive process that not only eliminated impurities but also broke down larger hydrocarbon molecules into smaller, more valuable products. Today, both hydrotreaters and hydrocrackers are deployed across various configurations to process a wide spectrum of petroleum fractions. Hydrotreaters handle vacuum gas oil, gas oil, kerosene, and even heavier residue fractions, effectively reducing sulfur, nitrogen, and other impurities. Conversely, hydrocrackers are designed to accommodate feedstocks such as coker gas oil, visbreaker gas oil, vacuum gas oil, and heavy cycle oil, characterized by boiling points ranging from 650 °F to 1050 °F (343 °C to 566 °C) (Robinson and Dolbear 2006).

Hydroprocessing encompasses a range of catalytic reactions occurring in the presence of hydrogen. The process of saturating double bonds in molecules by adding hydrogen catalytically in a reactor at specific temperatures and pressures is referred to as "hydrogenation." A triglyceride is composed of three fatty acid chains: linoleic, oleic, and stearic acid. Complete hydrogenation of this molecule involves converting all unsaturated fatty acids into saturated ones, resulting in three stearic acid chains.

Furthermore, another commonly used term for modifying a triglyceride molecule is "hydrotreatment." In hydrotreatment, hydrogen is added to carbonyl group after hydrogenation, and three additional reactions can occur simultaneously depending on the process's selectivity. Excess hydrogen addition after saturation can lead to the breakdown of the glycerol compound, forming propane and a chain of free fatty acids. The remainder of the carboxylic acid group must be eliminated in order to convert these free fatty acids into straight-chain alkanes, which can be done in the following ways: the hydrodeoxygenation (HDO) pathway, which reacts with hydrogen to produce a hydrocarbon with the same amount of carbon atoms as the fatty acid chain and two moles of water; and the decarboxylation (DCOX) route, which generates a hydrocarbon with one carbon atom lesser than the fatty acid chain and one mole of CO and H_2O.

Depending on the composition of the final n-alkanes produced, they may require further processing such as isomerization, cracking, and cyclization to enhance the combustion properties, yielding isoalkanes, lighter hydrocarbons along with aromatics. When hydroprocessing most vegetable oils, the end products typically involve organic liquid products, water, and gases (C_3H_8, H_2, CO, H_2S, CO_2, CH_4), and other hydrocarbons. A comprehensive understanding of these chemical reactions presents a valuable opportunity for their application in the synthesis of jet biofuels. One critical aspect of this method is the fact that the hydrogen requirement for these reactions varies from vegetable oil to vegetable oil.

The processes of hydrodeoxygenation and hydrodecarboxylation are demonstrated using a saturated molecule; palmitic triglyceride.

$$\mathrm{HDO}: C_{51}H_{98}O_6 + 12H_2 \rightarrow 3C_{16}H_{34} + C_3H_8 + 6H_2O$$

$$\mathrm{DCOx}: C_{51}H_{98}O_6 + 3H_2 \rightarrow 3C_{15}H_{32} + C_3H_8 + 3CO_2$$

$$\mathrm{DCO}: C_{51}H_{98}O_6 + 6H_2 \rightarrow 3C_{15}H_{32} + C_3H_8 + 3CO + H_2O$$

The HDO (hydrodeoxygenation) reaction consumes dodeca moles of hydrogen for every mole of triglyceride needing treatment. In contrast, the DCOx (decarboxylation) and DCO (decarbonylation) reactions require only three and six moles of hydrogen/mole triglyceride, respectively (plus one more hydrogen mole/double bond in the vegetable oil for saturation). The CO and CO_2 generated during hydrodecarboxylation reactions can be transformed into methane through a methanation reaction. To achieve this conversion, additional hydrogen would be necessary.

$$CO + 3H_2 \rightarrow CH_4 + H_2O$$

$$CO_2 + 4H_2 \rightarrow CH_4 + 2H_2O$$

When compared to the hydrodeoxygenation method, the hydrodecarboxylation approach will need three more hydrogen molecules. The higher the initial saturation of the feedstock, the more desirable it is since less hydrogen is required during the hydrogenation phase. The iodine value (IV) is commonly used to quantify the degree of unsaturation of fatty acids in a feedstock, with a higher IV indicating a greater number of double bonds present. Biodiesel is often made from feedstocks with higher levels of IV. However, depending on the carbon atom length of the intended product, one of these reactions will be more selective. All of this emphasizes the importance of selecting the vegetable oil when doing a feasibility analysis. This choice depends not only on its availability but also on its degree of saturation, considering the hydrogen requirements and how to ensure an adequate supply of hydrogen for the process (Vásquez et al. 2017). The hydro-isomerization and hydrocracking are subsequently followed by a fractionation step aimed at segregating the blends into desired products, viz., paraffinic diesel, naphtha, kerosene, and light gases (Wang and Tao 2016). For the classical mechanism of ideal hydrocracking for the conversion of an *n*-alkane on a bifunctional catalyst, one can refer to Weitkamp 2012.

To evaluate the efficiency of hydrocracking reactions, hydrocracking conversion is a useful metric. Hydrocracking conversion (%) (Bezergianni et al. 2009) is determined as the proportion of the heavier fraction of the feedstock that has undergone transformation into lighter products during the hydrocracking process:

$$\mathrm{Conversion}\,(\%) = \left[\left(\mathrm{Feed}_{360+} - \mathrm{Product}_{360+}\right) / \mathrm{Feed}_{360+}\right] \times 100$$

where $Feed_{360+}$ and $Product_{360+}$ are the wt% of the feed and product, respectively, which have a boiling point higher than 360 °C.

Additionally, to assess the hydrocracking process's efficiency in generating a specific product while minimizing the production of other by-products, selectivity is used. Selectivity can be quantified for various products (e.g., diesel, gasoline) depending on their defined boiling point ranges. For instance, for a product with specified initial and final boiling points, denoted as A and B, respectively, selectivity can be defined as

$$\text{Product selectivity}(\%) = \left[\left(\text{Product}_{A-B}\text{-Feed}_{A-B}\right) / \left(\text{Feed}_{360+}\text{-Product}_{360+}\right)\right] \times 100$$

where $Feed_{360+}$ and $Product_{360+}$ are the wt% of the feed and product, respectively, which have a boiling point higher than 360 °C (i.e., heavy molecules of feed and product) and Feed AB and Product AB are the wt% of the feed and product, respectively, which have a boiling point range between A and B degrees Celsius. Selectivity can be defined for diesel (180–360 °C), kerosene/jet (170–270 °C) and naphtha (40–200 °C) (Bezergianni and Kalogianni 2009).

When dealing with feeds having significant quantities of polyunsaturated fatty acids, efficient hydrogenation is critical. This is due to the proclivity of unsaturated fatty acids to undergo oligomerization and Diels–Alder reactions, each of which impair catalyst performance by generating coke and increasing hydrogen consumption in cracking processes (Žula et al. 2022).

7.4 Catalyst

Creating a hydrocracking catalyst typically consists of two primary phases: firstly, producing a support material with acidic properties, like zeolite (act as acidic support), and then applying metal nanoparticles onto this support. For a comprehensive examination of catalyst synthesis, please refer further to Saab et al. (2020). Nickel–molybdenum (NiMo) and nickel–tungsten (NiW) catalysts are extensively employed in hydroprocessing. Specifically, the NiMo/γ-Al_2O_3 catalyst has shown impressive performance in hydrotreating activities, encompassing hydrodesulfurization (HDS) and hydrodenitrogenation (HDN). Conversely, NiW/Y-zeolite catalysts exhibit encouraging hydrocracking capabilities and can selectively target the desired product outcome (Peng et al. 2018). The influence of precursor and promoter substances also plays a crucial role in catalytic reactions. In the petroleum industry, various types of promoters and active metals are employed for processes such as refining, hydrodesulfurization (HDS), and hydrodeoxygenation (HDO) (Ameen et al. 2017).

Hydrocracking technology has gained increasing importance in recent times, with various types of reactors utilized for upgrading heavy oil. These reactors, namely fixed-bed, ebullated-bed, moving-bed, and slurry-phase reactors, share fundamental principles but vary in technical aspects and their ability to handle impurities. Typically, fixed-bed reactors are well-suited for hydrotreating middle distillates or high API gravity feeds, while moving-bed or ebullated-bed reactors are employed for more complex feedstocks. In fixed-bed reactors, catalysts need periodic replacement, whereas moving-bed reactors introduce fresh catalysts at the top and remove deactivated ones at the bottom, showcasing catalyst expansion that can help reduce pressure drop to some extent. In cases where feed quality is insufficient for a fixed-bed reactor, combining moving-bed reactors in series or integrating ebullated-bed reactors with fixed-bed reactors can be effective solutions.

For hydrocracking heavy feeds in fixed-bed or ebullated-bed reactors, supported metal catalysts are generally favored. Catalyst activity depends on the active metals and supports, which include materials like zeolites, alumina and mixed alumina oxide, and mesoporous substances. These supported catalysts are typically prepared using wet or incipient wetness impregnation methods, followed by calcination, reduction, and utilization in hydrocracking reactions. Prior to the reaction, catalysts are pretreated with a sulfur agent to convert them into their sulfide form.

Fixed-bed, ebullated-bed, and moving-bed reactors encounter challenges related to mass transfer, pressure drop, feed diffusion, and intra-particle mass transfer between liquid and solid phases. Factors such as particle size and agitation speed amplitude must also be taken into account. Diffusion issues can be mitigated by employing high agitation speeds (>300 rpm) and ensuring an optimal mixture of reactants. However, these challenges limit the commercial feasibility of hydrocracking processes in these reactors. To address these issues and make the process more viable for industrial applications, slurry-phase hydrocracking is considered a promising alternative (Sahu et al. 2015).

7.5 Jet Fuel Specifications

Along with establishing target compositions, the specifications and requirements for jet fuel are primarily defined in terms of essential performance properties. These specifications are crucial to ensure a safe and efficient operation of aircraft. Some of the key specifications for jet fuels include minimum energy density by mass; high allowable freeze point temperature (specifies the coldest temperature at which the fuel should remain liquid); high allowable deposits in standard heating tests (ensures that the fuel does not leave harmful deposits when heated); high allowable viscosity (defines the fuel's resistance to flow, ensuring it can be pumped and atomized effectively); high allowable sulfur and aromatics content (limits the presence of sulfur and aromatic compounds, which can be detrimental to engine performance and emissions); high allowable amount of wear in standardized test (determines the level of wear and tear the fuel can cause to engine component); high acidity and mercaptan concentration (sets limits on acidic and mercaptan compounds, which can corrode fuel system components); minimum aromatics content (specifies the minimum amount of aromatic compounds for certain performance characteristics); minimum fuel electrical conductivity (ensures that the fuel can dissipate static electricity safely); and minimum allowable flash point (sets a minimum temperature at which the fuel can ignite).

Certification of aviation fuel is typically carried out according to specific standards, including

I. ASTM D1655: The American Society for Testing and Materials (ASTM) standard for aviation turbine fuels, covering various performance and quality criteria.
II. International Air Transport Association Guidance Material (Kerosene Type): Provides guidance on fuel quality and performance for aviation.
III. United Kingdom Ministry of Defence, Defence Standard (Def Stan): A British standard that sets requirements for aviation fuel used by the military (Wang and Tao 2016).

ASTM Specification D7566 applies to alternative jet fuels. This standard defines the qualities and criteria required to manage the generation and the standard of sustainable aviation fuels while also addressing safety issues. D7566 expands on D1655 (Lin and Wang 2020) to include specifications for synthetic paraffinic kerosene (SPK) blend stocks. One critical consideration for jet fuel is its high flash point, which is essential for safety as it reduces the risk of fire. While major fuel properties are consistent across different standards, some variations exist to accommodate specific requirements. For example, ensuring good cold flow properties, such as a lower freezing point, is crucial to guarantee that the fuel remains fluid at high altitudes, where temperatures can drop significantly (Wang and Tao 2016).

7.6 Factors Affecting Hydrocracking Process

The effectiveness of hydrocracking reactions is influenced by a combination of intrinsic (related to the catalyst) and extrinsic (related to the process) factors. Here is a breakdown of these factors:

Catalyst characteristics: The properties of the catalyst, including its morphology, shape selectivity, porosity, structure, and composition, have a great impact on hydrocracking performance. Catalysts with larger surface areas and improved accessibility to acid sites, such as nanosized and mesoporous zeolites, tend to exhibit enhanced conversion and reaction rates.

Incorporation of carbon nanotubes (CNTs): The introduction of carbon nanotubes (CNTs) during zeolite synthesis can create hierarchical zeolites having both the micro- and mesoporosity. This can boost catalytic activity, but the amount of CNTs added must be carefully controlled to avoid negatively impacting Bronsted acid sites and selectivity.

Variation in pore sizes: Adjusting the pore sizes of zeolites can either facilitate or hinder the accessibility of molecules to reaction sites. This allows for different reaction pathways based on molecular size.

Influence of binders: Adding binders to the catalyst can alter its mechanical and thermal properties, acidity, coke buildup, and porosity. This can often have positive effects on hydrocracking performance.

Choice of catalyst type: Different types of catalyst are used depending on the specific hydrocracking application. For example, catalyst based on nickel supported by zeolite beta can be suitable for aromatics while catalyst based on Ni–Mo supported over zeolite Y can be used for heavy vacuum gas oil. Catalyst based on Ni–Mo and supported over alumina can be employed for vegetable oil.

Catalyst deactivation: Evaluating catalyst deactivation is crucial for assessing its performance. Some catalysts, like nanosized β zeolite loaded with Ni2P, have demonstrated superior stability and resistance to coke formation compared to others.

Si/Al ratio: Enhancing the Si/Al ratio in protonic zeolites enhances stability during reactions and prevents de-alumination.

Feedstock type: The nature of the feedstock significantly influences the hydrocracking reaction pathway. Different feedstocks result in various mechanisms and product outcomes.

Process conditions: Factors like temperature, pressure, and the hydrogen-to-feed ratio are critical in determining hydrocracking performance. Elevated temperatures accelerate cracking but can also lead to issues like coking and catalyst deactivation. Selecting the right process parameters depends on the desired product goals.

Temperature: High value of hydrocracking temperature is beneficial for both conversion and overall biofuels' yield. Higher temperatures enhance cracking activity and improve the transformation of feedstock into biofuels. However, the choice of temperature depends on the desired biofuel product. Moderate temperatures are preferable if diesel production is the main goal. In contrast, higher temperatures are much suitable when gasoline synthesis is also a priority.

Liquid hourly space velocity (LHSV): Decreasing LHSV is advantageous for conversion and biofuels' yield. Lower LHSV values allow for more extensive cracking reactions to occur within the same time frame. This parameter is closely linked to the reaction kinetics.

Heteroatom removal: The heteroatoms (nitrogen, sulfur, oxygen) removal increases as the hydrocracking temperature rises. Deoxygenation, in particular, is a favorable reaction. Higher temperatures promote the efficient removal of these impurities, enhancing the quality of the biofuels produced.

Saturation: Saturation reactions are not favored at higher temperatures. This suggests the need for a pretreatment step before hydrocracking to saturate double bonds before addressing heteroatom removal. Ensuring full saturation is important for improving the properties of the resulting biofuels.

Catalyst deactivation: Over time, catalyst deactivation is observed as DOS increases. The effectiveness of the catalyst decreases with prolonged use, affecting various reaction mechanisms differently. Sulfur and nitrogen removal are affected earlier in the process, while saturation reactions are impacted only after reaching the maximum DOS studied. Conversion and oxygen removal are also affected, with their loss of effectiveness occurring more rapidly than other reaction mechanisms.

Effective hydrocracking relies on a combination of carefully tailored catalyst characteristics and precise control of process parameters to optimize the reaction for the desired biofuel production (Bezergianni and Kalogianni 2009; Saab et al. 2020).

7.7 Challenges in WCO Biofuels

Challenges in the generation of biofuels from WCO encompass several key aspects. Firstly, the by-product glycerol purification is energy-intensive, thereby increasing production costs. This expense may be partly compensated by the selling value of glycerol. To meet stringent aviation fuel specifications, bio-aviation fuel must have high flash point, along with excellent cold flow properties. Achieving this requires the hydrocracking and hydro-isomerization of normal paraffins generated during deoxygenation to produce a synthetic paraffinic kerosene (SPK) with carbon chains ranging from C9 to C15. However, hydrocracking reactions, though exothermic, are slow and predominantly occur in the latter part of the reactor. Overcracking can lead to diminished yields of jet fuel-range alkanes and increased production of lighter substances, such as C1–C4 and naphtha (C5–C8), which fall outside the aviation fuel range and possess low economic value compared to diesel or jet fuel. Moreover, pricing and availability challenges of glycerin, a significant by-product, pose economic and environmental considerations. Additionally, the substantial investments required for large-scale biodiesel production units may deter smaller-scale producers or regions with limited resources (Wang and Tao 2016).

One of the primary hurdles for the commercialization of biofuels lies in their economic competitiveness relative to conventional fuels. Despite recent price reductions, biojet fuel remains twice as expensive as fossil jet fuel, raising concerns about its economic viability. Fuel prices are intricately tied to the fluctuating costs of crude oil and biomass, introducing significant uncertainty into production costs. A potential solution for WCO biofuels is the implementation of subsidies to incentivize adequate disposal and biofuel adoption. For instance, the United States generates a substantial daily volume of WCO but offers a sale subsidy of $0.50 per gallon for WCO. The United States and Japan have stringent regulatory policies to ensure proper WCO disposal. Japan employs WCO-derived biodiesel for garbage trucks and supports biodiesel sales with consumption taxes, while Brazil and Korea mandate blending of WCO biodiesel with conventional biodiesel. However, research suggests that China's WCO policies focus more on regulation and administration, lacking market-oriented initiatives and adequate funding, resulting in limited biofuel consumption. While regulatory enforcement can improve WCO disposal, only a few countries combine law enforcement with subsidies, indicating room for enhancement in fully realizing the potential of WCO biofuels. In the European Union (EU), regulatory frameworks have driven WCO utilization and recovery, establishing a viable market while encouraging efficient WCO collection. Nevertheless, challenges remain as WCO biofuels in the EU are highly sensitive to established double counting rules and recent caps on WCO biofuels that may affect their attractiveness in member states.

The commercialization of WCO biofuels faces a significant challenge due to a limited supply of WCO. Inefficient recovery, particularly in smaller restaurants with cost and space constraints, contributes to recovery rates often below 40%. Globally, inadequate government oversight, the absence of penalties, high disposal fees, and insufficient incentives lead to improper WCO disposal. To increase recovery rates, suggested measures include subsidies for biofuel companies, infrastructure upgrades, stricter regulations, and improved biofuel sales policies.

WCO biofuel production currently tends to be limited to small-scale operations, primarily due to logistical challenges associated with large-scale production. Challenges include limited collection centers, improper facilities, farther distances between collection to production points, and lesser space constraints. Even if economically feasible, the varying properties of WCO present challenges for commercialization. Techno-economic analyses have shown that feedstock prices have a substantial influence on biojet fuel costs. While waste materials like grease have economic benefits, the pretreatment process adds complexity and increases production costs. Nonetheless, the usage of WCO as a source of commercial bio-aviation fuel faces challenges, particularly with regard to regulations. Furthermore, the difference in fatty acid saturation in WCO directly impacts the hydrogenation reaction's hydrogen requirements. Therefore, a more selective approach to WCO collection, separating it into bio-aviation fuel feedstocks, may be necessary to address these challenges effectively (Goh et al. 2020).

7.8 Conclusion

The global population is growing, leading to increased demand for energy. However, concerns are rising about resource scarcity, including food and water, and the depletion of fossil fuels. The aviation industry, a major energy consumer, is expected to see significant growth, causing a surge in jet fuel demand and greenhouse gas emissions. To address these challenges, there is a need for clean alternative fuels in aviation. Biojet fuel derived from renewable sources, like WCO, is a promising solution. WCO is generated from frying processes and includes various vegetable oils exposed to high temperatures. It is a valuable resource, with a growing market. Improper disposal of WCO can lead to environmental issues, but it can be converted into biodiesel and, potentially, bio-aviation fuel, reducing its impact. Various countries generate significant amounts of WCO, but more than 60% of it is not properly managed. WCO has potential applications beyond fuel, including biolubricants, asphalt modification, and animal feed. Repurposing WCO through chemical or biochemical processes can yield valuable products, contributing to sustainability in multiple industries. Challenges in WCO biofuel production include energy-intensive glycerol purification, slow and overcracking-prone hydrocracking reactions, and economic concerns related to glycerin. Biofuels face economic competitiveness issues, with biojet fuel costing twice as much as fossil jet fuel. Subsidies can incentivize proper disposal and adoption. Supply scarcity stems from inadequate recovery rates due to oversight issues and inappropriate disposal. Commercialization tends to be limited to small-scale operations due to logistical challenges. A selective approach to WCO collection may be needed to overcome regulatory hurdles and address these challenges effectively.

References

Ameen M, Azizan MT, Yusup S, Ramli A, Yasir M (2017) Catalytic hydrodeoxygenation of triglycerides: an approach to clean diesel fuel production. Renew Sust Energ Rev 80:1072–1088. https://doi.org/10.1016/j.rser.2017.05.268

Asiedu A, Barbera E, Naurzaliyev R, Bertucco A, Kumar S (2019) Waste cooking oil to jet-diesel fuel range using 2-propanol via catalytic transfer hydrogenation reactions. Biofuels:1–14. https://doi.org/10.1080/17597269.2018.1532754

Awogbemi O, Onuh EI, Inambao FL (2019) Comparative study of properties and fatty acid composition of some neat vegetable oils and waste cooking oils. Int J Low Carbon Technol 14(3):417–425. https://doi.org/10.1093/ijlct/ctz038

Awogbemi O, Kallon DVV, Aigbodion VS, Panda S (2021) Advances in biotechnological applications of waste cooking oil. Case Stud Chem Environ Eng 4:100158. https://doi.org/10.1016/j.cscee.2021.100158

Bezergianni S, Kalogianni A (2009) Hydrocracking of used cooking oil for biofuels production. Bioresour Technol 100(17):3927–3932. https://doi.org/10.1016/j.biortech.2009.03.039

Bezergianni S, Voutetakis S, Kalogianni A (2009) Catalytic hydrocracking of fresh and used cooking oil. Ind Eng Chem Res 48(18):8402–8406. https://doi.org/10.1021/ie900445m

Doliente SS, Narayan A, Tapia JFD, Samsatli NJ, Zhao Y, Samsatli S (2020) Bio-aviation fuel: a comprehensive review and analysis of the supply chain components. Front Energy Res 8. https://doi.org/10.3389/fenrg.2020.00110

El-Sawy MS, Hanafi SA, Ashour F, Aboul-Fotouh TM (2020) Co-hydroprocessing and hydrocracking of alternative feed mixture (vacuum gas oil/waste lubricating oil/waste cooking oil) with the aim of producing high quality fuels. Fuel 269:117437. https://doi.org/10.1016/j.fuel.2020.117437

Goh BHH, Chong CT, Ge Y, Ong HC, Ng J-H, Tian B, Ashokkumar V, Lim S, Seljak T, Józsa V (2020) Progress in utilisation of waste cooking oil for sustainable biodiesel and biojet fuel production. Energy Convers Manag 223:113296. https://doi.org/10.1016/j.enconman.2020.113296

Khodadadi MR, Malpartida I, Tsang C-W, Lin CSK, Len C (2020) Recent advances on the catalytic conversion of waste cooking oil. Mol Catal 494:111128. https://doi.org/10.1016/j.mcat.2020.111128

Lin C-H, Wang W-C (2020) Direct conversion of glyceride-based oil into renewable jet fuels. Renew Sust Energ Rev 132:110109. https://doi.org/10.1016/j.rser.2020.110109

Negash YA, Amare DE, Bitew BD, Dagne H (2019) Assessment of quality of edible vegetable oils accessed in Gondar City, Northwest Ethiopia. BMC Res Notes 12(1). https://doi.org/10.1186/s13104-019-4831-x

Peng C, Cao Z, Du Y, Zeng R, Guo R, Duan X, Fang X (2018) Optimization of a pilot hydrocracking unit to improve the yield and quality of jet fuel together with heavy naphtha and tail oil. Ind Eng Chem Res 57(6):2068–2074. https://doi.org/10.1021/acs.iecr.7b04981

Robinson PR, Dolbear GE (2006) Hydrotreating and hydrocracking: fundamentals. In: Practical advances in petroleum processing. Springer New York, New York, pp 177–218

Saab R, Polychronopoulou K, Zheng L, Kumar S, Schiffer A (2020) Synthesis and performance evaluation of hydrocracking catalysts: a review. J Ind Eng Chem 89:83–103. https://doi.org/10.1016/j.jiec.2020.06.022

Sahu R, Song BJ, Im JS, Jeon Y-P, Lee CW (2015) A review of recent advances in catalytic hydrocracking of heavy residues. J Ind Eng Chem 27:12–24. https://doi.org/10.1016/j.jiec.2015.01.011

Vásquez MC, Silva EE, Castillo EF (2017) Hydrotreatment of vegetable oils: a review of the technologies and its developments for jet biofuel production. Biomass Bioenergy 105:197–206. https://doi.org/10.1016/j.biombioe.2017.07.008

Wang W-C, Tao L (2016) Bio-jet fuel conversion technologies. Renew Sust Energ Rev 53:801–822. https://doi.org/10.1016/j.rser.2015.09.016

Wang M, Dewil R, Maniatis K, Wheeldon J, Tan T, Baeyens J, Fang Y (2019) Biomass-derived aviation fuels: challenges and perspective. Prog Energy Combust Sci 74:31–49. https://doi.org/10.1016/j.pecs.2019.04.004

Weitkamp J (2012) Catalytic hydrocracking-mechanisms and versatility of the process. ChemCatChem 4(3):292–306. https://doi.org/10.1002/cctc.201100315

Why ESK, Ong HC, Lee HV, Gan YY, Chen W-H, Chong CT (2019) Renewable aviation fuel by advanced hydroprocessing of biomass: challenges and perspective. Energy Convers Manag 199:112015. https://doi.org/10.1016/j.enconman.2019.112015

Žula M, Grilc M, Likozar B (2022) Hydrocracking, hydrogenation and hydro-deoxygenation of fatty acids, esters and glycerides: mechanisms, kinetics and transport phenomena. Chem Eng J 444:136564. https://doi.org/10.1016/j.cej.2022.136564

Chapter 8
Techno-Economic Analysis of Biojet Fuel Production

Y. Lalitha Kameswari, Samakshi Verma, and Sonu Kumar

Abstract The creation of biojet fuel (BJF) has become a viable alternative as the aviation sector looks to lessen its carbon impact and dependency on fossil fuels. Utilizing a process-based model to assess several biojet fuel production methods, such as feedstock cultivation, conversion technologies, and refining processes, the research includes both technical and economic elements. A thorough cost analysis is carried out, taking into account capital expenses, operating costs, and income streams while taking market dynamics and feedstock price uncertainty into account. Key findings show that, in some circumstances, the manufacture of biojet fuel has promising potential. The cost-competitiveness of biojet fuel in comparison to conventional jet fuel is highlighted as being significantly influenced by the type of feedstock, the effectiveness of the conversion technology, and economies of scale. Sensitivity analysis shows how policy incentives and changing feedstock costs affect the overall techno-economic viability. Additionally, this analysis takes into account broader ramifications, like the potential decrease in greenhouse gas emissions and the diversification of energy sources for the aviation industry. The findings offer insightful information for stakeholders seeking to promote the development of sustainable aviation fuels, including politicians and investors. This techno-economic analysis concludes by highlighting the significance of a comprehensive assessment

Y. L. Kameswari
Department of Electronics and Communication Engineering, Koneru Lakshmaiah Education Foundation (Deemed to be University), Vaddeswaram, Guntur, Andhra Pradesh, India

Department of Electrical and Electronics Engineering, MLR Institute of Technology, Hyderabad, Telangana, India

S. Verma (✉)
Department of Biotechnology, Faculty of Engineering and Technology, Rama University, Kanpur, Uttar Pradesh, India

S. Kumar
Department of Electronics and Communication Engineering, K.S.R.M. College of Engineering, Kadapa, Andhra Pradesh, India

Manipur International University, Imphal, Manipur, India

A. Kuila (ed.), *Biojet Fuel: Current Technology and Future Prospect*, Clean Energy Production Technologies, https://doi.org/10.1007/978-981-99-8783-2_8

when determining the viability of producing biojet fuel. Even though there are still difficulties, improvements in conversion technology and favorable market circumstances may spur the commercialization of biojet fuel and help the aviation sector become more ecologically conscious. This chapter gives a thorough techno-economic analysis (TEA) of the manufacturing of biojet fuel, evaluating its viability, economics, and potential environmental advantages.

Keywords Aviation industry · Sustainability · Economic viability · Techno-economic analysis of biojet fuel, etc.

8.1 Introduction

The aviation industry, a crucial element of international transportation, is under increasing pressure to reduce its negative environmental effects, especially its role in greenhouse gas emissions (Aakko-Saksa et al. 2023). The demand for sustainable aviation fuels increases as climate change worries intensify. The development of biojet fuel from renewable biological feedstocks offers a possible way to lessen the carbon footprint of the aviation sector and increase its long-term profitability. This study undertakes a thorough techno-economic analysis (TEA) of the production of biojet fuel in an effort to evaluate the viability, feasibility, and probable difficulties of switching from conventional jet fuel to biojet fuel (Detsios et al. 2023). This analysis offers a comprehensive knowledge of the opportunities and challenges in the way of sustainable aviation by integrating technological, economic, and environmental issues (Ng et al. 2021). This chapter's main goal is to assess the techno-economic elements of large-scale biojet fuel production. To do this, the analysis takes into account several steps in the manufacturing of biojet fuel, such as feedstock selection, cultivation, conversion technologies, and refinement (Escalante et al. 2022). This study quantifies the inputs, outputs, and expenses linked to each production chain using a process-based model, enabling a thorough cost analysis. The evaluation of the economic viability of the manufacturing of biojet fuel is a vital component of this investigation. The goal of the study is to compare the cost-competitiveness of biojet fuel to conventional jet fuel by taking into account capital expenses, operating costs, and prospective revenue streams (Elkelawy et al. 2022). Sensitivity assessments are used to provide a more thorough knowledge of potential future scenarios by taking into account uncertainties in feedstock costs, technology improvements, and regulatory frameworks (Van Schoubroeck et al. 2021). This study also recognizes the wider effects of biojet fuel production. Beyond its viability economically, the environmental advantages of biojet fuel including fewer carbon dioxide emissions and less reliance on fossil fuels are investigated (Tiwari et al. 2023). The analysis also takes into account policy incentives and regulatory frameworks that may have an impact on the uptake of biojet fuel.

Techno-economic analysis adds to the expanding body of knowledge about environmentally friendly aviation fuels. This analysis provides insights into the difficulties and possibilities of shifting to more environmentally conscious aircraft practices by evaluating the technical and financial viability of producing biojet fuel (Michaga et al. 2022). The study's ensuing sections go into its methodology, data sources, and specific findings in order to give light on how biojet fuel might affect how people travel in the future.

8.2 Feedstock's Composition and Their Properties for Biojet Fuel Production

The phrase "feedstock's composition" in a techno-economic analysis (TEA) of the manufacture of biojet fuel refers to the chemical and elemental composition of the raw materials used to make biojet fuel (Baral et al. 2019). Given that it has a direct impact on the effectiveness of conversion processes, yields of valuable products, and total costs, feedstock composition is a crucial consideration when assessing the viability and economics of the manufacture of biojet fuel. Depending on the source material, the feedstock used to produce biojet fuel might have a very different makeup (Escalante et al. 2022). The manufacturing of biojet fuel frequently uses a variety of biomass as feedstocks, including waste fats, algae, plant oils, and lignocellulosic materials. With different concentrations of carbon, hydrogen, oxygen, nitrogen, sulfur, and other components, each feedstock has a distinct composition (Kusenberg et al. 2022). Additionally, feedstocks could include non-biomass components or contaminants that must be taken into account during the conversion process.

The composition of the feedstock affects a number of significant elements of the techno-economic analysis (Kumar et al. 2020). The energy content and chemical composition of various feedstocks can have an impact on how well they can be transformed into biojet fuel. Higher yields of biojet fuel may result from feedstocks with higher energy densities or better chemical characteristics for conversion processes (Goh et al. 2020). Distribution of the final product as well as its yield and composition can be influenced by the feedstock's makeup, as can the composition of the biojet fuel and other byproducts produced during the conversion process (Adeniyi et al. 2018). While some feedstocks may produce more desired byproducts than desired ones, such as jet fuel-suitable hydrocarbons, others may provide higher yields of desired products (Gutierrez-Antonio et al. 2017). The structure of the feedstock can affect how the conversion process is designed and set up. To attain the best conversion yields, certain feedstocks could need unique processing conditions, catalysts, or technology. The composition of the feedstock affects the expenses involved with its acquisition, transportation, and processing (Wright et al. 2010). A more cost-effective production process might be enabled by feedstocks with higher energy content or better compatibility with conversion technology. The environmental effects of producing biojet fuel can also depend on the composition of the feedstock (Escalante et al. 2022). For instance, when made into biojet fuel, feedstocks with higher carbon content might produce fewer greenhouse gas emissions.

To effectively model and simulate the complete manufacturing process in a techno-economic analysis, it is critical to characterize the composition of the feedstock accurately (Zhang et al. 2021). Understanding the feedstock's chemical makeup, elemental composition, contaminants, and other important characteristics is necessary for this. To optimize their composition before conversion, certain feedstocks could need various pre-processing procedures (Ramos et al. 2022). The composition of the feedstock is ultimately a key factor that influences the technical viability, economic viability, and sustainability of production pathways for biojet fuel (Goh et al. 2022). Within the TEA, proper consideration of feedstock composition provides more precise cost projections, performance forecasts, and environmental assessments. The practicality and effectiveness of the conversion process are greatly influenced by the composition and characteristics of the various feedstocks utilized in the manufacturing of biojet fuel (Goh et al. 2022).

Overview of some typical feedstocks and their pertinent characteristics for the manufacturing of biojet fuel is given below:

1. ***Oil/Lipid extract from plants such as camelina, jatropha, and soybean.***
   (a) *Composition*: Triglycerides, which are fatty acids (both saturated and unsaturated) connected to a glycerol backbone, make up the majority of plant oils. Plant oils often have significant levels of energy, different unsaturation levels, and distinctive fatty acid compositions (Jadhav and Annapure 2023). These characteristics affect whether they are appropriate for conversion processes.
   (b) *Algae*: Proteins, carbohydrates, and lipids (oil) can all be found in algae. Lipids play a crucial role in the production of biojet fuel. Algal lipids can include a lot of lipids; however, different strains of algae might have varied fatty acid profiles and lipid structures (Reddy et al. 2023).
   (c) *Fatty and oil waste*: Animal fats and used cooking oils are examples of waste fats and oils, which are obtained from industrial sources or during the cooking process (Lopes et al. 2020).
   (d) *Properties*: Waste fats and oils can have a wide range of contaminants, free fatty acids, and breakdown products in their composition. Consistent conversion could require pre-processing (Ghadge et al. 2022).

2. ***Wood chips and other lignocellulosic biomass, such as agricultural residues.***

   Lignocellulosic biomass is made up of three different compounds: cellulose, hemicellulose, and lignin. Lignin, another component of lignocellulosic biomass, is a complex polymer that adds rigidity and resistance to enzymatic degradation, making the biomass recalcitrant (Deng et al. 2023).

   Properties: To convert lignocellulosic feedstocks into sugars for future fermentation or conversion to biofuels, more complicated conversion procedures (such as biochemical or thermochemical techniques) are needed (Awasthi et al. 2023).

3. ***Organic waste and municipal solid waste.***

   Composition: Organic trash, including food scraps and yard clippings, is mixed with municipal solid waste (MSW) (Yakah et al. 2023).

   Properties: The quality and availability of potential biofuel precursors are impacted by the content of MSW, which can vary greatly. Sorting and pre-processing are frequently required (Yaashikaa et al. 2020).

4. ***Agriculture and forestry leftovers.***

   Composition: These are made up of sawdust, wood chips, and crop wastes.

   Qualities: Depending on the type of residue, the composition and qualities will vary. Pre-processing may be needed on some feedstocks to get rid of contaminants and make them more suitable for conversion (Velusamy et al. 2022).

   Important characteristics to take into account while assessing feedstocks are:

1. Energy content: In general, biojet fuel with higher energy content has better potential yields (Zhu et al. 2022).
2. Fatty acid profile: The kinds of fatty acids contained in lipid-based feedstocks might affect the characteristics of the resulting biojet fuel (Wang et al. 2022).
3. Moisture content: High moisture content might make drying more labor-intensive (Hiloidhari et al. 2023).
4. Ash content: High ash content might cause equipment for conversion to foul up and erode (Tobio-Perez et al. 2022).
5. Oxygen content: High oxygen content may affect the biojet fuel's energy density (Lim et al. 2021).
6. Impurities: Metals, sulfur, nitrogen, and other impurities can affect catalysts and subsequent processes (Neves et al. 2020).

It's crucial to remember that a feedstock's applicability depends on the particular conversion pathway being studied because different technologies are better suited for different feedstock types (Dahiya 2020). A complete evaluation of the qualities of the feedstock in relation to the selected conversion process should be included in a full techno-economic analysis, taking into account things like conversion efficiency, yields, and overall costs (Alkasrawi et al. 2020).

8.3 Opportunities and Barriers for Adopting BJF and Their Feedstocks

8.3.1 Opportunities

1. The ability of biojet fuel to dramatically lower greenhouse gas emissions as compared to traditional fossil-based jet fuel is the main driver behind its adoption. Biojet fuels can help the aviation sector reduce its environmental effect and work towards carbon neutrality (Zhu et al. 2022).
2. A lot of nations and international aviation organizations are putting laws and rules into place to encourage the use of biojet fuels and other sustainable aviation

fuels. Adoption of biojet fuel can be facilitated by regulatory incentives like blending requirements and carbon pricing (Harahap et al. 2023).
3. The manufacturing of biojet fuel is possible with a variety of feedstocks, including algae, waste oils, and agricultural leftovers, which offers flexibility and lessens reliance on particular raw materials. The supply chain's resilience is increased by this diversification (Ahmed et al. 2023).
4. The efficiency of the production processes for biojet fuel is always being improved through ongoing research and development (Zhang et al. 2020). Cost savings and increased scalability may result from improvements in feedstock choices, conversion technology, and refining methods.
5. Through the production and processing of feedstock, the biojet fuel industry's expansion can promote economic growth in rural areas. Additionally, it can generate employment in the manufacturing, research, and development fields related to the generation of biofuels (Thanigaivel et al. 2022).

8.3.2 Barriers

1. Finding reliable and reasonably priced feedstock sources can be difficult when there is a shortage of viable feedstocks combined with rivalry from other businesses (such as food production) (Usman et al. 2023).
2. Growing energy crops for the manufacture of biojet fuel might lead to worries about changing land uses, deforestation, and potential conflicts with the production of food. It is essential to ensure sustainable land use practices (Ahmed et al. 2023).
3. At the moment, biojet fuels are frequently more expensive than regular jet fuel. Cost parity can only be attained by technology improvements, scale economies, and supportive legislation (Tiwari et al. 2023).
4. Because biojet fuels differ from conventional jet fuels in their characteristics, current aviation infrastructure and engines must be modified. There may be compatibility problems, necessitating certification and retrofitting investments (Zhang et al. 2020).

Technical constraints include effective feedstock conversion, catalyst development, and process optimization due to the complexity of biojet fuel production processes, particularly for advanced feedstocks like lignocellulosic biomass (Walls and Rios-Solis 2020). From small-scale to commercial-scale biojet fuel production, it is necessary to make significant financial expenditures and ensure that the product is ready for the market (Larson et al. 2020). It is crucial to increase public acceptance of biojet fuels and debunk common misconceptions about their use, food competitiveness, and sustainability. The market dynamics of biojet fuels may be impacted by the regulatory environment (Zhang et al. 2023). Future policy, incentive, and subsidy uncertainty can have an impact on project planning and investor confidence.

Biojet fuels have a lot of potential for lowering aviation emissions and diversifying energy sources, but there are a lot of opportunities and challenges associated with their use (Rony et al. 2023). To improve the sustainability and viability of biojet fuel generation, these issues must be addressed holistically through technological innovation, helpful legislation, stakeholder collaboration, and ongoing research (Hiloidhari et al. 2023).

8.3.3 Adoption of Biojet Fuel (BJF) and Their Feedstocks: Opportunities and Challenges

The main reason for using biojet fuel is that it has the potential to produce much fewer greenhouse gas emissions than regular jet fuel. This supports international initiatives aimed at reducing climate change and achieving environmental objectives (Tiwari et al. 2023). By having access to a variety of feedstock resources, such as plant oils, algae, used cooking oil, and lignocellulosic materials, flexibility is increased and reliance on a single resource is reduced (Escalante et al. 2022). The supply chain for biojet fuel is more resilient as a result of this diversification. To power conversion operations and improve sustainability overall, biojet fuel production can be integrated with renewable energy sources like solar or wind (Chong et al. 2022). By reusing waste materials for the manufacture of useful fuel, using waste fats and agricultural leftovers as feedstocks promotes the principles of the circular economy. By generating jobs in agriculture, feedstock production, refining, and research, the rise of the biojet fuel business can promote economic growth (Ahmed et al. 2023). Local feedstock production and farming can boost rural economies and cut emissions associated with transportation.

Obtaining sufficient and reliable feedstock is a significant hurdle. Algae as a feedstock could have scaling issues, while other feedstocks could compete with food production or cause land use issues (Sarwer et al. 2022). Growing biofuel feedstock on the same area as food production could cause issues with food security and morality. Various feedstocks need for distinct conversion processes, each of which has unique technical difficulties (Lynd et al. 2022). It might be challenging to create efficient and affordable conversion procedures for each feedstock.

Due to technical, financial, and logistical issues, moving from small- to large-scale commercial production can be difficult (Goh et al. 2020). Depending on the price of feedstock, the effectiveness of the conversion technology, and the scale involved, it may now be more expensive to produce biojet fuel than conventional jet fuel (Tiwari et al. 2023). Due to the special characteristics of biojet fuels, modifications to the current aviation infrastructure, engines, and fuel delivery systems may be required (Dahal et al. 2021). Public acceptability of biojet fuels, feedstocks, and their environmental advantages depends on increasing public awareness of and comprehension of these topics (Kumar et al. 2023). Investment and adoption may be hampered by a lack of consistent and encouraging policy frameworks, including mandates, mandated programs, and regulatory standards (Zetzsche and Anker-Serensen 2022). Each feedstock faces unique obstacles, such as the complexity of growing algae, the difficulty of converting lignocellulosic biomass, and the consistency of obtaining waste oil. For new biojet fuel types and feedstocks, meeting the safety and quality requirements set by the aviation sector can be a barrier to entrance (Usman et al. 2023). Governments, businesses, researchers, and stakeholders must work together to create long-lasting, all-encompassing solutions to these opportunities and barriers (Kumar et al. 2023). Technology breakthroughs, enabling legislation, public involvement, and the capacity to address feedstock-specific difficulties are all necessary for the successful deployment of biojet fuels (Priya et al. 2023).

8.4 Specification for Biojet Fuel Production

To ensure safe and effective usage in aviation while minimizing environmental impact, the requirements for biojet fuel manufacturing specify the quality and performance parameters that biojet fuel must achieve (Goh et al. 2020). These requirements are necessary to keep biojet fuel compatible with the current aircraft infrastructure and engines. The ASTM D7566 standard outlines the specifications for biojet fuels and other aviation turbine fuels containing synthetic hydrocarbons (Ng et al. 2021). This standard includes information on performance, characteristics, and composition. To achieve consistent and dependable conversion processes, the feedstock used to produce biojet fuel must adhere to strict purity criteria. The characteristics of the final biojet fuel can be influenced by the fatty acid composition, oxygen content, and impurity levels of the feedstock (Misra et al. 2023). In order to retain the performance and range of an aeroplane, biojet fuel should have similar energy content to conventional jet fuel. To provide optimum handling, storage, and flow properties, biojet fuel should have density and viscosity values comparable to those of conventional jet fuel (Lahijani et al. 2022). Biojet fuel needs to have a low enough freezing point to avoid gelling in cold weather. Low sulfur concentration is essential for reducing emissions and avoiding engine component damage. In order to reduce microbial development, corrosion, and icing problems, biojet fuel should have low water content (Johnson et al. 2022). The cetane number is significant for ignition quality and combustion performance if the biojet fuel contains hydrocarbons in the diesel range. While the oxygen content in some biojet fuels is important for the oxygenate function, it should nevertheless fall below certain bounds to ensure engine compatibility (Lahijani et al. 2022). The greenhouse gas emissions linked to the manufacture and consumption of biojet fuel are measured by carbon intensity. It is a crucial factor to consider when evaluating the environmental advantages of biofuels. The infrastructure used in aviation, such as fuel storage tanks, pipelines, and aircraft fuel systems, should be compatible with biojet fuel (Dahal et al. 2021). In terms of ignition quality, combustion efficiency, and emissions, biojet fuel ought to function similarly to conventional jet fuel in engines (Sundararaj et al. 2019). To ensure adherence to aviation safety and quality standards, biojet fuel needs go through certification and approval procedures. In order to produce biojet fuel blends that meet the required quality standards, specifications may additionally include the permissible blending ratios with conventional jet fuel (Goh et al. 2022). These regulations guarantee that biojet fuel complies with strict safety, performance, and environmental standards. They facilitate the seamless integration of biojet fuel into current aviation operations and support the sustainability objectives of the aviation sector.

8.5 Pathways for Synthesis of BJF

There are multiple techniques to make biojet fuel (BJF), and each one uses various feedstocks and conversion processes. The availability of feedstock, the level of technological development, cost-effectiveness, and sustainability all play a role in the decision-making process (Alkaraan et al. 2023). The following are some typical methods for producing BJF.

8.5.1 Hydroprocessed Esters and Fatty Acids (HEFA)

Plant oils, leftover cooking oil, and animal fats are used as feedstock.

Process: Triglycerides are transesterified to create fatty acid methyl esters (FAME), which are then hydroprocessed to create hydrocarbons like regular jet fuel (Huang et al. 2022).

Benefits: Makes use of already-existing refinery infrastructure and produces biojet fuel with characteristics similar to those of conventional jet fuel.

Challenges include the scarcity of feedstock and competition from the food industry.

8.5.2 Synthesis Through Fischer-Tropsch (FT)

Syngas (a combination of hydrogen and carbon monoxide) generated from biomass is the feedstock.

Process: Synthetic paraffinic kerosene (SPK) is created through the catalytic conversion of syngas into hydrocarbons in the FT synthesis process, which also creates biojet fuel (Goh et al. 2022).

Benefits: Can use a range of feedstocks, including agricultural wastes and waste biomass.

High capital expenses and complicated gasification and syngas production procedures are challenges.

8.5.3 ATJ (Alcohol-to-Jet)

Alcohols generated from biomass, such as butanol or ethanol, are the feedstock.

Process: Alcohols are dehydrated and oligomerized to create long-chain hydrocarbons appropriate for aviation fuel (Domenech et al. 2022).

Benefits: Can use current infrastructure for alcohol production; may result in lower production costs.

Challenges include a process that requires a lot of energy and a limited supply of feedstock.

8.5.4 Hydrothermal Catalysis (CH)

Waste oils or lignocellulosic biomass are used as feedstock.

Process: High-pressure hydrogen and heat are used to process biomass in order to disassemble complicated structures into simpler hydrocarbons (Ke et al. 2022).

Benefits: Wide range of feedstocks can be used; greater yields may be possible.

Process optimization and catalyst development face difficulties.

8.5.5 (Lipid-CHTC) Lipid-Catalytic Hydrothermal Conversion

Feedstock: Lipid-rich feedstocks like algae.

Process: Algal lipids are transformed into hydrocarbons by a catalytic hydrothermal conversion process that takes place at extremely high temperatures and pressures (Ravichandran et al. 2022).

Advantages: Possibility of high yields; ability to use lipid-rich feedstocks like algae.

Process improvement needed for algae harvesting and culture problems; process improvement needed.

8.5.6 Biochemical Transformation

Lignocellulosic biomass or other complex sugars are used as feedstock.

Process: Biomass is transformed into biojet fuel precursors by enzymatic or microbiological processes, which are then upgraded to hydrocarbons (Keasling et al. 2021).

Benefits include the use of non-food biomass and the possibility for high efficiency.

Complex metabolic processes and process efficiency face difficulties.

Each route has its own benefits, difficulties, and particular technical requirements. The decision of the pathway is influenced by regional variables, feedstock availability, technological readiness, and economic reasons (Julio et al. 2021). A multidisciplinary approach is used in the development of BJF, including feedstock production, conversion technologies, process optimization, and sustainability analyses (Cervi et al. 2021).

8.6 Production Routes for Biojet

Here are some typical production methods for biojet fuel (BJF), each of which makes use of various feedstocks and conversion procedures.

8.6.1 Hydroprocessed Esters and Fatty Acids (HEFA) Pathway

Plant oils, leftover cooking oil, and animal fats are used as feedstock.

Process: The feedstock is transesterified to produce fatty acid methyl esters (FAME), which are then hydroprocessed to yield hydrocarbons that are similar to traditional jet fuel (Chong et al. 2022).

Benefits: Produces biojet fuel with characteristics that are very similar to those of conventional jet fuel and may make use of existing refinery infrastructure.

Challenges include the scarcity of feedstock and competition for those resources from the food and other industries.

8.6.2 Synthesis Through Fischer-Tropsch (FT)

Syngas (a combination of hydrogen and carbon monoxide) generated from biomass is the feedstock.

Process: Synthetic paraffinic kerosene (SPK), a biojet fuel, is produced through the catalytic conversion of syngas into hydrocarbons in the FT synthesis process (Emmanouilidou et al. 2023).

Advantages: Offers flexibility in manufacturing and can employ a variety of feedstocks, including waste biomass.

High capital expenses and complicated gasification and syngas production procedures are challenges.

8.6.3 Pathway from Alcohol-to-Jet (ATJ)

Alcohols generated from biomass, such as butanol or ethanol, are the feedstock.

Process: Alcohols are dehydrated and oligomerized to create long-chain hydrocarbons appropriate for aviation fuel (Domenech et al. 2022).

Benefits: Reduced greenhouse gas emissions, potential integration with the infrastructure already in place for the manufacture of alcohol.

Challenges include a process that requires a lot of energy and a limited supply of feedstock.

8.6.4 Pathway for Catalytic Hydrothermolysis (CH)

Waste oils or lignocellulosic biomass are used as feedstock.

Process: Heat and high-pressure hydrogen are used to convert biomass from complicated structures into simpler hydrocarbons (Ke et al. 2022).

Wide variety of feedstock alternatives, potential for better yields, and decreased pre-processing of feedstock are benefits.

Process optimization and catalyst development face difficulties.

8.6.5 Pathway for Lipid-Catalytic Hydrothermal Conversion (Lipid-CHTC)

Feedstock: Lipid-rich feedstocks like algae.

Process: Algal lipids are transformed into hydrocarbons by a catalytic hydrothermal conversion process that takes place at extremely high temperatures and pressures (Ravichandran et al. 2022).

Benefits include the use of lipid-rich feedstocks like algae, the possibility for large yields, and the ability to remediate wastewater.

Algae cultivation problems and effective lipid extraction are issues.

8.6.6 Pathway of Biochemical Conversion

Lignocellulosic biomass or other complex sugars are used as feedstock.

Process: Biomass is transformed into biojet fuel precursors by enzymatic or microbiological processes, which are then upgraded to hydrocarbons (Keasling et al. 2021).

Benefits include the use of biomass other than food, the potential for high efficiency, and the potential for co-products such biochemicals.

Challenges: Availability of feedstock, efficiency of enzymes and microbes, and complex metabolic pathways.

The availability of feedstock, technological readiness, economic viability, and environmental considerations are only a few examples of the variables that influence the production route decision (Julio et al. 2021). Additionally, improvements in these pathways are constantly being made with the goals of boosting sustainability, cutting costs, and improving efficiency.

8.7 Sensitivity Detection

Sensitivity analysis in the manufacturing of biojet fuel entails determining the effects of changes in a variety of input parameters and variables on the final results, costs, and efficiency of the production process (Fitriasari et al. 2023). This study aids in determining the variables that have the biggest impact on the outcomes and can help with risk assessment, process optimization, and decision-making. Identify the factors and variables that have the biggest impact on the technological, economic, and environmental aspects of producing biojet fuel (Julio et al. 2021). Costs of feedstock, conversion effectiveness, capital expenses, energy usage, and governmental incentives are a few examples (Manikandan et al. 2023). Set ranges for each parameter that has been chosen. Consider a range of probable feedstock prices that represent market volatility, for instance, when analyzing feedstock costs. Change one parameter at a time, while maintaining the other values constant, and track how the changes impact important performance measures like production costs, greenhouse gas emissions, and net present value (Chen and Quinn 2021). Change several parameters at once to see how they interact. By doing so, interactions and nonlinear effects can be found. To simulate the full generation of biojet fuel, use mathematical models or process simulation tools (Kroyan et al. 2022). You may evaluate how parameter changes affect the entire process chain using these tools.

Consider the sensitivity analysis's findings. Determine which variables have the biggest effects on the outcomes and which have the least. To see these effects, utilize sensitivity plots, tornado diagrams, or correlation matrices (Senova et al. 2023). Rank the criteria according to their degree of sensitivity. Give top priority to those that will have the biggest impact on the project's overall success, economic viability, and environmental performance. Utilize scenario analysis to investigate multiple "what-if" situations by combining the values of various parameters. Understanding the variety of possible outcomes and hazards is aided by this. Sensitivity analysis can highlight weak spots and potential threats (Uludere Aragon et al. 2023). Utilize risk management techniques to deal with uncertainties, such as diversifying the feedstocks used or creating backup plans. Locate areas for optimization based on the results of the sensitivity analysis. Identify the variables that can be changed to improve project resilience, reduce environmental impact, and ensure economic viability (Hiloidhari et al. 2023). Take into account how the viability and allure of the manufacturing process for biojet fuel might be affected by the susceptibility to policy changes, such as carbon pricing or biofuel mandates (Chong and Ng 2021). Sensitivity analysis is a helpful tool for producing biojet fuel when making decisions. It offers perceptions into the stability of the manufacturing process, the susceptibility of financial indicators to changes in the market, and the possibility for efficiency and cost-effectiveness gains (Vela-Garcia et al. 2020).

8.8 Various Potential Assessments

Life Cycle Assessment (LCA): Throughout every stage of the manufacture and use of biojet fuel, including feedstock cultivation, processing, transportation, and burning, LCA assesses the environmental impact. It evaluates elements including resource depletion, energy use, and greenhouse gas emissions (Hiloidhari et al. 2023).

Qualities of Biojet Fuel (BJF): Examining the density, viscosity, flash point, freezing point, energy content, and combustion characteristics of biojet fuel is one way to determine its qualities (Why et al. 2022). The compatibility with aviation engines and infrastructure is determined by these features. A yield evaluation determines how much biojet fuel can be produced from a specified amount of feedstock. It takes into account variables including feedstock quality variations, processing losses, and conversion efficiency.

***Recovery Cost*: **The costs connected with removing and separating biojet fuel from the conversion process are referred to as recovery costs. It comprises expenses for byproduct management, catalysts, and separation technologies (Romero-Izquierdo et al. 2022).

***Input Costs*: **The costs associated with purchasing the feedstock, raw materials, and other inputs required for the manufacturing of biojet fuel are referred to as input costs (Umenweke et al. 2023). Depending on the supply of feedstock and the state of the market, these costs may change.

***Conversion Cost*: **The costs related to the actual conversion process, such as energy use, catalysts, equipment upkeep, and labor, are included in the conversion cost (Wang et al. 2022).

***Operational Cost*: **Operational costs include continuing costs associated with running the facility that produces biojet fuel. This covers labor costs, utilities, upkeep, and other ongoing costs (Tanzil et al. 2021).

***Production Cost*: **The total cost of production includes all input costs, conversion costs, operating costs, and other costs incurred during the complete manufacturing of biojet fuel (Vela-Garcia et al. 2020).

Together, these diverse analyses offer a thorough understanding of the manufacturing of biojet fuel's viability, economics, and effects on the environment. They enable stakeholders to take well-informed decisions, streamline procedures, and pinpoint problem areas in order to promote the growth of a competitive and sustainable biojet fuel business.

8.9 Environmental Significance of BJF Production

The manufacture of biojet fuel (BJF) is significant for the environment since it has the potential to lessen the harmful effects of aviation on the environment (Cervi et al. 2021). The ability of BJF production to reduce greenhouse gas emissions is one of the most important advantages. BJF may emit fewer greenhouse gases than traditional jet fuel, particularly if it is made from feedstocks with low carbon intensity or through the use of cutting-edge conversion techniques (Rajpoot et al. 2023). BJF participates in global initiatives to mitigate climate change and achieve carbon neutrality in the aviation sector by lowering the aviation sector's dependency on fossil fuels and causing fewer emissions (Fathi et al. 2023). From feedstock cultivation to fuel burning, emissions are taken into account throughout the entire supply chain in the life cycle study of BJF production (Julio et al. 2021). This thorough evaluation aids in calculating the overall carbon footprint decrease. BJF provides a sustainable and renewable option to fossil-based jet fuels, assisting in reducing reliance on non-renewable resources and slowed-down fossil fuel reserve depletion (Dahal et al. 2021). BJF produces less sulfur and particulate matter than traditional jet fuel, which can enhance the local air quality near airports and flight paths.

8.9.1 Innovations in Conversion Technologies, Refining Processes, and Sustainable Feedstock Cultivation

The research and development efforts targeted at BJF production lead to innovations in these fields. These developments have broader effects on bio-based goods and sustainable energy (Gunasekaran et al. 2021). BJF production promotes the use of a variety of feedstocks, which can lessen strain on particular land resources and minimize harmful land-use changes like deforestation. By employing waste fats, oils, or agricultural byproducts as feedstocks, BJF manufacturing can be integrated with waste management systems (Sharno and Hiloidhari 2022). Reusing waste materials for beneficial purposes encourages a circular economy. The demand for BJF may influence the creation of enabling laws, technology, and financial investments in processes for sustainable forestry, agriculture, and conversion (Hiloidhari et al. 2023).

BJF production can change people's perceptions of sustainable aviation and nudge travelers to think about how their travel decisions will affect the environment (Larsson et al. 2020). Projects like the International Civil Aviation Organization's (ICAO) Carbon Offsetting and Reduction Scheme for International Aviation (CORSIA) are examples of how the environmental significance of BJF production fosters international collaboration on sustainability goals within the aviation industry (Atmowidjojo et al. 2021). By lowering aviation's carbon footprint, supporting renewable energy sources, and pushing technological advancements that go beyond the aviation industry, BJF production has the potential to greatly contribute to environmental sustainability (Cervi et al. 2021).

8.10 Future Prospects of BJF

The prospects for biojet fuel (BJF) in the future are positive due to rising environmental concerns, aviation industry objectives, regulatory backing, technological improvements, and expanding sustainability consciousness (Hooda and Yadav 2023).

The airline industry is under pressure to lower its carbon emissions as global climate targets become increasingly ambitious. By offering a less-carbon-intensive substitute for standard jet fuel, BJF provides a useful option to accomplish these objectives (Gray et al. 2021). Governments and international organizations are putting policies and rules into place that encourage the use of BJF and other sustainable aviation fuels. These regulations may improve the business climate and encourage investments in the production of BJF. To develop BJF production, airlines, aircraft makers, and fuel producers are working together (Julio et al. 2021). Research, development, and commercialization initiatives are being driven by partnerships between aviation stakeholders and renewable energy companies. Ongoing studies are aimed at enhancing feedstock utilization, conversion efficiency, and BJF production technology (Escalante et al. 2022). Higher yields, lower manufacturing costs, and enhanced fuel characteristics could result from this. As technology develops and economies of scale are realized, BJF production may become more competitive in terms of price. A feasible choice for airlines, BJF might become more affordable as production volumes rise (Dahal et al. 2021). For the production of BJF, researchers are looking into new feedstock sources, such as algae, trash, and non-food crops (Goh et al. 2020). These feedstocks might be more readily available, more environmentally friendly, and less competitive with food production. The need for more environmentally friendly travel options is being driven by growing public awareness of climate change and environmental sustainability. Airlines that use BJF can set themselves apart in a cutthroat industry. By creating clear, globally accepted certification and quality standards for BJF, it will be possible to increase passenger and airline confidence that the fuel satisfies all safety and performance requirements (Kumar et al. 2023). To assist BJF production infrastructure, research, and innovation, governments and private investors are funding and offering incentives.

The production of BJF can work in harmony and with greater sustainability with other renewable energy sectors, including bioenergy, biorefineries, and waste-to-energy processes (Julio et al. 2021). To demonstrate the viability and performance of BJF, airlines are undertaking successful demonstration flights. These flights serve to illustrate how BJF is used in real-world situations and can encourage interest and adoption (Lyu et al. 2023). International partnerships, like the Carbon Offsetting and Reduction Scheme for International Aviation (CORSIA) run by ICAO, show a commitment to lowering aviation emissions and establish a conducive environment for the implementation of the BJF (Atmowidjojo et al. 2021). BJF has the potential to transform the aviation sector by offering a cleaner substitute for traditional jet fuel. Future prospects of BJF are projected to be characterized by higher acceptance, improved sustainability, and a smaller carbon footprint for aviation as technology develops, costs fall, and regulatory backing rises (Peters et al. 2023).

8.11 Conclusion

A crucial tool for assessing the viability, sustainability, and economics of switching to renewable and more ecologically friendly aviation fuels is the techno-economic analysis of the generation of biojet fuel. This thorough analysis covers a thorough investigation of numerous variables that affect the cost, performance, and method of producing biojet fuel. Stakeholders may make well-informed decisions that support the development of the biojet fuel industry and its favorable effects on the aviation industry and the environment by integrating technical, economic, and environmental concerns. The techno-economic analysis offers insights into the potential difficulties and opportunities related to the production of biojet fuel through the evaluation of feedstock availability, conversion technologies, operational costs, and market dynamics. It allows for a complete understanding of the variables influencing production costs, enabling the identification of potential areas for cost- and efficiency-saving optimization. The analysis also considers how producing biojet fuel affects the environment, including how it may help achieve global sustainability goals by lowering greenhouse gas emissions and improving air quality. Stakeholders are given a clearer understanding of the environmental advantages provided by biojet fuel by quantifying the life cycle impacts and contrasting them with those of conventional jet fuel. Techno-economic analysis's insights help the aviation sector make decisions about where to spend, how to build policies, and what research to prioritize. This is because the sector wants to comply with regulations and targets for carbon reduction. The economic viability and competitiveness of the production of biojet fuel are projected to be further improved by ongoing improvements in feedstock diversity, conversion technology, and process efficiency. The techno-economic analysis emphasizes the significance of a balanced strategy that takes into account both environmental stewardship and economic viability. The aviation sector can move towards a more sustainable and resilient future while supporting international efforts to slow down climate change by encouraging innovation, cooperation, and investment in the development of biojet fuel.

References

Aakko-Saksa PT, Lehtoranta K, Kuittinen N, Järvinen A, Jalkanen JP, Johnson K et al (2023) Reduction in greenhouse gas and other emissions from ship engines: current trends and future options. Prog Energy Combust Sci 94:101055

Adeniyi OM, Azimov U, Burluka A (2018) Algae biofuel: current status and future applications. Renew Sust Energ Rev 90:316–335

Ahmed M, Alam MN, Abdullah A, Ahmad Z (2023) Bio-jet fuel: an overview of various feedstock and production routes. In: AIP conference proceedings, vol. 2785, no. 1. AIP Publishing

Alkaraan F, Elmarzouky M, Hussainey K, Venkatesh VG (2023) Sustainable strategic investment decision-making practices in UK companies: the influence of governance mechanisms on synergy between industry 4.0 and circular economy. Technol Forecast Soc Chang 187:122187

Alkasrawi M, Rajangam AS, Tawalbeh M, Kafiah F, Al-Othman A, Al-Asheh S, Sun Q (2020) Techno-economic analysis and a novel assessment technique of paper mill sludge conversion to bioethanol toward sustainable energy production. Int J Energy Res 44(15):12602–12613

Atmowidjojo A, Rianawati E, Chin BLF, Yusup S, Quitain AT, Assabumrungrat S, et al (2021, December). Supporting clean energy in the ASEAN: policy opportunities from sustainable aviation fuels initiatives in Indonesia and Malaysia. In: IOP conference series: earth and environmental science, vol. 940, no. 1. IOP Publishing, p 012031

Awasthi MK, Sar T, Gowd SC, Rajendran K, Kumar V, Sarsaiya S et al (2023) A comprehensive review on thermochemical, and biochemical conversion methods of lignocellulosic biomass into valuable end product. Fuel 342:127790

Baral NR, Kavvada O, Mendez-Perez D, Mukhopadhyay A, Lee TS, Simmons BA, Scown CD (2019) Techno-economic analysis and life-cycle greenhouse gas mitigation cost of five routes to bio-jet fuel blendstocks. Energy Environ Sci 12(3):807–824

Cervi WR, Lamparelli RAC, Gallo BC, de Oliveira Bordonal R, Seabra JEA, Junginger M, van der Hilst F (2021) Mapping the environmental and techno-economic potential of biojet fuel production from biomass residues in Brazil. Biofuels Bioprod Biorefin 15(1):282–304

Chen PH, Quinn JC (2021) Microalgae to biofuels through hydrothermal liquefaction: open-source techno-economic analysis and life cycle assessment. Appl Energy 289:116613

Chong CT, Ng JH (2021) Biojet fuel in aviation applications: production, usage and impact of biofuels. Elsevier

Chong JW, Chemmangattuvalappil NG, Thangalazhy-Gopakumar S (2022) Aviation biofuels: conversion routes and challenges. In: Sustainable Technologies for the oil palm Industry: latest advances and case studies. Springer Nature Singapore, Singapore, pp 33–85

Dahal K, Brynolf S, Xisto C, Hansson J, Grahn M, Grönstedt T, Lehtveer M (2021) Techno-economic review of alternative fuels and propulsion systems for the aviation sector. Renew Sust Energ Rev 151:111564

Dahiya A (ed) (2020) Bioenergy: biomass to biofuels and waste to energy. Academic Press

Deng W, Feng Y, Fu J, Guo H, Guo Y, Han B et al (2023) Catalytic conversion of lignocellulosic biomass into chemicals and fuels. Green Energy Environ 8(1):10–114

Detsios N, Theodoraki S, Maragoudaki L, Atsonios K, Grammelis P, Orfanoudakis NG (2023) Recent advances on alternative aviation fuels/pathways: a critical review. Energies 16(4):1904

Doménech P, Pogrebnyakov I, Nielsen AT, Riisager A (2022) Catalytic production of long-chain hydrocarbons suitable for jet-fuel use from fermentation-derived oxygenates. Green Chem 24(9):3461–3474

Elkelawy M, Bastawissi HAE, Radwan AM, Ismail MT, El-Sheekh M (2022) Biojet fuels production from algae: conversion technologies, characteristics, performance, and process simulation. In: Handbook of algal biofuels. Elsevier, pp 331–361

Emmanouilidou E, Mitkidou S, Agapiou A, Kokkinos NC (2023) Solid waste biomass as a potential feedstock for producing sustainable aviation fuel: a systematic review. Renew Energy

Escalante J, Chen WH, Tabatabaei M, Hoang AT, Kwon EE, Lin KYA, Saravanakumar A (2022) Pyrolysis of lignocellulosic, algal, plastic, and other biomass wastes for biofuel production and circular bioeconomy: a review of thermogravimetric analysis (TGA) approach. Renew Sust Energ Rev 169:112914

Fathi M, Zio E, Pardalos PM (eds) (2023) Handbook of smart energy systems. Springer Nature

Fitriasari EI, Won W, Liu JJ (2023) Sustainability assessment of biojet fuel produced from pyrolysis oil of woody biomass. Sustainable Energy Fuels

Ghadge R, Nagwani N, Saxena N, Dasgupta S, Sapre A (2022) Design and scale-up challenges in hydrothermal liquefaction process for biocrude production and its upgradation. Energy Convers Manage X 14:100223

Goh BHH, Chong CT, Ge Y, Ong HC, Ng JH, Tian B et al (2020) Progress in utilisation of waste cooking oil for sustainable biodiesel and biojet fuel production. Energy Convers Manag 223:113296

Goh BHH, Chong CT, Ong HC, Seljak T, Katrašnik T, Józsa V et al (2022) Recent advancements in catalytic conversion pathways for synthetic jet fuel produced from bioresources. Energy Convers Manag 251:114974

Gray N, McDonagh S, O'Shea R, Smyth B, Murphy JD (2021) Decarbonising ships, planes and trucks: an analysis of suitable low-carbon fuels for the maritime, aviation and haulage sectors. Adv Appl Energy 1:100008

Gunasekaran M, Kumar G, Karthikeyan OP, Varjani S (2021) Lignocellulosic biomass as an optimistic feedstock for the production of biofuels as valuable energy source: techno-economic analysis, environmental impact analysis, breakthrough and perspectives. Environ Technol Innov 24:102080

Gutiérrez-Antonio C, Gómez-Castro FI, de Lira-Flores JA, Hernández S (2017) A review on the production processes of renewable jet fuel. Renew Sust Energ Rev 79:709–729

Harahap F, Nurdiawati A, Conti D, Leduc S, Urban F (2023) Renewable marine fuel production for decarbonised maritime shipping: pathways, policy measures and transition dynamics. J Clean Prod 137906

Hiloidhari M, Sharno MA, Baruah DC, Bezbaruah AN (2023) Green and sustainable biomass supply chain for environmental, social and economic benefits. Biomass Bioenergy 175:106893

Hooda SK, Yadav S (2023) Green finance for sustainable aviation: stakeholder perspectives and systematic review. Int J Prof Bus Rev 8(5):31

Huang X, Dong S, Yang X (2022) Refining lipid for aviation biofuel at the molecular level. Renew Energy 201:148–159

Jadhav HB, Annapure US (2023) Triglycerides of medium-chain fatty acids: a concise review. J Food Sci Technol 60(8):2143–2152

Johnson JU, Carpenter M, Williams C, Pons JF, McLaren D (2022) Complexities associated with nucleation of water and ice from jet fuel in aircraft fuel systems: a critical review. Fuel 310:122329

Julio AAV, Batlle EAO, Trindade AB, Nebra SA, Reyes AMM, Palacio JCE (2021) Energy, exergy, exergoeconomic, and environmental assessment of different technologies in the production of bio-jet fuel by palm oil biorefineries. Energy Convers Manag 243:114393

Ke L, Wu Q, Zhou N, Xiong J, Yang Q, Zhang L et al (2022) Lignocellulosic biomass pyrolysis for aromatic hydrocarbons production: pre and in-process enhancement methods. Renew Sust Energ Rev 165:112607

Keasling J, Garcia Martin H, Lee TS, Mukhopadhyay A, Singer SW, Sundstrom E (2021) Microbial production of advanced biofuels. Nat Rev Microbiol 19(11):701–715

Kroyan Y, Wojcieszyk M, Kaario O, Larmi M (2022) Modeling the impact of sustainable aviation fuel properties on end-use performance and emissions in aircraft jet engines. Energy 255:124470

Kumar A, Saini K, Bhaskar T (2020) Hydochar and biochar: production, physicochemical properties and techno-economic analysis. Bioresour Technol 310:123442

Kumar T, Basakran GK, Marsuki MZ, Wash AM, Mohsin R, Majid ZA, Ghafir MFA (2023) Exploring socioeconomic and political feasibility of aviation biofuel production and usage in Malaysia: a thematic analysis approach using expert opinion from aviation industry. Green Energy and Intell Transp 100111

Kusenberg M, Zayoud A, Roosen M, Thi HD, Abbas-Abadi MS, Eschenbacher A et al (2022) A comprehensive experimental investigation of plastic waste pyrolysis oil quality and its dependence on the plastic waste composition. Fuel Process Technol 227:107090

Lahijani P, Mohammadi M, Mohamed AR, Ismail F, Lee KT, Amini G (2022) Upgrading biomass-derived pyrolysis bio-oil to bio-jet fuel through catalytic cracking and hydrodeoxygenation: a review of recent progress. Energy Convers Manag 268:115956

Larson ED, Kreutz TG, Greig C, Williams RH, Rooney T, Gray E et al (2020) Design and analysis of a low-carbon lignite/biomass-to-jet fuel demonstration project. Appl Energy 260:114209

Larsson J, Matti S, Nässén J (2020) Public support for aviation policy measures in Sweden. Clim Pol 20(10):1305–1321

Lim JHK, Gan YY, Ong HC, Lau BF, Chen WH, Chong CT et al (2021) Utilization of microalgae for bio-jet fuel production in the aviation sector: challenges and perspective. Renew Sust Energ Rev 149:111396

Lopes M, Miranda SM, Belo I (2020) Microbial valorization of waste cooking oils for valuable compounds production–a review. Crit Rev Environ Sci Technol 50(24):2583–2616

Lynd LR, Beckham GT, Guss AM, Jayakody LN, Karp EM, Maranas C et al (2022) Toward low-cost biological and hybrid biological/catalytic conversion of cellulosic biomass to fuels. Energy Environ Sci 15(3):938–990

Lyu C, Liu X, Wang Z, Yang L, Liu H, Yang N et al (2023) An emissions inventory using flight information reveals the long-term changes of aviation CO2 emissions in China. Energy 262:125513

Manikandan S, Vickram S, Sirohi R, Subbaiya R, Krishnan RY, Karmegam N et al (2023) Critical review of biochemical pathways to transformation of waste and biomass into bioenergy. Bioresour Technol 128679

Michaga MFR, Michailos S, Akram M, Cardozo E, Hughes KJ, Ingham D, Pourkashanian M (2022) Bioenergy with carbon capture and storage (BECCS) potential in jet fuel production from forestry residues: a combined techno-economic and life cycle assessment approach. Energy Convers Manag 255:115346

Misra P, Alvarez-Majmutov A, Chen J (2023) Isomerization catalysts and technologies for biorefining: opportunities for producing sustainable aviation fuels. Fuel 351:128994

Neves RC, Klein BC, da Silva RJ, Rezende MCAF, Funke A, Olivarez-Gómez E et al (2020) A vision on biomass-to-liquids (BTL) thermochemical routes in integrated sugarcane biorefineries for biojet fuel production. Renew Sust Energ Rev 119:109607

Ng KS, Farooq D, Yang A (2021) Global biorenewable development strategies for sustainable aviation fuel production. Renew Sust Energ Rev 150:111502

Peters MA, Alves CT, Onwudili JA (2023) A review of current and emerging production technologies for biomass-derived sustainable aviation fuels. Energies 16(16):6100

Priya A, Hu Y, Mou J, Du C, Wilson K, Luque R, Lin CSK (2023) Introduction: an overview of biofuels and production technologies. In: Handbook of biofuels production, pp 3–24. https://doi.org/10.1533/9780857090492.1.3

Rajpoot AS, Choudhary T, Chelladurai H, Verma TN, Pugazhendhi A (2023) Sustainability analysis of spirulina biodiesel and their bends on a diesel engine with energy, exergy and emission (3E's) parameters. Fuel 349:128637

Ramos A, Monteiro E, Rouboa A (2022) Biomass pre-treatment techniques for the production of biofuels using thermal conversion methods–a review. Energy Convers Manag 270:116271

Ravichandran SR, Venkatachalam CD, Sengottian M, Sekar S, Kandasamy S, Subramanian KPR et al (2022) A review on hydrothermal liquefaction of algal biomass on process parameters, purification and applications. Fuel 313:122679

Reddy MMS, Tiwari S, Chauhan VS (2023) Algal lipids, lipidomics, and biosurfactants. In: Algae materials. Academic Press, pp 313–342

Romero-Izquierdo AG, Gutiérrez-Antonio C, Gómez-Castro FI, Hernández S (2022) Synthesis and intensification of a biorefinery to produce renewable aviation fuel, biofuels, bioenergy and chemical products from Jatropha Curcas fruit. IET Renew Power Gener 16(14):2988–3008

Rony ZI, Mofijur M, Hasan MM, Ahmed SF, Almomani F, Rasul MG et al (2023) Unanswered issues on decarbonizing the aviation industry through the development of sustainable aviation fuel from microalgae. Fuel 334:126553

Sarwer A, Hussain M, Al-Muhtaseb AAH, Inayat A, Rafiq S, Khurram MS et al (2022) Suitability of biofuels production on commercial scale from various feedstocks: a critical review. ChemBioEng Rev 9(5):423–441

Senova A, Tobisova A, Rozenberg R (2023) New approaches to project risk assessment utilizing the Monte Carlo method. Sustainability 15(2):1006

Sharno MA, Hiloidhari M (2022) Life cycle carbon, energy, water, and landuse footprints of agricultural residue-based biojet fuel in Bangladesh. Bioresour Technol Rep 20:101234

Sundararaj RH, Kumar RD, Raut AK, Sekar TC, Pandey V, Kushari A, Puri SK (2019) Combustion and emission characteristics from biojet fuel blends in a gas turbine combustor. Energy 182:689–705

Tanzil AH, Brandt K, Wolcott M, Zhang X, Garcia-Perez M (2021) Strategic assessment of sustainable aviation fuel production technologies: yield improvement and cost reduction opportunities. Biomass Bioenergy 145:105942

Thanigaivel S, Priya AK, Dutta K, Rajendran S, Vasseghian Y (2022) Engineering strategies and opportunities of next generation biofuel from microalgae: a perspective review on the potential bioenergy feedstock. Fuel 312:122827

Tiwari R, Mishra R, Choubey A, Kumar S, Atabani AE, Badruddin IA, Khan TY (2023) Environmental and economic issues for renewable production of bio-jet fuel: a global prospective. Fuel 332:125978

Tobio-Perez I, Alfonso-Cardero A, Diaz-Dominguez Y, Pohl S, Piloto-Rodriguez R, Lapuerta M (2022) Thermochemical conversion of Sargassum for energy production: a comprehensive review. Bioenergy Res 15(4):1872–1893

Uludere Aragon NZ, Parker NC, VanLoocke A, Bagley J, Wang M, Georgescu M (2023) Sustainable land use and viability of biojet fuels. Nat Sustain 6(2):158–168

Umenweke GC, Pace RB, Santillan-Jimenez E, Okolie JA (2023) Techno-economic and life-cycle analyses of sustainable aviation fuel production via integrated catalytic deoxygenation and hydrothermal gasification. Chem Eng J 452:139215

Usman M, Cheng S, Boonyubol S, Cross JS (2023) The future of aviation soars with HTL-based SAFs: exploring potential and overcoming challenges using organic wet feedstocks. Sustain Energy Fuels 7(17):4066–4087

Van Schoubroeck S, Thomassen G, Van Passel S, Malina R, Springael J, Lizin S et al (2021) An integrated techno-sustainability assessment (TSA) framework for emerging technologies. Green Chem 23(4):1700–1715

Vela-Garcia N, Bolonio D, Mosquera AM, Ortega MF, Garcia-Martinez MJ, Canoira L (2020) Techno-economic and life cycle assessment of triisobutane production and its suitability as biojet fuel. Appl Energy 268:114897

Velusamy S, Subbaiyan A, Murugesan SR, Shanmugamoorthy M, Sivakumar V, Velusamy P et al (2022) Comparative analysis of agro waste material solid biomass briquette for environmental sustainability. Adv Mater Sci Eng 2022

Walls LE, Rios-Solis L (2020) Sustainable production of microbial isoprenoid derived advanced biojet fuels using different generation feedstocks: a review. Front Bioeng Biotechnol 8:599560

Wang WC, Liu YC, Nugroho RAA (2022) Techno-economic analysis of renewable jet fuel production: the comparison between Fischer-Tropsch synthesis and pyrolysis. Energy 239:121970

Why ESK, Ong HC, Lee HV, Chen WH, Asikin-Mijan N, Varman M, Loh WJ (2022) Single-step catalytic deoxygenation of palm feedstocks for the production of sustainable bio-jet fuel. Energy 239:122017

Wright MM, Daugaard DE, Satrio JA, Brown RC (2010) Techno-economic analysis of biomass fast pyrolysis to transportation fuels. Fuel 89:S2–S10

Yaashikaa PR, Kumar PS, Saravanan A, Varjani S, Ramamurthy R (2020) Bioconversion of municipal solid waste into bio-based products: a review on valorisation and sustainable approach for circular bioeconomy. Sci Total Environ 748:141312

Yakah N, Samavati M, Akuoko Kwarteng A, Martin A, Simons A (2023) Prospects of waste incineration for improved municipal solid waste (MSW) Management in Ghana—a Review. Clean Technol 5(3):997–1011

Zetzsche DA, Anker-Sørensen L (2022) Regulating sustainable finance in the dark. Eur Bus Organ Law Rev 23(1):47–85

Zhang L, Butler TL, Yang B (2020) Recent trends, opportunities and challenges of sustainable aviation fuel. In: Green energy to sustainability: strategies for global industries, pp 85–110. https://doi.org/10.1002/9781119152057.ch5

Zhang Z, Delcroix B, Rezazgui O, Mangin P (2021) Simulation and techno-economic assessment of bio-methanol production from pine biomass, biochar and pyrolysis oil. Sustain Energy Technol Assess 44:101002

Zhang W, Luckert M, Qiu F (2023) Asymmetric price transmission and impulse responses from US crude oil to jet fuel and diesel markets. Energy 283:128425

Zhu X, Xiao J, Wang C, Zhu L, Wang S (2022) Global warming potential analysis of bio-jet fuel based on life cycle assessment. Carbon Neutrality 1(1):25

Chapter 9
Different Applications of Bio-Jet Fuel

Ankita Kumari, Depak Kumar, Priyanka Sati, Sudesh Kumar, Ashok Kumar Yadav, and Ajay Singh Verma

Abstract Bio-jet fuel, referred to as aviation biofuel in academic literature, is a form of sustainable fuel obtained from biomass sources, including plants, algae, and waste materials. The sustainable nature of alternative jet fuel, as opposed to traditional jet fuel sourced from fossil fuels, has garnered significant interest. This focus is mostly owing to its ability to mitigate greenhouse gas emissions and minimise reliance on finite resources. The integration of contemporary technology advancements and the production of alternative jet fuels are pivotal factors that possess the potential to mitigate the presence of greenhouse gases (GHGs), namely carbon dioxide (CO_2) emissions. The emissions of greenhouse gases (GHGs) generated by conventional jet fuel have become a significant cause for concern on a worldwide level. In comparison to traditional aviation fuel, biofuel is widely perceived as possessing greater renewable attributes and exhibiting reduced environmental pollution. Bio-based jet fuels have the potential to serve as a viable alternative. This chapter offers a thorough analysis of the various bio-jet fuel varieties, their production procedures, the legal environment in which they are used, and their effects on the environment. The following processes are often employed for the production of bio-jet fuel using both edible and inedible feedstock. Second-generation biofuels, which possess both environmental benefits and sophisticated technical features, provide a highly favourable option. The potential of biomass jet fuel as a viable substitute for conventional jet fuel is significant, as it caters to the requirements of both commercial and military aircraft. This chapter will address the contemporary

A. Kumari · P. Sati · S. Kumar
Department of Chemistry, Banasthali Vidyapith, Newai, Rajasthan, India

D. Kumar (✉) · A. K. Yadav
Department of Chemical Engineering, Banasthali Vidyapith, Newai, Rajasthan, India
e-mail: kdepak@banasthali.in

A. S. Verma
Division of Research & Innovation, School of Applied and Life Sciences, Uttaranchal University, Dehradun, Uttarakhand, India

A. Kuila (ed.), *Biojet Fuel: Current Technology and Future Prospect*, Clean Energy Production Technologies, https://doi.org/10.1007/978-981-99-8783-2_9

technology, significant concerns, practicality, and obstacles in the emerging sector that are being presented.

Keywords Bio-jet production · Aviation fuels standards · Feedstocks · Applications · Environmental aspects

9.1 Introduction

Bio-jet fuel, commonly referred to as aviation biofuel, serves as a viable substitute for conventional jet fuel, as it is sourced from renewable biological resources. In contrast to traditional jet fuels that are predominantly obtained from crude oil, bio-jet fuel is produced from biomass sources such as plants, algae, and waste materials. The proposed solution is a more sustainable and ecologically conscious option for fuelling aircraft. Bio-jet fuel is a form of renewable energy that may be derived from a diverse range of organic sources, such as agricultural crops (e.g. sugarcane, corn, and soy), algae, and waste oils. This characteristic renders it a viable and environmentally friendly substitute for fossil fuels, therefore mitigating reliance on diminishing oil sources (Abdullah et al. 2019). Bio-jet fuel has been found to have a lower greenhouse gas (GHG) emission profile when compared to conventional jet fuel. The reduction of the carbon footprint of the aviation industry, which is widely recognised for its substantial contributions to greenhouse gas (GHG) emissions, serves as a means to combat climate change.

Bio-jet fuel has been specifically formulated to ensure compatibility with the current engines and infrastructure utilised in the aviation industry. The integration of this fuel with traditional jet fuel does not necessitate any alterations to aircraft engines or fuelling systems, hence facilitating a seamless transition within the aviation sector. Continual research and development endeavours are driving advancements in the field of bio-jet fuel producing technology. Ongoing research endeavours are focused on the advancement of methodologies aimed at enhancing the efficiency of biofuel production processes, so rendering them more economically feasible and capable of being scaled up. International organisations and governments are actively promoting the adoption of bio-jet fuels within the aviation sector. Numerous nations have established specific objectives and regulatory measures with the aim of augmenting the utilisation of biofuels within the aviation sector, therefore fostering a more environmentally conscious trajectory for the future of air transportation. Bio-jet fuel has the potential to be derived from a diverse array of feedstock sources, encompassing non-food crops as well as waste products (Agusdinata et al. 2011). The availability of a diverse range of feedstock alternatives serves to mitigate potential conflicts with food crops and encourages the utilisation of underutilised resources. Many airlines and aviation companies are investing in bio-jet fuel as part of their corporate sustainability initiatives. By using bio-jet fuel for their flights, these companies are demonstrating their commitment to reducing their environmental impact.

Although bio-jet fuel has significant potential, there are several obstacles that must be overcome, including the availability of feedstock, manufacturing costs, and scalability (Cantarella et al. 2015). Continuous research and collaborative efforts by governmental bodies, companies, and research institutions play a pivotal role in surmounting these obstacles and establishing bio-jet fuel as a feasible choice for wider implementation.

9.2 History of Bio-Jet Fuel

The origins of bio-jet fuel may be traced back to the early 2000s, when extensive research and development endeavours were undertaken to explore viable and environmentally friendly substitutes for conventional jet fuel. In February 2008, Virgin Atlantic successfully executed the inaugural commercial aircraft flight with biofuel, marking a significant milestone in aviation history. In one of the aircraft's engines, a mixture of coconut and babassu oil was utilised throughout the voyage from London to Amsterdam. ASTM International, formerly known as the American Society for Testing and Materials, has certified the specification for aviation turbine fuel containing synthetic hydrocarbons (Chiaramonti and Horta Nogueira 2017). This approval signifies a significant milestone in the certification and potential commercialisation of bio-jet fuels in the aviation industry.

The inauguration of the initial commercial-scale bio-jet fuel manufacturing facility in Geismar, Louisiana, marks a significant milestone for Dynamic Fuels, a collaborative effort between Tyson Foods and Syntroleum. The production facility generated sustainable diesel and aviation fuel by using animal fats, greases, and vegetable oils. In June 2012, Lufthansa achieved the distinction of becoming the inaugural airline to successfully execute a scheduled transatlantic passenger flight with bio-jet fuel. The air travel route between Frankfurt and Washington, D.C., employed a biofuel mixture sourced from recycled cooking oil. A number of carriers, such as Cathay Pacific, United Carriers, and Alaska Airlines, have initiated regular commercial flights use mixes of bio-jet fuel. This development signifies a notable step towards the incorporation of biofuels into conventional aviation practices.

Fulcrum Bio Energy has just established a bio-jet fuel production plant in Reno, Nevada. This facility possesses the capability to transform municipal solid waste into transportation fuels with low carbon emissions, namely bio-jet fuel (Dincer and Acar 2014).

The research and development endeavours persisted in their concentration on enhancing the efficacy of bio-jet fuel production methods, investigating novel sources of feedstock, and tackling obstacles associated with scalability and cost-efficiency.

Over the course of recent years, the aviation sector, together with its diverse range of stakeholders, has achieved notable progress in the advancement of bio-jet fuel technology. Production routes for bio-jet fuel is shown in Fig. 9.1. Ongoing research, substantial expenditures, and collaborative efforts are important for the advancement of sustainable bio-jet fuel alternatives, with the aim of achieving economic feasibility and widespread use within the aviation industry (Gonzalez et al. 2011).

9.3 Types of Bio-Jet Fuel

Bio-jet fuel, often known as aviation biofuel, may be derived from a diverse range of feedstock sources and can be synthesised using numerous methodologies. Bio-jet fuel types are essentially classified according to the feedstock used and the technology applied throughout the production process. Here are the main types.

9.3.1 Feedstock-Based Classification

9.3.1.1 Plant-Based Bio-Jet Fuel

Derived from various plant oils, such as soybean, palm, jatropha, and camelina (Ail and Dasappa 2016). These oils can be converted into bio-jet fuel through processes like esterification and hydrogenation.

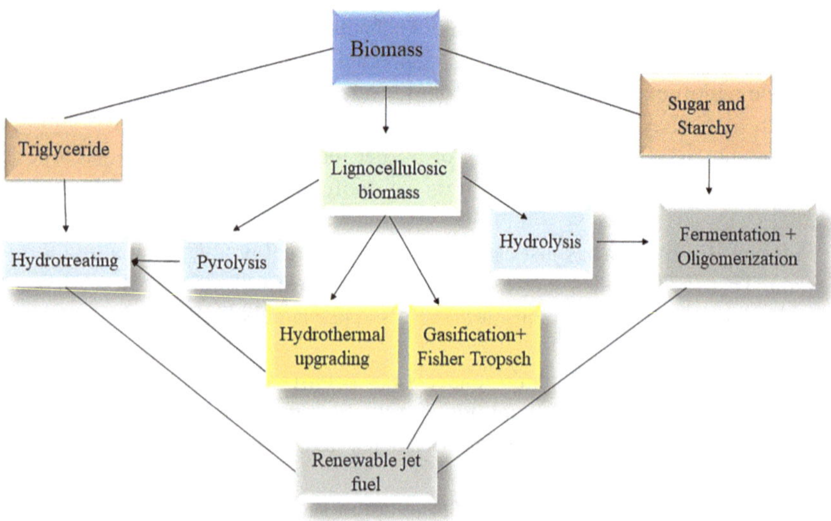

Fig. 9.1 Production routes for bio-jet fuel

9.3.1.2 Algae-Based Bio-Jet Fuel

Produced from algae, which have high oil content and can be cultivated rapidly. Algae-based bio-jet fuel is in the research and development stage and holds great promise due to its potential for high yields.

9.3.1.3 Waste-Based Bio-Jet Fuel

Created from waste and residue materials, including used cooking oil (UCO), animal fats, agricultural residues, and municipal solid waste. These feedstocks are considered sustainable as they repurpose waste materials into valuable energy sources (Han et al. 2013).

9.3.2 Process-Based Classification

9.3.2.1 Hydroprocessed Esters and Fatty Acids (HEFA)

HEFA bio-jet fuel is derived from vegetable oils and animal fats through a hydroprocessing method. This process involves hydrogenation and refining to convert triglycerides into hydrocarbons suitable for aviation use. HEFA is the most common type of bio-jet fuel in commercial use.

9.3.2.2 Fischer-Tropsch Synthesis (FT)

FT bio-jet fuel is produced through a chemical reaction known as Fischer-Tropsch synthesis. Syngas, a mixture of hydrogen and carbon monoxide, is transformed into hydrocarbons through this process. FT synthesis can utilise various feedstocks, including biomass and coal, to produce aviation biofuel (Khatun et al. 2017).

9.3.2.3 Alcohol-to-Jet (ATJ)

ATJ bio-jet fuel is synthesised from alcohols, such as ethanol and butanol, through dehydration and chemical processes. These alcohols are derived from biomass sources and can be converted into hydrocarbons suitable for jet engines.

9.3.2.4 Biomass-to-Liquid (BTL)

BTL bio-jet fuel is produced through gasification of biomass, followed by Fischer-Tropsch synthesis to convert the syngas into liquid hydrocarbons. This method allows for the use of a wide range of feedstocks, including wood, agricultural residues, and organic waste (Liu et al. 2013).

9.3.3 Blending and Certification

The aviation industry frequently uses blends of conventional jet fuel and bio-jet fuel, such as Jet A-1/bio-jet fuel. The bio-jet fuel blending ratio can change and is frequently decided in accordance with certification standards established by aviation authorities. These standards are put in place to ensure that the bio-jet fuel is compatible with the current aircraft and infrastructure.

It is important to acknowledge that continuous research and development endeavours are being conducted in the realm of bio-jet fuels, hence facilitating the study of novel feedstocks and manufacturing. These developments aim to enhance the sustainability, efficiency, and economic viability of bio-jet fuels, making them more widely adopted in the aviation industry (Murphy et al. 2015).

9.4 Advantages of Bio-Jet Fuel

Bio-jet fuel, which is produced from sustainable biomass feedstocks, has several benefits, rendering it a compelling substitute for conventional jet fuel. Here are some key advantages of bio-jet fuel.

9.4.1 Reduced Greenhouse Gas Emissions

Carbon-Neutral or Low-Carbon Footprint: Because the carbon dioxide (CO_2) absorbed during the growth of the feedstock balances the CO_2 emitted during combustion, bio-jet fuel is regarded as carbon-neutral. This leads to a net reduction in greenhouse gas emissions, mitigating climate change (Mohammed et al. 2019).

9.4.2 *Energy Security and Diversification*

Reduced Dependence on Fossil Fuels: By utilising renewable feedstocks, bio-jet fuel reduces reliance on fossil fuels, enhancing energy security and decreasing vulnerability to oil price fluctuations and supply disruptions.

9.4.3 *Compatibility with Existing Infrastructure*

Effortless Integration: Bio-jet fuel is made to work with current aircraft engines and fuelling systems. It can be used without modifying aircraft or refuelling systems by blending it with regular jet fuel.

9.4.4 *Job Creation and Economic Development*

Rural Development: Bio-jet fuel production creates jobs and stimulates economic growth, especially in rural areas where feedstocks like crops and algae are cultivated and harvested.

9.4.5 *Waste Utilisation and Sustainability*

Utilisation of Waste Materials: Bio-jet fuel can be produced from various waste materials, including forestry waste, agricultural residues, and municipal solid waste. This promotes the efficient use of waste, reducing pollution, and promoting sustainability (Pham et al. 2010).

9.4.6 *Promotion of Innovation and Research*

Technological Advancements: Research and development in bio-jet fuel technology drive innovation in biotechnology, chemistry, and engineering, leading to more efficient production processes and novel feedstock options.

9.4.7 Corporate Sustainability and Environmental Responsibility

Corporate Initiatives: Many companies and airlines invest in bio-jet fuel to align with their sustainability goals and demonstrate environmental responsibility, enhancing their corporate image and social responsibility efforts (Pirker et al. 2016).

9.4.8 Compliance with Regulations and International Agreements

Alignment with Environmental Goals: The use of bio-jet fuel aligns with international agreements and initiatives aimed at reducing greenhouse gas emissions, such as the Paris Agreement. Countries and airlines investing in bio-jet fuel contribute to global efforts to combat climate change.

9.4.9 Air Quality Improvement

Reduced Air Pollutants: Bio-jet fuel has the potential to reduce emissions of air pollutants, such as sulphur oxides and particulate matter, leading to improvements in air quality and public health.

9.4.10 Technological Advancements and Scale-Up Potential

Ongoing Research: Continuous research and development efforts lead to technological advancements, addressing challenges related to feedstock availability, production costs, and scalability, making bio-jet fuel a more viable option for the aviation industry. Bioprocessing to make bio-jet fuel is shown in Fig. 9.2.

These advantages make bio-jet fuel an increasingly viable and sustainable alternative to traditional jet fuels, driving its adoption in the aviation sector and contributing to a more environmentally friendly future for air travel.

9.5 Disadvantages of Bio-Jet Fuel

While bio-jet fuel offers numerous advantages, it also has its share of challenges and disadvantages. Here are some of the key disadvantages associated with bio-jet fuel.

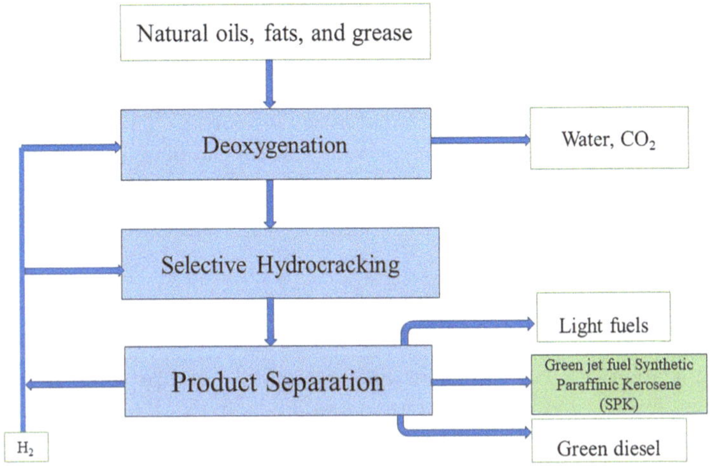

Fig. 9.2 Bioprocessing to make jet fuel

9.5.1 Limited Feedstock Availability

Competition with Food Production: Some biofuel crops can compete with food crops for agricultural resources, leading to concerns about food security and increased food prices. Striking a balance between food and fuel production is a significant challenge (Rathmann et al. 2010).

9.5.2 Land Use and Deforestation

Land Use Change: Converting natural habitats or forests into biofuel crop cultivation can lead to deforestation, disrupting ecosystems, reducing biodiversity, and contributing to habitat loss and greenhouse gas emissions.

9.5.3 Water Usage and Impact on Water Resources

High Water Requirements: Certain biofuel crops, especially those grown in arid regions, can require significant amounts of water, leading to increased stress on local water resources and potential conflicts with agriculture and communities.

9.5.4 Energy Intensive Production

Energy Input: The production processes for bio-jet fuel can be energy-intensive, potentially offsetting some of the environmental benefits. It's crucial to develop methods that minimise energy inputs and maximise output efficiency (Shah et al. 2019).

9.5.5 Impact on Soil Quality

Soil degradation can result from the intensive cultivation of biofuel crops, which lowers soil fertility and long-term agricultural productivity.

9.5.6 Greenhouse Gas Emissions

Indirect Emissions: The entire lifecycle of bio-jet fuel, including cultivation, processing, and transportation, can still result in greenhouse gas emissions, especially if fossil fuels are used in these processes.

9.5.7 Economic Viability and Scale-Up Challenges

Production Costs: Bio-jet fuel production can be expensive compared to conventional jet fuel, making it economically challenging for widespread adoption without subsidies or incentives.

Limited Scale-Up: Large-scale production of bio-jet fuel faces challenges related to feedstock availability, technology scalability, and investment, hindering rapid deployment.

9.5.8 Technological and Infrastructure Challenges

Infrastructure Compatibility: While efforts have been made to ensure compatibility with existing aircraft and infrastructure, full integration still faces challenges, including storage, transportation, and distribution logistics (Samsatli and Samsatli 2018b).

9.5.9 *Food Security Concerns*

Diversion of Agricultural Resources: Large-scale cultivation of biofuel crops can divert agricultural resources away from food production, potentially impacting food security in certain regions. Table 9.1 shows benefits and drawbacks of bio-jet fuel. Addressing these challenges is crucial for the sustainable development and deployment of bio-jet fuel. Continued research, innovation, and international collaboration are essential to overcoming these disadvantages and making bio-jet fuel a truly sustainable and widely adopted alternative to traditional jet fuels.

9.6 Growth of Aviation Industry

With a compound annual growth rate (CAGR) of 18% by 2020, the Indian aviation industry would have the fastest growth. The profitability of smaller airlines is impacted by India's high fuel prices and taxes, if not the highest in the world. In the past 5 years, both domestic and international air traffic in India has increased by about 20%. Carbon dioxide (CO_2), nitrogen oxides (NO_X), water vapour (H_2O), and particulate matter (PM) are the primary greenhouse gas (GHG) emissions produced by air travel during flight, which contribute to global air pollution. As an alternative to fossil fuel-based jet A-1 fuel, green fuel is urgently needed by the aviation industry. Second-generation biofuels are an option that can be combined with fossil fuel-based Jet A-1 to partially meet the demand for aviation fuel while reducing the net emissions of carbon dioxide (CO_2), nitrogen oxides (NO_X), water vapour (H_2O), and particulate matter (PM). These fuels can be blended with aviation fuel in part using the current refuelling infrastructure, which makes the global supply chain simple. Bio-jet fuel production process is shown in Fig. 9.3.

Carbon dioxide (CO_2) emissions are decreased by 3.16 kg for every kilogramme of fuel saved (Suresh et al. 2018). However, compared to other modes of transportation like cars, there are a number of difficulties in applying biofuels for aviation

Table 9.1 Benefits and drawbacks of bio-jet fuel

Advantages	Disadvantages
Theoretically limitless supply of feedstock.	The lack of biodiversity and pest susceptibility that come with monocultures.
Long-term risk reduction in the event of fuel spillage.	If energy crops become more profitable for farmers than food crops, there could be a conflict with the supply of food.
Depending on the production process, can burn with lower net CO_2 emissions.	Changing land use in a negative way, such as clearing vegetation from the area, using fertilisers to the point where the water becomes eutrophic, and using water and energy to cultivate the land.
Use as a "drop-in" replacement for current engines.	Boundaries of space and time, for example, feedstock may not be grown year-round or at all in some areas if certain conditions are needed.

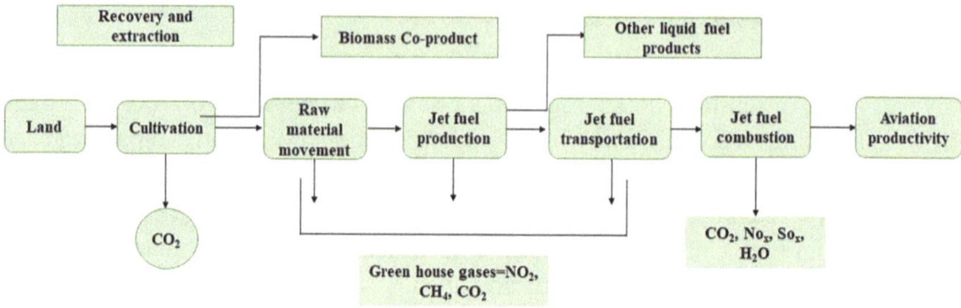

Fig. 9.3 Bio-jet fuel production process

purposes. Meeting stricter fuel specifications than those for automobiles is one of the biggest obstacles to the use of biofuels in aviation. Here are some of the primary requirements for environmentally friendly alternative jet fuels:

- Can be blended with regular jet fuel,
- Able to utilise the same infrastructure for supply and don't need to modify engines or aircraft (drop-in fuel),
- Meet the same requirements as traditional jet fuel, especially in terms of resistance to cold (Jet A: 40 °C, Jet A-1: 47 °C),
- A lot of energy (minimum 42.8 MJ/kg),
- Meet sustainability requirements like reducing lifecycle carbon emissions, requiring less freshwater, not competing with food production, and not deforestation. Any alternative fuel must unquestionably adhere to the strict requirements for aviation fuel set by the current aircraft industry.

9.7 Various Applications of Bio-Jet Fuels

Bio-jet fuel has various applications and benefits, including the following.

9.7.1 First-Generation Biofuels (Edible Crops)

These are biofuels made from various edible plants, such as soybean, rapeseed, and various derivatives of palm oil. This sort of fuel falls under the category of first-generation biofuel (Oladosu and Msangi 2013). Soybean and rapeseed oil are typically used in some regions of America and Europe to produce biodiesel and its by-products (Tao et al. 2014). Palm oil was used as a feedstock in the production of biodiesel in the southern region of Asia. The claim that using first-generation feedstocks for biodiesel synthesis results in a significant amount of GHG emissions into the atmosphere is one of the biggest issues facing academics and scientists (Woytiuk et al. 2017).

9.7.2 Second-Generation Biofuels (Lignocellulose, Non-Edible Crops, Animal Feedstock)

Non-edible plants such as jatropha and camelina are employed as feedstock for the production of second-generation biofuels. Camelina has robust growth in regions characterised by high temperatures. The cultivation of this particular plant species is characterised by its ease of growth and its ability to provide a substantial quantity of oil seeds. The *Jatropha curcas* plant, which exhibits a preference for a humid climate, has the capability to be cultivated throughout the whole year. According to a study conducted by researchers (Zhu et al. 2018), it has been shown that jatropha has a higher oil production capacity compared to other plant species. The use of bio-jet fuel generated from camelina exhibited superior fuel efficiency and shown less environmental impact. Camelina has the potential to yield an annual production of 800 million metric gallons of oil. Jatropha exhibits higher energy levels, ranging from 18 to 19 MJ kg^{-1}, and possesses a notable seed dry weight of up to 55% (Zhang et al. 2018). The aforementioned ethanol is often regarded as the most superior renewable fuel source on a global scale, exhibiting commendable environmental benefits.

9.7.3 Third-Generation Biofuels (Algal Feedstock)

A biofuel source from the third generation of biofuels is algae feedstock. Algae can be collected using a variety of techniques, including flotation procedures, centrifugal sedimentation techniques, flocculation methods, and filtering techniques. The predominant post-harvest technique employed is drying, sometimes denoted as dehydration. Moreover, it provides additional commercial benefits, including enhanced simplicity and cost-effectiveness in comparison to similar technologies (Yang et al. 2019). When compared to other fuels, third-generation or advanced biofuels regularly demonstrate superior cost-effectiveness and performance. In comparison to preceding generations of feedstock, algae have been seen to have a greater year-round productivity. Algal farming employs brackish, saline, and wastewater as alternative sources of water, as opposed to freshwater, hence necessitating a greater water demand compared to conventional food crops. There are many noteworthy proteins that may be obtained as by-products from the process of oil extraction, which have the potential to be repurposed and utilised as fertilisers. Algae possesses a significant quantity of oil. Utilising photobiology for hydrogen production, algae demonstrate a reduced emission of greenhouse gases compared to alternative forms of oil refineries. This methodology employs reduced energy consumption and incurs lower financial expenses in the production of algae. Table 9.2 shows bio-jet fuel production technologies.

Table 9.2 Bio-jet fuel production technologies

Technologies	Production processes
Alcohol-to-jet	1. N-butanol-to-jet 2. Ethanol-to-jet 3. Methanol-to-jet 4. Iso-butanol-to-jet
Oil-to-jet	1. Catalytic hydrothermolysis 2. Hydroprocessed renewable jet 3. Hydrotreated depolymerised cellulosic jet
Gas-to-jet	1. Gas fermentation 2. Fischer-Tropsch synthesis
Sugar-to-jet	1. Catalytic upgrading 2. Direct sugar to hydrocarbons

Two limitations of this approach are low production and the requirement for a significant amount of land (Trivedi et al. 2015). Closed photo-bioreactor systems have a higher energy input and entail a greater financial cost, up to $22.4 per gallon.

Commercial Aviation: One of the primary applications of bio-jet fuel is in commercial aviation. Airlines can use bio-jet fuel to power their aircraft, reducing their carbon footprint and contributing to a more sustainable aviation industry.

Military Aviation: Military aviation operations can also benefit from bio-jet fuel. Armed forces of various countries have started exploring the use of bio-jet fuel to enhance their energy security and reduce environmental impact.

Climate Change Mitigation: Bio-jet fuel helps mitigate climate change by reducing carbon dioxide (CO_2) emissions. Unlike fossil fuels, bio-jet fuel is derived from renewable sources, and the CO_2 released during its combustion is part of the natural carbon cycle, making it a more sustainable option.

Energy Security: Bio-jet fuel production can enhance energy security by diversifying the sources of aviation fuel. It reduces dependency on imported fossil fuels, which can be affected by geopolitical tensions and price fluctuations.

Rural Development: The production of feedstocks for bio-jet fuel, such as certain crops and algae, can create economic opportunities in rural areas. This can lead to the development of local economies and job creation.

Waste Utilisation: Various organic waste products, such as agricultural residues, forestry waste, and municipal solid waste, can be used to make bio-jet fuel. This promotes the efficient use of waste materials and reduces the environmental impact of landfilling (Tapia and Samsatli 2020).

Innovation and Research: The development and application of bio-jet fuel technology drive innovation in the fields of biotechnology, chemistry, and engineering. Research in this area can lead to more efficient production processes and novel feedstock options.

Corporate Sustainability: Many companies and organisations are investing in bio-jet fuel to demonstrate their commitment to sustainability. Using bio-jet fuel for corporate travel and transportation aligns with their environmental goals and corporate social responsibility initiatives.

Reduction of Air Pollutants: Bio-jet fuel has the potential to reduce emissions of air pollutants, such as sulphur oxides and particulate matter, which can have harmful effects on air quality and human health.

International Agreements: The adoption of bio-jet fuel aligns with international agreements and initiatives aimed at reducing greenhouse gas emissions, such as the Paris Agreement. Countries and airlines that invest in bio-jet fuel contribute to global efforts to combat climate change.

9.7.4 Fourth-Generation Feedstocks

Several options, including non-biological resources and genetically modified organisms, are included in ATAG's assessment of the portfolio of feedstocks for sustainable aviation fuels. These resources are classified as fourth-generation (4-G) feedstocks, as identified by ATAG. Genetically modified organisms, such as cyano bacteria, microalgae fungus, and yeast, have been subject to genetic alterations that result in increased oil and/or sugar production, as well as the ability to sequester carbon dioxide. However, it is important to note that these advancements are still in the early stages of scientific investigation. Despite the promise of biofuels, further research is required to investigate the possible health and environmental concerns associated with these creatures. Additionally, studies are needed to explore ways for containment and mitigation when these organisms are introduced into global supply chains. Non-biological feedstocks, such as CO_2, water, renewable power, and sunshine, have the potential to be a more ecologically friendly alternative, particularly when industrial plant exhaust gases are utilised. One potential approach is the power-to-liquid (PtL) pathway, which entails the electrolysis of water using renewable electricity to generate hydrogen and oxygen. Subsequently, the hydrogen is reacted with CO_2/CO to synthesise biomass-derived aviation fuel (BAF). Nevertheless, in the long run, the environmental advantages of PtL fuels, such as their near carbon neutrality and minimal demands for water and land, may surpass the economic considerations and external effects associated with CJF. An alternative approach involves harnessing concentrated solar energy to facilitate the electrolysis of water and the decomposition of CO_2, resulting in the creation of syngas as a precursor for biomass-derived aviation fuel (BAF) synthesis. Although these pathways are currently in the preliminary phases of investigation, Richter et al. (2018) have found two European endeavours, namely Sunfire and SOLAR-JET, which have successfully showcased the ability to produce jet fuel with carbon dioxide, water, and solar energy. Regarding the research conducted on the supply chains of 4-G feedstocks, it is worth noting that the existing studies in this area have been rather few (Samsatli and Samsatli 2018a). However, Mesfun et al. (2017) have made a significant contribution by employing a spatiotemporal Mixed Integer Linear Programming (MILP) model to examine the integration of power-to-gas (PtG) and power-to-liquid technologies within the context of an Alpine energy supply system. Once these technologies reach a state of commercial maturity, it is anticipated that biomass-derived alternative fuels (BAF) sourced from fourth generation feedstocks would offer the highest level of sustainability. These fuels have the potential to achieve zero carbon emissions and facilitate the integration of electricity, heating, and aviation sectors.

9.7.5 Potential Source of Microalgae for Bio-Jet Fuel

Microalgae have been utilised as promising sources of biofuel for several decades. Microalgae have similarities to terrestrial plants in that they possess chlorophyll, a crucial photosynthetic pigment, enabling them to convert carbon dioxide and water into sugar through the process of photosynthesis (Richter et al. 2018). In general, microalgae have a greater oil content in comparison to alternative sources of biofuel. Microalgae have demonstrated comparative benefits over other sources of biofuel, mostly attributed to their ability to achieve significant levels of biomass productivity and oil content within a very brief timeframe. Based on the findings of Behera and Varma (2016), it has been shown that the annual microalgal oil output is significantly higher than that of soybeans and palm oil, with approximate ratios of 92–215 times and 11–26 times, respectively. According to Trent (Quarton and Samsatli 2018), it has been demonstrated that microalgae have the potential to yield a substantial amount of fuel, ranging from 2000 to 5000 gallons per acre (equivalent to 18,708 to 46,770 L/ha) annually. This output surpasses that of palm oil, the second most prolific source, which yields from 1400 to 4400 gallons per acre (equivalent to 13,096 to 41,157 L/ha). According to previous research, microalgae have demonstrated a much higher energy yield per hectare in comparison to other types of terrestrial crops, with estimates ranging from 30 to 100 times greater. In a comparative study, it was shown that microalgae yielded an annual biofuel production of 94,000 L/ha, but maize only yielded 560 L/ha per year. According to the research conducted by Chisti (2007), it was observed that the production of microalgae often exhibited a twofold rise within a span of 1 day. However, several types of microalgae were shown to have a doubling time of 3.5 h, with oil deposition exceeding 80% of the dry biomass weight. Microalgae have the capacity to produce advantageous by-products, including carbohydrates, proteins, and residual biomass subsequent to the extraction of oil. The use of microalgae co-products has been seen in many applications such as animal feed, fertiliser, and ethanol production via fermentation (Perkis and Tyner 2018). The utilisation of existing techniques for microalgae cultivation can provide a consistent provision of feedstock for the generation of bio-jet fuel. Microalgae exhibit a greater capacity for lipid production in comparison to alternative sources of biofuel feedstocks. Lipids found in microalgae may be categorised into two distinct forms: (a) non-polar or neutral lipids, which function as energy stores, and (b) polar lipids, which are utilised as food supplements and act as elements of organelles and membranes. Microalgae possess the capacity to acquire lipid content ranging from 30% to 70%. According to Marcilla et al. (2013), the thermochemical conversion techniques of hydrothermal liquefaction and catalytic pyrolysis have been identified as viable methods for the manufacture of biofuels from microalgae. The experimental findings indicate that the maximum bio-oil yield (43%) is achieved at a temperature of 350 °C, accompanied with a heating value of 39 MJ/kg. Additionally, dote et al. found that 64% of the bio-oil in their sample had a higher heating value (HHV) of 45.9 MJ/kg. In contrast, Fong et al. conducted a study on the impact of several catalysts (HZSM-5 zeolite, limestone (LS), bifunctional HZSM-5/LS) on the catalytic pyrolysis of *Chlorella vulgaris*. The findings of this study indicate that the use of bifunctional HZSM-5/LS catalyst in catalytic pyrolysis has considerable potential as a thermochemical route for the synthesis of biofuels, when compared to non-catalytic pyrolysis. The use of the bifunctional HZSM-5/LS catalyst led to a reduction in the average activation energy (EA) (133.26 kJ/mol) and enthalpy (ΔH) (128 kJ/mol), suggesting a decrease in the energy demands for the biofuel generation process. Furthermore, the inclusion of carbohydrates derived from microalgae serves as an additional chemical constituent in the manufacture of biofuels. Carbohydrates such as starch, glucose, cellulose, paramylon, and laminarin play a significant role as energy sources. According to Arun et al. (Oladosu and Msangi 2013), the authors reported that bioethanol and biobutanol may be produced by extracting the carbohydrate content of microalgae. In the field of biofuel generation, *Saccharomyces cerevisiae* is commonly employed for fermentation processes to generate bioethanol, whilst *Clostridium acetobutylicum* yeast is utilised for the creation of biobutanol. Microalgae have some benefits over carbohydrates derived from other sources due to their absence of lignin and hemicelluloses. In addition, the carbohydrate constituents present in microalgae can be used for the generation of methane gas. The study conducted by Passos et al. (2016) examined several thermochemical pre-treatment conditions, including acids and alkaline, to determine their effectiveness in enhancing biogas generation from microalgae biomass. The findings indicate that the use of thermochemical pre-treatment including a 0.5% acid and 0.5% alkaline solution resulted in significantly increased methane production. Specifically, the methane output observed in this pre-treatment condition was found to be 82% and 86% greater compared to the control condition. While the alkaline pre-treated condition yields the maximum amount of methane, it is not considered an optimal environment for methane production because to its longer lag phase (1.20 day) and lower rate of methane generation compared to the acid pre-treated condition.

In brief, the use of bio-jet fuel encompasses a wide range of significant implications, including but not limited to the reduction of carbon emissions and the mitigation of climate change. Additionally, it contributes to the advancement of rural development and serves as a catalyst for innovation within the energy sector. The use of bio-jet fuel is anticipated to have a substantial impact on the future of sustainable aviation as technology progresses.

9.8 Current Policies on Bio-Jet Fuels

The distribution and advancement of conversion technologies for biofuels can be significantly influenced by government legislation. Renewable fuel standards aim to assure the cost-competitiveness of alternative jet fuel by providing financial assistance for the establishment of new technology production plants and delivering agricultural incentives. The objective of this initiative was to provide the necessary infrastructure and logistics for the delivery of biofuel. The European Union (EU) has established a programme aimed at the advancement of biofuel energy development within the timeframe of 2020 to 2030, with the objective of enhancing aviation fuel policy (Mesfun et al. 2017). According to projections, the demand for alternative jet fuel in Indonesia is expected to increase by 5% by the year 2025, compared to the levels observed in 2018. In 2018, Indonesia had a requirement of 2%. Furthermore, the People's Republic of China has implemented a comprehensive 5-year energy strategy. It is projected that there will be a substantial increase in the utilisation of renewable energy sources, such as biomass-derived fuels, in the coming years [89]. During the CAEP/11 cycle (2016–2019), the ICAO Task Force on different Fuels (AFTF) conducted an assessment on the efficacy of different policy instruments in facilitating the commercial utilisation of jet fuel. There are some technologies presently undergoing production, such as the FT and HEFA. Nevertheless, several sophisticated technologies require more development prior to their industrialisation. The concept of sustainable aviation fuel certifications was initially proposed by ASTM. In 2009, fuels generated using the Fischer-Tropsch (FT) technique were officially sanctioned. Consequently, the petrol manufactured by HEFA received clearance from the American Society for Testing and Materials (ASTM) in the year 2011. In 2014, SIP Fuel achieved its third ASTM certification, which was obtained using the hydroprocessed fermentation method that produces Isoparaffin. In order to fulfil the necessary criteria, it is vital to validate additional methodologies such as cellulosic jet, alcohol-to-jet, and paraffinic kerosene jet. Between the years 2011 and 2015, a total of 2500 commercial passenger flights were analysed. Within this sample, it was observed that 22 different airlines utilised a blend of bio-jet fuel, constituting around 50% of the overall fuel composition (Liu et al. 2013). Furthermore, it is worth mentioning that in the year 2016, the global production of sustainable aviation fuel exceeded 4.5 million litres, which represents a significant increase of over 100% compared to the production levels seen in 2015. Aireg (Germany), Biofuel Net (Canada), Plan de Vuelo (Mexico), CAAFI (US), Sustainable Aviation (UK), Ubrabio (Brazil), NISA (Nordic countries), Bioport Holland (The Netherlands), Bioqueroseno (Spain), and AISAF (Australia) are now engaged in the development of plans pertaining to the distribution of jet fuel. Presently, there are ongoing initiatives of this nature in Japan, China, Qatar, the United Arab Emirates, and Israel. When considering the accessibility of data pertaining to biomass feedstock, previous research has indicated a more substantial influence on the production of bio-jet fuel. These data have been compiled to facilitate the evaluation process. China possesses significant energy potentials in both lignocellulosic and sugar feedstocks, estimated at 6.72 EJ/yr and 0.41 EJ/yr, respectively. According to the projections made by the US Energy Information Administration, there is an anticipated increase in the global consumption of jet fuel, reaching a total of 22.88 exajoules by the year 2040 (Ganguly et al. 2018).

9.9 Challenges and Future Outlook

Due to alternative biofuels' superior environmental performance compared to conventional petrol, which is primarily explained by the environment's excessive emission of greenhouse gases (GHGs) and carbon dioxide (CO_2), the use of alternative biofuels is essential. Several obstacles are still being experienced. The aviation industry stands to gain several environmental advantages from the utilisation of sustainable bio-jet fuel. These benefits encompass the ready availability of feedstock, the formation of novel sectors, as well as the promotion of both economic and environmental sustainability. The achievement of sustained success over an extended period of time may be impeded by the choice of feedstock, which should possess attributes such as a high yield while minimising any adverse impact on food production. Considering the imperative nature of large-scale manufacturing, the requirement for a feedstock that is economically viable has equal significance (Zaher et al. 2015). The creation of feedstock and the establishment of a formula for the widespread production of alternative fuels necessitate minimal resource requirements in terms of water quality, land and fertiliser availability. In order to mitigate the price disparity and facilitate the procurement of fuels by airlines, as well as foster a market-oriented approach, it is imperative to implement incentive programmes or compensation mechanisms that promote environmental well-being by encouraging the use of bio-jet fuels. This strategy is expected to attract potential investors and mitigate the perceived level of risk. The manufacture of bio-jet fuel is influenced by several factors, including the cost of extracted fuels and by-products, operational expenses, product yield, conversion efficiency, and feedstock prices. Hence, in order to achieve cost reduction, it is imperative to enhance feedstock productivity, maximise the value of co-products, optimise equipment distribution, establish ideal reaction conditions, utilise low-cost catalysts, and optimise recovery processes, among several other factors. The enhanced competitiveness of alternative jet fuels can be facilitated by reduced market price in comparison to traditional jet fuel. This, in turn, can result in more backing and funding for the research and development of bio-jet production, as well as advancements in technology. In recent decades, extensive research and development efforts have been dedicated to exploring and establishing various production pathways. The properties of the fuel exhibit similarities to conventional aviation petrol, but with a few notable distinctions, such as a lower proportion of aromatic compounds that can lead to fuel leakage (Xue et al. 2017). Nevertheless, appropriate chemical compounds are introduced to compensate for this limitation. The success and profitability of the F-T synthesis, similar to HRJ, may be attributed to its ability to utilise a wide range of feedstocks, hence enhancing its adaptability. One of the notable limitations of this approach is its protracted and intricate procedure, which is contingent upon many factors, rendering it among the costliest alternatives for producing presently available. Conversely, the use of less complex and cost-effective production methods is limited by the limiting supply of raw materials, necessitating a dependence on proximity for accessibility. The accessibility of affordable hydrogen is an additional essential element in the manufacturing of jet fuel from biomass. Small-scale businesses are unaffected, but large-scale manufacturers must take the availability of hydrogen as a fuel source into account.

9.10 Conclusions

The imperative for sustainable aviation is indisputable, given the projected escalation of the sector's prominence within the realm of global transportation. The investigation yielded results indicating that second-generation biofuels are widely recognised for their high level of technological sophistication, economic feasibility, and environmental sustainability. Jatropha possesses negligible nutritional value, although it exhibits remarkable potential for energy generation. The primary findings of this research suggest that oils and lipids may be converted into bio-jet fuels by the use of intricate and advanced techniques such as hydrogenation, decarboxylation, hydrodeoxygenation, hydrodecarbonylation, and/or isomerisation. A variety of methodologies have been devised for the conversion of biomass; however, the HEFA process has gained significant prominence and recognition as the most often utilised and esteemed approach for the production of biofuels, owing to its advantageous attributes of cost-effectiveness and superior energy efficiency. Despite the higher initial investment required, Fischer-Tropsch (FT) technologies are considered commercially established and effective methods for reducing greenhouse gas (GHG) emissions. Lignocellulose has superior characteristics as a feedstock for biofuel production; yet, its conversion into bio-jet fuel poses challenges due to the intricacies involved in the conversion process. The process involves many sequential stages and exhibits inherent accessibility for the production of bio-jet fuel. The technological and economic viability of bio-jet production methods has been demonstrated in laboratory settings. However, in order to transition these technologies into large-scale production, more advancements in catalyst and feedstock are necessary. In order to achieve a substantial reduction in greenhouse gas (GHG) and carbon dioxide (CO_2) emissions within the aviation sector, it is imperative to streamline the processes associated with GHG and CO_2 emissions. It is imperative for researchers and scientists to prioritise efforts towards bridging this disparity.

References

Abdullah B, Syed Muhammad SAF, Shokravi Z, Ismail S, Kassim K, Mahmood AN et al (2019) Fourth generation biofuel: a review on risks and mitigation strategies. Renew Sust Energ Rev 107:37–50

Agusdinata DB, Zhao F, Ileleji K, Delaurentis D (2011) Life cycle assessment of potential biojet fuel production in the United States. Environ Sci Technol 45:9133–9143

Ail SS, Dasappa S (2016) Biomass to liquid transportation fuel via FischerTropsch synthesis – technology review and current scenario. Renew Sust Energ Rev 58:267–286

Behera BKVarma A, (2016) From algae to liquid fuels. In: Microbial resources for sustainable energy. Springer 123–80

Cantarella H, Nassar AM, Cortez LAB, Junior RB (2015) Potential feedstock for renewable aviation fuel in Brazil. Environ Dev 15:52–63

Chiaramonti D, Horta Nogueira LA (2017) Aviation biofuels: processes, opportunities, constraints, and perspectives. In: Singh R, Kumar S (eds) Biofuels production and processing technology. CRC Press, pp 295–318

Chisti Y (2007) Biodiesel from microalgae. Biotechnol Adv 25(3):294–306

Dincer I, Acar C (2014) Review and evaluation of hydrogen production methods for better sustainability. Int J Hydrog Energy 40:11094–11111

Ganguly I, Pierobon F, Bowers TC, Huisenga M, Johnston G, Eastin IL (2018) 'Woods-to-wake' life cycle assessment of residual woody biomass-based jet-fuel using mild bisulfite pretreatment. Biomass Bioenergy 108:207–216

Gonzalez R, Treasure T, Wright J, Saloni D, Phillips R, Abt R, Jameel H (2011) Exploring the potential of eucalyptus for energy production in the southern United States: financial analysis of delivered biomass. Part, I. Biomass Bioenergy 35:755–766

Han SH, Cho DH, Kim YH, Shin SJ (2013) Biobutanol production from 2-year-old willow biomass by acid hydrolysis and acetonebutanol-ethanol fermentation. Energy 61:13–17

Khatun R, Reza MIH, Moniruzzaman M, Yaakob Z (2017) Sustainable oil palm industry: the possibilities. Renew Sust Energ Rev 76:608–619

Liu G, Yan B, Chen G (2013) Technical review on jet fuel production. Renew Sust Energ Rev 25:59–70

Marcilla A, Catal´a L, García-Quesada JC, Vald´es FJ, Hern´andez MR (2013) A review of thermochemical conversion of microalgae. Renew Sustain Energy Rev 27:11–9

Mesfun S, Sanchez DL, Leduc S, Wetterlund E, Lundgren J, Biberacher M et al (2017) Power-to-gas and power-to-liquid for managing renewable electricity intermittency in the Alpine region. Renew Energy *107*:361–372

Mohammed IY, Abakr YA, Mokaya R (2019) Integrated biomass thermochemical conversion for clean energy production: process design and economic analysis. J Environ Chem Eng 7:103093

Murphy HT, O'Connell DA, Raison RJ, Warden A, Booth TH, Her A et al (2015) Biomass production for sustainable aviation fuels: a regional case study in Queensland. Renew Sust Energ Rev 44:738–750

Oladosu G, Msangi S (2013) Biofuel-food market interactions: a review of modeling approaches and findings. Agriculture 3:53–71

Passos F, Felix L, Rocha H, Pereira JdO, de Aquino S (2016) Reuse of microalgae grown in full-scale wastewater treatment ponds: thermochemical pretreatment and biogas production. Bioresour Technol 209:305–12

Perkis DF, Tyner WE (2018) Developing a cellulosic aviation biofuel industry in Indiana: a market and logistics analysis. Energy 142:793–802

Pham V, Holtzapple M, El-Halwagi M (2010) Techno-economic analysis of biomass to fuel conversion via the MixAlco process. J Ind Microbiol Biotechnol 37:1157–1168

Pirker J, Mosnier A, Kraxner F, Havlík P, Obersteiner M (2016) What are the limits to oil palm expansion? Glob Environ Change 40:73–81

Quarton CJ, Samsatli S (2018) Power-to-gas for injection into the gas grid: what can we learn from real-life projects, economic assessments and systems modelling? Renew Sust Energ Rev 98:302–316

Rathmann R, Szklo A, Schaeffer R (2010) Land use competition for production of food and liquid biofuels: an analysis of the arguments in the current debate. Renew Energy 35:14–22

Richter S, Braun-Unkhoff M, Naumann C, Riedel U (2018) Paths to alternative fuels for aviation. CEAS Aeronaut J 9:389–403

Samsatli S, Samsatli NJ (2018a) A general mixed integer linear programming model for the design and operation of integrated urban energy systems. J Clean Prod 191:458–479

Samsatli S, Samsatli NJ (2018b) A multi-objective MILP model for the design and operation of future integrated multi-vector energy networks capturing detailed spatio-temporal dependencies. Appl Energy 220:893–920

Shah Z, Veses RC, Vaghetti JCP, Amorim V, da Silva R (2019) Preparation of jet engine range fuel from biomass pyrolysis oil through hydrogenation and its comparison with aviation kerosene. Int J Green Energy 16:350–360

Suresh P, Malina R, Staples MD, Lizin S, Olcay H, Blazy D et al (2018) Life cycle greenhouse gas emissions and costs of production of diesel and jet fuel from municipal solid waste. Environ Sci Technol 52:12055–12065

Tao L, Tan ECD, McCormick R, Zhang M, Aden A, He X et al (2014) Techno-economic analysis and life-cycle assessment of cellulosic isobutanol and comparison with cellulosic ethanol and n-butanol. Biofuel Bioprod Biorg 8:30–48

Tapia JFD, Samsatli S (2020) Integrating fuzzy analytic hierarchy process into a multi-objective optimisation model for planning sustainable oil palm value chains. Food Bioprod Process 119:48–74

Trivedi P, Malina R, Barrett SRH (2015) Environmental and economic trade-offs of using corn Stover for liquid fuels and power production. Energy Environ Sci 8:1428–1437

Woytiuk K, Campbell W, Gerspacher R, Evitts RW, Phoenix A (2017) The effect of torrefaction on syngas quality metrics from fluidized bed gasification of SRC willow. Renew Energy 101:409–416

Xue C, Liu M, Guo X, Hudson EP, Chen L, Bai F et al (2017) Bridging chemical- and bio-catalysis: high-value liquid transportation fuel production from renewable agricultural residues. Green Chem 19:660–669

Yang J, Xin Z, He QS, Corscadden K, Niu H (2019) An overview on performance characteristics of bio-jet fuels. Fuel 237:916–936

Zaher FA, Hussein AES, Hassan BH, Hamed FS (2015) Potential of castor oil as a feedstock for the production of bio-fuel via catalytic hydrocracking. Curr Sci Int 4:443–449

Zhang B, Wu J, Yang C, Qiu Q, Yan Q, Li R et al (2018) Recent developments in commercial processes for refining bio-feedstocks to renewable diesel. Bioenergy Res 11:689–702

Zhu Z, Chu F, Dolgui A, Chu C, Zhou W, Piramuthu S (2018) Recent advances and opportunities in sustainable food supply chain: a model oriented review. Int J Prod Res 56:5700–5722

Chapter 10
Sustainability of Biojet Fuel

Nikita Bhati and Arun Kumar Sharma

Abstract The unavoidable detrimental impacts on the environment due to continuous dependency on traditional jet fuels have urged global initiatives in the direction of alternate possibilities for the aviation sector. The lack of possibilities for decarbonization of fossil fuels has made the adoption of biojet fuels (BJF) a success because of their critical contribution to the aviation sector as a means to reduce greenhouse gas (GHG) emissions. The long lifespan and substantial capital expenses of aircraft make the rapid substitution with carbon-neutral technologies a less favorable choice. Therefore, "drop-in" solutions that can be installed seamlessly in the engines of current aircraft may be needed. The usage of lignocellulosic biomass in the Fischer-Tropsch production pathway has the highest probability of reducing GHG emissions and could possibly be useful for the mid- to long-range objectives of the airline sector, but because of its restricted technological development and higher capital expenditures, more study and optimization are needed before it can be implemented on a large scale. Practically, the "optimum" raw materials and advancements in logistics management are significantly reliant on spatiotemporal parameters. Furthermore, most studied factors are connected to one another, and the strategies that are operative in the mitigation of GHG emissions are mostly expensive. Therefore, guidelines must be rationalized via the constituents of logistics management to aid the economic and long-term use of BJF.

Keywords Greenhouse gas emissions · Biojet fuel · Conventional jet fuel · Life-cycle assessment

---

N. Bhati · A. K. Sharma (✉)
Department of Bioscience and Biotechnology, Banasthali Vidyapith, Newai, Rajasthan, India

A. Kuila (ed.), *Biojet Fuel: Current Technology and Future Prospect*, Clean Energy Production Technologies, https://doi.org/10.1007/978-981-99-8783-2_10

10.1 Introduction

The Intergovernmental Panel on Climate Change (IPCC) anticipates that earth's temperature will upsurge by 2.5–7.8 °C until the year 2100, compared to the normal for the time frame between 1850 and 1900 (Bernstein et al. 2008). The globalization of trade and travel has led to a significant increase in air transport, which contributes about 2% of all industrial sector total carbon emissions. Biofuel or electrical power are not practical aviation fuel substitutes for the foreseeable future because of the requisite energy density, though jet fuel can be employed in fuel-cells to generate electrical energy remotely for dispersed purposes. Present methods to produce jet fuel utilize fossil fuel as a feedstock and necessitate large amenities that are complicated to operate.

Mitigation of carbon release in the airline sector is necessary to overcome climate change. The airline sector has made major strides towards mitigating emissions, involving enhancements to air traffic management and jet fuel efficiency, with the goal of ensuring reliable, efficient, and sustainable flights. But less than 15% mitigation of GHG emissions was achieved (The global aviation industry 2010). The utilization of substitute aviation fuel is thought to be the most effective strategy to attain zero-carbon flight operations since it can reduce carbon emissions by an additional 50–80% in contrast to various non-fuel approaches. So as to accomplish a substantial decrease in GHG emissions, it is crucial to address the fuel source utilized by the airlines. BJF is a non-depletable alternate fuel to traditional petroleum-based aviation gas that must be permitted by the American Society for Testing and Materials (ASTM) aviation fuel norms. It can be generated from both biological and non-biological sources (Ng et al. 2021). The most widespread technique for BJF production is hydroprocessing technology, which has been accepted by the majority of test flights (Gutiérrez-Antonio et al. 2017).

Conventional jet fuel (CJF), made from petroleum, is a mixture of several paraffin hydrocarbons (HCs). Aviation fuel has a HC length that falls between that of gasoline and diesel. Aviation fuel (or paraffin) is the intermediate distillate in a traditional refinery, accounting for about 10% of the petroleum portion, while gasoline and petrodiesel are in the majority. Jet fuel is preferred as an aviation fuel over petrol as it has low volatility and density; in contrast to fossil diesel, aviation fuel is also light and less susceptible to wax at lower temperatures (Yang et al. 2019). An alternative jet fuel (AJF) must resemble CJF on account of its physical and chemical characteristics. So as to meet the high energy requirements of long-haul flights, an ideal jet fuel should have good cold stabilizing properties at temperatures between −47 °C and 40 °C and altitudes above 30,000 feet (The Engineering ToolBox 2003; Wilbrand 2018).

Jet A and Jet A-1 are the two major kerosene-based CJFs used in the industry. Jet A-1 possesses a low melting point of −4 °C, thus becoming a good option for overseas flights. A jet fuel should have a desired composition of 75–85 vol% paraffins, cycloparaffins, and iso-paraffins, with the residual 15–25 vol% being aromatics and olefins. Further crucial features involve worldwide accessibility, reasonable costs, good flow behavior, and effective combustion properties. As a "drop-in" fuel, AJF may therefore be simply incorporated into prevailing structures, enabling a smooth conversion (Tiwari et al. 2023). In comparison to CJF, which normally has ecological footprints of around 4 metric tonnes per metric tonne of fuel, a BJF should have reduced environmental impacts during its whole life-cycle (Kargbo et al. 2021).

BJF (commonly referred to as bio-aviation fuel, BAF, or sustainable aviation fuel, SAF), a form of biofuel for the aviation industry, is viewed as a brief- to mid-range approach for the complete mitigation of the air sector's GHG pollution. Yang et al. (2019) reported the normative conditions for both CJF and BJF to which producers are required to adhere. It states that the subsequent emissions characteristics of a jet operating on BJF would resemble those of Jet A-1 (Tiwari et al. 2023). However, atmospheric carbon sequestration during biomass growth establishes a closed carbon cycle, which is liberated at the termination of its life process as BJF, resulting in overall lower GHG pollution in contrast to CJF (Doliente et al. 2020). Although this renders BJF an appealing AJF choice, numerous problems emerged during its execution. It failed to gain adequate funding because of a lack of administrative and industrial support (Gegg et al. 2014), a shortage of raw materials, the indeterminate commercial viability of the production paths, and insufficient logistics management approval. The employment of different feedstocks, like lignocellulosic biomass (Cheng and Brewer 2017), vegetable oils (Vasquez et al. 2017), municipal and agrarian residues (Jiménez-Díaz et al. 2017), and microalgae (Bwapwa et al. 2018), for potential BJF production has been reported.

10.2 Biojet Fuel

BJF is a synthesized paraffinic kerosene (SPK) blend made from biomass that is added to conventional aviation fuel made from petrol (Wang et al. 2019). A list of five SPK varieties for combining (in a certain amount percent) with CJF as approved by ASTM D7566-19a (ASTM 2019) and their production pathways, comprising an overview of the method on the basis of which SPK are categorized, is shown in Table 10.1.

The hydroprocessed esters and fatty acids (HEFA) production pathway, an aviation fuel generation path from oil, creates HEFA-SPK by deoxygenation of fats and oils with subsequent hydroprocessing (Wang et al. 2019). Other oil-to-jet platforms include hydrous pyrolysis of plant or algal oil and rapid pyrolysis of cellulose, with subsequent aviation fuel advancement (Wang and Tao 2016). The gas-to-jet technology uses the Fischer-Tropsch (FT) production pathway to transform syngas from biomass gasification into paraffinic and olefinic HCs, which are then hydroprocessed to create FT-SPK. Gas-to-jet platforms can also be used to create FT-SPK/A (Fischer-Tropsch synthetic paraffinic kerosene with aromatics) with the incorporation of alkylated and bio-derived aromatics (de Medeiros et al. 2022). In the alcohol-to-jet (ATJ) synthesis path, biomass is disintegrated to acquire soluble sugars; sugar fermentation for the production of iso-butanol and ethanol; and subsequent

Table 10.1 An overview of production pathways of SPK (Yang et al. 2019; Wang and Tao 2016)

Production pathway	SPK	Outline of the method	Airline companies
Oil-to-jet	HEFA-SPK	Deoxygenation of fats and oils with subsequent hydroprocessing	Virgin Blue, Lufthansa, GE Aviation, Boeing, Virgin Atlantic, Rolls-Royce, Romanian Air, Alaska Airlines, Interjet, Air New Zealand, Continental, JAL, CFM, Pratt & Whitney, Air France, Thomson Airways, Air China, Porter Airlines
Gas-to-jet	FT-SPK	Gasification of biomass to produce syngas ($CO + H_2$); FT to synthesize paraffins and olefins, with subsequent hydroprocessing	United Airlines, British Airways, Airbus, Qatar Airways
Gas-to-jet	FT-SPK/A	Besides FT-SPK, the aromatic concentration is purposely improved by addition of alkylated & bio-based aromatics	N/A
Alcohol-to-jet	ATJ-SPK	Breakdown of biomass to produce simple sugars; sugar fermentation for production of iso-butanol and ethanol, with subsequent dehydration, oligomerization, hydrogenation, and fractionation	Virgin Atlantic, United Airlines, Boeing, Continental Airlines
Sugar-to-jet	SIP-SPK	Breakdown of biomass to produce simple sugars; sugar fermentation to obtain farnesene, with subsequent hydroprocessing and fractionation	GE, Boeing, Azul Airlines, Trip Airlines, Embraer

dehydration, oligomerization, hydrogenation, and fractionation to obtain ATJ-SPK (Ng et al. 2021). The sugar-to-jet (SIP) or direct sugar-to-hydrocarbon (DSHC) aviation fuel production comprises the breakdown of biomass to obtain simple sugars and farnesene production from sugar fermentation, followed by hydroprocessing and fractionation to obtain DSHC-SP (Doliente et al. 2020). Other SIP platforms include direct sugar to HCs and catalytic transformation of sugar or sugar intermediary products through biochemical or chemical methods, with subsequent progression to aviation fuel by aqueous phase reforming (Wang and Tao 2016).

Figure 10.1 summarizes the benefits and drawbacks of BJF; however, in order for it to be a really sustainable substitute, emissions reductions are needed during all manufacturing steps, including extraction, refining, and shipping. Additional potential benefits include a rise in employment, energy security, and cost stability. The use of BJF increases the potential development of rural areas in the form of increased employment in crop cultivation and harvesting and enhanced yields from marginal

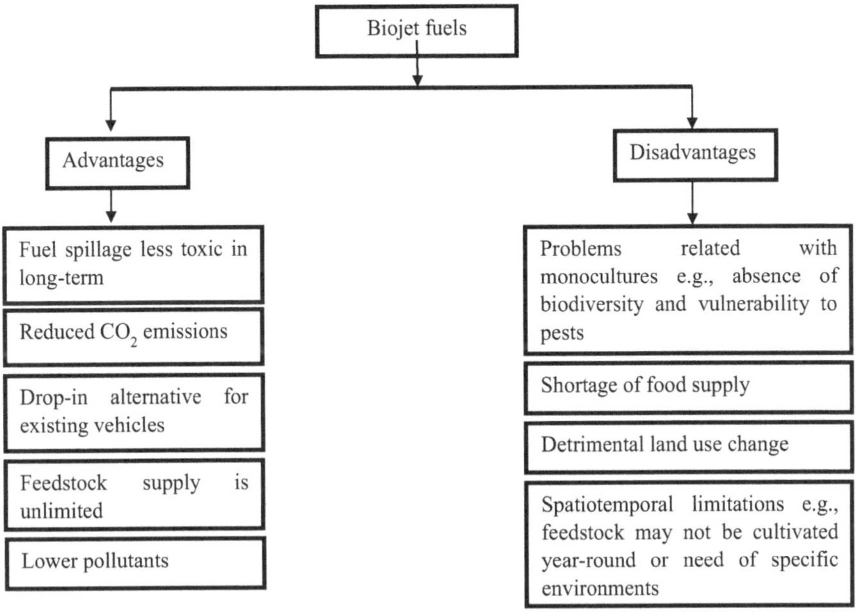

Fig. 10.1 Biojet fuels—advantages and disadvantages (Doliente et al. 2020)

non-agricultural land. Regardless of its financial advantages, BJF implementation has not received enough funding (Gegg et al. 2014). For the production pathways to be economically competitive with crude refinery production, funds in the form of incentives and governmental assistance are required (Hendricks et al. 2011).

BJF faces issues comparable to those faced by clean energy fuels overall, with the primary issue being how to guarantee that the raw materials, derived from organic debris or other renewable sources, are safe, eco-friendly, commercially viable, and sufficiently accessible to meet spatiotemporal needs (Su et al. 2015; Hendricks et al. 2011). As the airline sector, as well as the heating, chemicals, transportation, and electrical industries, work to transition from reliance on non-renewable fuels to bioenergy, their needs for similar raw materials lead to a fresh inventory competitiveness (de Jong et al. 2017).

10.3 Raw Materials for SPK Derived from Biomass

Raw materials required for BJF production can be grouped into following categories: 1st generation (1G), 2nd generation (2G), 3rd generation (3G), and 4th generation (4G) as shown in Table 10.2. A crucial parameter in selection of a raw material is its accessibility. In case of agricultural raw materials, their obtainability and potential yield are interconnected.

Table 10.2 Raw materials for biojet fuel production (Alalwan et al. 2019; Staples et al. 2018; Hari et al. 2015)

Categories	Feedstocks
1st generation (1G)	Starch and sugar crops: Potato, sweet sorghum, wheat, corn Oil seed crops: Rapeseed, soybean, sunflower, camelina, oil palm
2nd generation (2G)	Wood energy crops: Poplar, eucalyptus, willow Oil-seed energy crops: Castor beans, jatropha Grass energy crops: Miscanthus, Napier grass, switch grass Food and municipal waste: Biogenic fraction of municipal solid waste, used cooking oil, animal fats Agricultural and forestry residues: Sorghum straw, wheat straw, corn stover, wood harvesting/processing remains
3rd generation (3G)	Algae: Microalgae
4th generation (4G)	Genetically modified organisms Non-biological raw materials: Renewable electricity, water, CO_2

10.3.1 1st Generation (1G) Raw Materials

The 1G group includes consumable agricultural crops like sugarcane, corn, sugar beets, wheat, and oil palm (Lee and Lavoie 2013). These crops are harvested for their fat/oil content, sugar, and starch. The well-known HEFA technique may quickly transform these fats or oils into aviation fuel. The newly developed DSCH technology can be used for the processing of starch or sugar. The USA is highly interested in ATJ, another new technology, because they have an abundant supply of 1G ethanol made from corn (Radich 2015). Although corn utilizes water effectively, the total amount that needs to be grown will lead to a high water demand and more fertilizer consumption.

Growing agriculture can put a burden on a nation's water supply and can lead to eutrophication and other water-related problems like scarcity. These are the major disadvantages of using 1G raw materials because the majority of agricultural crops have higher nutrient and water requirements. Conflict for water, space, and energy sources with food production is another major obstacle to the production of 1G raw materials (Moioli et al. 2018). The practical solution to the shortage of land resources has been to expand into forestland, but this has resulted in deforestation and the loss of biodiversity (Paschalidou et al. 2016). The cultivation of oil palm, a dependable agricultural crop and potential BJF raw material, has also remained connected to these negative effects (Ayompe et al. 2021; Meijaard et al. 2020; Vijay et al. 2016).

10.3.2 2nd Generation (2G) Raw Materials

The nourishment versus energy source conundrum of 1G raw materials can be overcome by non-edible 2G biomass resources (Liu et al. 2021). They are categorized into two primary categories: waste biomass and biofuel crops. Further, waste feedstock is divided into food and municipal wastes as well as agricultural and forestry residues. Despite the categorization, 2G raw materials are either sugar- or oil-rich in nature. However, the sugars of 2G feedstocks, in contrast to 1G crops, are encased in the hard and resistant lignocellulosic network of cell walls of plants, necessitating pretreatment with thermochemical transformations, enzymes, and microorganisms for biofuel production (Cavelius et al. 2023; Yu et al. 2022).

The primary problems with the use of 2G feedstocks are the practical obstacles and excessive prices of these conversion processes (Alalwan et al. 2019). Nevertheless, lignocellulosic 2G feedstocks are a possible substitute for 1G crops due to their relative availability and low use competition (Nazari et al. 2021). Utilizing residual biomass also provides even better advantages, including the development of sustainable development, the elimination of waste, and conservation of the environment (Okolie et al. 2021; Richter et al. 2018). Up until this point, the generation of biofuels for land transportation from 2G raw materials still falls behind 1G raw materials (Doliente et al. 2020). Millinger et al. (2017) estimated that for land transportation in the long-term, liquid biofuels from 1G raw materials will be more economical compared to those from 2G raw materials, whereas gaseous biofuels produced from 2G raw materials for gas-operating transports are expected to be the more resource- and cost-competitive choice in the near future. However, as gaseous fuels are impractical for the aviation industry, liquid biofuels from 2G feedstocks may develop to be more significant (Millinger et al. 2017). However, it must be established that there is a sufficient, reliable, and cost-effective supply of 2G feedstocks.

Various wood energy crops and grass have been suggested as 2G raw materials for BJF production by biochemical and thermochemical methods (Wang et al. 2019). Grass energy crops are desirable for biofuel production because of their significant lignocellulosic percent and widely accessible harvesting technologies (Herr et al. 2016; Schorling et al. 2015). Wood may be a better feedstock source than grasses because of their higher biomass accessibility per area and cheaper transport expenses (Lu and El Hanandeh 2017; Murphy et al. 2015). Typically, short-rotation coppices are employed for the farming of woody energy crops for the production of biofuel. These trees grow quickly and are harvested after a cycle or rotation of about 10 years. Additionally, in times of drought, short-rotation coppices can replenish the limited availability of grass energy crops (Doliente et al. 2020).

Residual biomass may be a better alternative to energy crops as it requires fewer space (as they are generated from domestic, industrial, agro-forestry, and commercial sectors), very little economic significance, and less water footprints than cultivated plants (Rao and Rathod 2019; Mathioudakis et al. 2017; Caicedo et al. 2016). The primary category of waste feedstock includes a variety of agrarian and forest leftovers. Usually, these are lignocellulose-containing waste materials from agriculture, reaping, woodcutting, and postharvest processes like grinding, breaking, and processing wood (Doliente et al. 2020; Ali et al. 2019). There are existing practices accessible to turn this waste biomass into aviation fuels, for instance, butanol, pyrolysis oil, ethanol, and syngas (Devi et al. 2022; Ren et al. 2022; Karthick and Nanthagopal 2021; Pandiyan et al. 2019; de Corato et al. 2018). Initiatives to produce BJF from agro-forestry waste using direct sugar-to-farsenene and isobutanol-to-jet processes have been described (AviationPros 2015). The only waste stream with a viable purpose in the aviation sector at the moment is low-cost UCO (used cooking oil) (Roth et al. 2018). Jet fuel that is generated from or combined with UCO has been employed in a lot of experimental and passenger flights (Yang et al. 2019). Recently, Boeing flights in China have used jet fuel blends that contain UCO from homes and eateries that end up in the drains (Karmee 2017).

10.3.3 3rd Generation (3G) Raw Materials

Microalgae as a source of energy promises great output and the accessibility of fatty acids easily changeable to BJF through HEFA (Elkelawy et al. 2022; Martinez-Villarreal et al. 2022). Additionally, there is an increase in the development of thermochemical approaches using pyrolysis and hydrothermal liquefaction processes to streamline as well as expand the manufacturing routes (Ağbulut et al. 2023). Thus, microalgae are widely acknowledged for producing bioenergy on a huge scale. Despite substantial investment in algae biofuels, there are still many practical and technical problems (Martinez-Villarreal et al. 2022; Warshay et al. 2011).

Challenges faced in the production, processing, and extraction of oil processes that are currently ineffective and costly in terms of capital and resources, as well as suppressive environmental effects, hinder industrialization (Sudhakar et al. 2019; Muhammad et al. 2021; Goh et al. 2019). There have been several test and pilot microalgae production facilities, as well as demonstration flights that used jet fuel derived from algae, but as of now, no such production is still economically viable (Lim et al. 2021; Bwapwa et al. 2018; Martinez-Villarreal et al. 2022).

10.3.4 4th Generation (4G) Raw Materials

Cyanobacteria, microalgae, yeast, and fungi are a few examples of genetically modified organisms that have exaggeratedly increased oil and sugar yields and negative carbon capabilities that are still in the early phases of study (Mat Aron et al. 2020). Despite the fact that they have a promising future as biofuels, more research is required on the wellness and ecological concerns that these microbes can represent, as well as their control and reduction measures when they are introduced into global supply networks (Shokravi et al. 2021).

Non-living raw materials, such as renewable electricity, CO_2, sunlight, and water, may be a more eco-friendly choice, particularly when exhaust gases from power plants are used (ATAG 2017). One method is power-to-liquid (PtL), which includes breaking water into hydrogen and oxygen using an electrolyzer powered by renewable energy and then combining the hydrogen with CO_2/CO to make BJF. According to environmental and techno-economic assessments, in the near future, PtL fuels (influenced by the cost of green energy) will be costlier than CJF. Though, in the long-term, ecological profits of PtL fuels such as zero carbon and a lower requirement of space and water, as well as enhancements in financial profits, can possibly compete with the monetary profits of CJF (Dieterich et al. 2020).

Another method is to split water and CO_2 using concentrated solar energy to create syngas, which is a precursor to the production of BJF (Richter et al. 2018). Considering the case that both pathways are still in the initial stages of study, Richter et al. (2018) found two European programs, SOLAR-JET and Sunfire, that proved the generation of aviation fuel with solar energy, water, and CO_2. Even though research on supply networks of 4G raw materials has been restricted thus far, Mesfun et al. (2017) used a spatiotemporal MILP model for the combination of PtL and power-to-gas (PtG) processes in an Alpine energy supply. Relying on the cost of carbon and non-renewable fuels, the research established that green energy sources are more adaptable when combined with PtL and PtG processes, as they transform the additional erratic renewable energy into fuels as well as allow the usage of substantial quantities of captured CO_2 (0.20–15 million metric tonnes annually) via fuel generation. The commercial maturity of these technologies ensures that BJF from 4G feedstocks becomes the most eco-friendly technology with the possibility of zero GHG emissions and connecting the heating, power, and airline industries (Mesfun et al. 2017).

10.4 Production Pathways for SPK

There have been significant advancements in the study of BJF production pathways, and a few have been given approval for commercial employment. In the sections that follow, HEFA, FT, ATJ, and hydroprocessing of fermented sugars (HFS) are discussed and contrasted.

Life-cycle analysis (LCA) is a procedure that enables the evaluation of the ecological effects instigated both on the environmental and human wellness of an organization (Wang et al. 2019). The LCA system boundary determines the manufacturing procedures that are involved, and all feed-in and results from every method or stage are comprised (Pan et al. 2018). The fundamental processes covered by the LCA are demonstrated in Fig. 10.2.

10.4.1 HEFA

In order to produce aviation fuel, the HEFA method involves the use of pyrolysis oil, animal fats, waste cooking oil, algal oil, and vegetable oils in hydroprocessing. In the HEFA process, the fats and waste oils used are derived from sustainable sources. It is also possible to identify appropriate and sustainable feedstocks for specific nations based on their geographic and commercial characteristics.

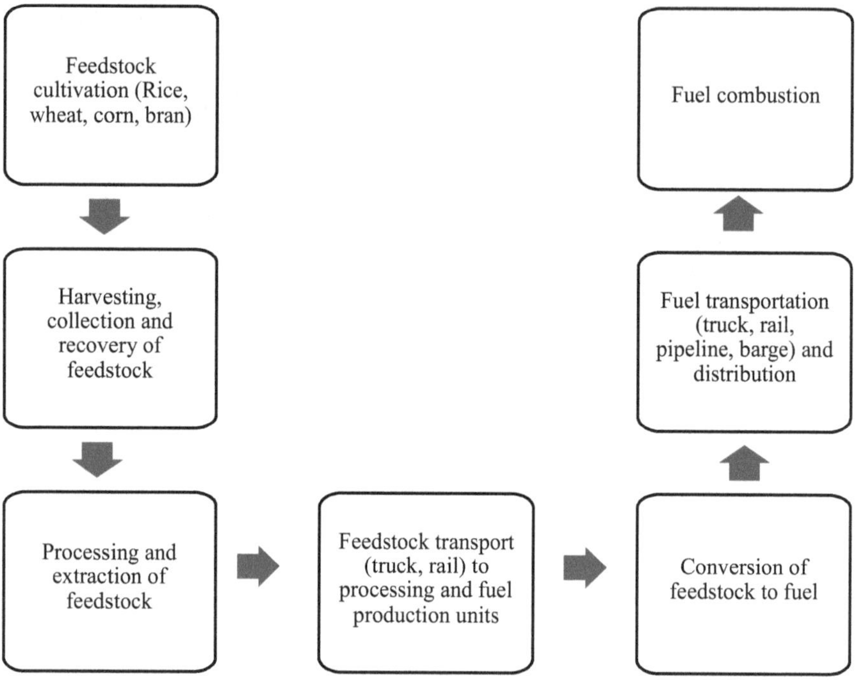

Fig. 10.2 Life-cycle steps (ICAO 2019)

The initial steps in the process involve oil extraction from oil-rich biomass. Unsaturated fatty acids and glycerides in the extracted oil must be converted to saturated triglycerides using a catalytic hydrogenation procedure in order to eliminate the double bond. The pressure of 0.7–4 bar in the presence of nickel as a catalytic agent at 150–220 °C is required for the hydrogenation process, whereas in the presence of palladium and platinum as catalysts, a low temperature of 80–120 °C can also be used. Triglyceride can be thermally hydrolyzed into 1 molecule of glycerol and 3 molecules of free fatty acids (FFA), and glycerol can then be further transformed into propane by adding hydrogen (Alenezi et al. 2010). The FFA undergoes either a hydrodeoxygenation (HDO) or a decarboxylation (DCO) process to remove the oxygen, resulting in the production of octadecane ($C_{18}H_{38}$) and heptadecane ($C_{17}H_{36}$), respectively. The key distinction among both processes is that the prior one needs 9 mol of hydrogen and produces water as a byproduct, whereas the other produces CO_2. The HDO process requires a substantial hydrogen expenditure rate at high pressure. A heterogeneous catalytic agent, for instance, sulfided NiMo and CoMo maintained on alumina, is commonly used in this process, which is normally conducted at temperatures between 300 °C and 600 °C (Seo et al. 2022; Huber et al. 2006).

An alternative treatment for eliminating the oxygen concentration in FFA is DCO, which rejects CO_2 rather than H_2O as in HDO. The benefit of DCO is that it performs effectively at low pressure, resulting in lesser hydrogen expenditure

(Rahmawati et al. 2023). Although straight-chain paraffins ($C_{18}H_{38}$ from HDO or $C_{17}H_{36}$ from DCO) are formed, the end outcomes fail to match the requirements for aviation fuel implementation regarding flash point, cloud point, and freeze point (Tao et al. 2017; Wang 2016). Thus, straight-chain paraffins undergo additional processing in a hydroisomerization process to produce branched-chain paraffins with the goal of reducing the freeze point to comply with the aviation fuel criteria (Wang 2016). SPK, which has a carbon chain length reaching from C_9 to C_{15}, is created by the hydrocracking reaction, which can happen either sequentially or simultaneously with hydroisomerization (Wang 2016). The corporations producing HEFA fuel include Dynamic Fuels, Neste Oil, UOP, and AltAir.

10.4.1.1 LCA

Numerous research teams have established an LCA for GHG emissions (Bailis and Baka 2010). The soybean yield, liming emissions, N_2O emissions from fertilizer, and H_2 supplies in the hydrotreating method all contribute to the soybean oil's GHG emissions, which range from 40% to 80% of those of traditional aviation fuel (89 gCO_2e/MJ for Jet A fuel) (Hileman et al. 2009). On account of land use change, there is a rise in GHG emissions. According to one study, 800% more emissions were obtained from low soybean yields in tropical rainforests compared to traditional aviation fuel. In comparison to traditional aviation fuel production methods, the pollutants of the palm oil to aviation fuel method are around 30–40%, as a result of the methane releases from palm oil-mill sewage management, H_2 necessities in the hydrotreating procedure, palm fresh-fruit-bunch yield per acre, hydroprocessing fuel yield, and farming energy. Upon consideration of land use change, emissions of GHG rise to the range of 40–800% for CJF (Stratton et al. 2010).

The pollutants from rapeseed oil are around 45–87% of those from CJF and rise to 87–147% when taking land use change into account. The pollutants from jatropha oil are around 36–52% of those from CJF, and N_2O emissions signify total emissions by more than 20%. The supposition that peripheral land will be exploited results in zero pollutants from land use change. The emissions for algae oil fall between 16% and 220% of those for CJF. In comparison to using conventional fuels, GHG emissions are cut by 45% when H_2 is produced from fossil gas and biochar is utilized to sustain the process energy. The GHG pollutants are decreased by 103% compared to CJF when H_2 is made from improved pyrolysis oil and biochar is used as fertilizer (Elgowainy et al. 2012).

10.4.2 FT

FT can be used in combination with a number of biomass conversion techniques, including pyrolysis, gasification, and liquefaction, to generate synthetic fuel. This chapter will emphasize the gasification-FT pathway, as it is the approved and marketable pathway for the production of aviation fuel. Gasification is the process of turning carbonaceous resources, like biomass, into syngas at a high temperature, usually exceeding 1000 °C. The main components of syngas are carbon monoxide and hydrogen, which are necessary for producing FT liquid (Gogulancea et al. 2023). FT synthesis creates a variety of HCs with diverse carbon chain lengths, for instance light HCs (C_1–C_4), which are light gases and can be employed straightly in gas turbines to produce heat and power or refined into LPG; naphtha (C_5–C_{10}) and kerosene (C_{10}–C_{16}), which can be mixed into petrol and aviation gas; distillate (C_{14}–C_{20}), which can be processed into diesel fuel; and waxes (C_{20}^+), which can be hydrocracked to form diesel (Kargbo et al. 2022).

The major key benefit of FT liquid is that it is entirely devoid of sulfur and comprises fewer aromatics than petrol and diesel, which leads to lower environmental contamination (Martinelli et al. 2020). This parameter, as well as the necessity to prevent catalyst poisoning, infers that the demand for the feedstock for FT synthesis is stricter. Thus, the syngas should be treated to eliminate any particles, tars, sulfur- and nitrogen-comprising complexes, and additional contaminants to prevent equipment fouling (Alcazar-Ruiz et al. 2022). Syngas cleanup continues to be a key challenge for the unified system of biomass gasification with FT synthesis, and further study is required to guarantee that the FT feed is cleaned to an acceptable quality while achieving significant cost reduction (dos Santos and Alencar 2020).

FT synthesis has been used for commercial purposes primarily by Shell, which uses syngas derived from natural gas, and Sasol, which uses syngas derived from coal (Ail and Dasappa 2016). The majority of biomass gasification-FT techniques remained at the experimental stage, for instance, the Syndi'ese-BtS project by CEA/Air Liquide and the Total in France and Velocys/Red Rock Biofuels projects in Austria and U.S. (Ng et al. 2021).

10.4.2.1 LCA

The following GHG emissions are taken into account in the syngas and fuel production methods (Marano and Ciferno 2001): (a) emissions of CO_2 from gasification, FT synthesis, traditional fuel burning, and emitting from fossil gas manufacture; (b) emissions of methane from fleeting plant and pipeline discharges, partial burning, and coal bed methane discharges; and (c) emissions of N_2O from fuel burning and biomass cultivation. The FT coal-to-liquid (CTL) and FT biomass-to-liquid (BTL) have substantial life-cycle GHG emissions (Xie et al. 2011). In comparison to traditional jet fuel with carbon capture, the FT CTL process emits GHGs at a rate that is 10% higher with carbon sequestration and 120% higher without carbon sequestration.

Due to the very low contribution of the raw materials for FT BTL, either wood-derived biomass or forest remains, emissions of GHG from the FT BTL procedure are between 92% and 95% lower than those of traditional jet fuel. This is also true because the biomass itself provides 48% of the energy needed for the transformation procedures, such as gasification or FT synthesis. The life-cycle GHG emissions from the usage of switchgrass, corn stover, and forestry remains are 2.0, 9.0, and

12.2 gCO_2e/MJ, respectively (with a soil carbon-change credit, the CO_2 capture resulting from land use change leads the emissions of GHG), which are around 2%, 10%, and 14% of those of CJF, respectively (Stratton et al. 2010).

Additionally, it is proposed that finding techniques and developments, for instance, carbon capture, joint power and fuel production, joint coal and biomass processing, and enhanced vehicle technique, will aid in lowering GHG and other pollutants from the FT method (Taylor et al. 2011). In comparison to FT BTL, hydroprocessed renewable jet fuel has about 62–92% higher GHG emissions because of the higher use of chemicals and fertilizer (Agusdinata et al. 2011). Studies have been done on the impact of carbon sequestration and storage on the environment, which is calculated based on the radiative forcing of the production of FT fuel and use chains based on GHG emissions (Holmgren and Hagberg 2009). It is determined that the climate effects of FT gas from peat are 30–40% less than from CJF without carbon sequestration and storage. The environmental effects of peat-based FT gases are 50–84% less than those of CJF with carbon capture and storage.

10.4.3 ATJ

In the ATJ process, shorter chain alcohols (like ethanol, butanol, and methanol) are converted into longer chain HCs (C_8–C_{16} alkane). There are two main methods for turning alcohol into aviation fuel: (1) methanol to olefins (MTO), and then Mobil's olefin to gasoline/distillate (MOGD); (2) processing butanol, ethanol, isobutanol, and other alcohols through dehydration, oligomerization, and hydrogenation. Biochemical processes like fermentation (Martinez Hernandez and Ng 2018) as well as thermochemical methods like gasification and pyrolysis (Ng and Sadhukhan 2011) can be used for the production of alcohol from biomass. A growing trend in emerging technology is the production of alcohol using microbial synthesis (Soleimani et al. 2017; Lan and Liao 2013).

Methanol can be transformed into aviation fuel through MTO and, subsequently, MOGD, technologies developed by ExxonMobil. Methane (1.4 wt%), C_2–C_4 paraffins (6.5 wt%), C_2–C_4 olefins (56.4 wt%), and C_5–C_{11} petrol (35.7 wt%) are produced from methanol when it is fed to the MTO fluidized bed reactor, which functions at 482 °C and 1 bar with ZSM-5 as a catalyst (Baliban et al. 2013). The olefin fractionation unit fractionates this produced slate from the MTO unit to produce light gases, petrol, and olefins. The product yield is enhanced by recycling light gases into the MTO unit. In the fractionation column, gasoline is separated as the only product. Olefins are further processed in the MOGD unit, a fixed-bed reactor running at 400 °C and 1 bar in the presence of a ZSM-5 catalyst (Baliban et al. 2013). Distillate (82 wt%), gasoline (15 wt%), and light gases (3 wt%) are the products from the MOGD unit. As a result of the integration of MTO and MOGD, the MOGD fractionation unit produces portions of light gases (C_1–C_4), petrol (C_5–C_{11}), aviation fuel (C_{11}–C_{13}), and diesel (C_{14}^{+}).

The production of jet fuels from alcohol can be accomplished through dehydration, oligomerization, and hydrogenation. Firstly, the alcohol is dehydrated to form alkenes at a temperature of 288–343 °C and a pressure of <14 bar (Ashok et al. 2019). The dehydration process can be aided by the use of acidic catalysts, for instance, γ-type zeolites, alumina-based catalysts, amberlyst acidic resins, and ZSM-5 zeolites (Geleynse et al. 2018). The next step is the oligomerization procedure, which combines alkene molecules to create longer-chain HCs like dimers, trimers, and tetramers at a temperature of 100 °C with the aid of a Nafion or Amberlyst-35 catalyst (Harvey and Quintana 2010). Mostly, dimers are recycled to produce trimers and tetramers with better yields, which provide C_{12}–C_{16} olefins for aviation fuel. In the final phase, hydrogenation, olefins are saturated to form paraffinic kerosene using an external hydrogen supply and a PtO_2 catalyst (Harvey and Quintana 2010).

The businesses that manufacture the ATJ gas are Gevo, UOP, Coskata, Cobalt/Navy, LanzaTech, and BRI. Lufthansa signed a pact with Gevo in 2014 to analyze and assess their ATJ gas for use in commercial flights. This demonstrates the expanding attention of ATJ to the aviation fuel sector.

10.4.3.1 LCA

LCA studies for the ATJ method include a strong emphasis on ethanol (Pereira et al. 2019), *n*-butanol (Li et al. 2016), and iso-butanol (Tao et al. 2014) production. The LCA can be classified into four groups for alcohol fuel production: (a) raw materials (land-use-change), (b) on-site enzyme production, (c) biorefinery process, and (d) biorefinery co-product credits. When alternate routes are taken into account, the performance of the thermochemical process and the biochemical process differ marginally in terms of fossil fuel usage, GHG emissions, and water use (Mu et al. 2010). Studies on the conversion of *n*-butanol and iso-butanol have concentrated on fossil fuel consumption, consumptive water usage, emissions, and potential global warming. In comparison to the iso-butanol production method, the *n*-butanol manufacturing procedure emits more direct emissions, including sulfur dioxide, carbon dioxide, and NO_2 (Tao et al. 2014).

Cellulase/enzyme production and cellulase seed fermentation both result in CO_2 production; however, the main source is combustion. Diatomic nitrogen in the combustion air undergoes a high-temperature oxidation reaction to produce nitrogen dioxide. The quantity of sulfuric acid utilized through the pretreatment procedure has an important influence on sulfur dioxide emissions. But *n*-butanol biorefining requires more water consumption than iso-butanol refining. Biomass feedstocks are accountable for the majority of the possibilities for global climate change and fossil fuel usage (Tao et al. 2014). In comparison to *n*-butanol conversion, iso-butanol conversion uses 5.15 MJ/GGE more natural gas. Future research should pay greater attention to the LCA of the ATJ fuel progression methods, as it is still unidentified.

10.4.4 HFS

The procedure comprises (a) a pretreatment stage for the separation of sugars from lignin; (b) sugars are converted into farnesene ($C_{15}H_{24}$) via enzymatic hydrolysis and fermentation; (c) recovery of farnesene from solid-liquid separation; and (d) hydroprocessing to farnesene ($C_{15}H_{32}$), the BJF. This method, marketed by Amyris and Total, employs a *S. cerevisiae* strain (PE-2) in the fermentation procedure to produce farnesene through the mevalonate pathway (Ng et al. 2021; Jiménez-Díaz et al. 2017). Farnesene can be produced with a yield of around 16.8% and a productivity of 16.9 $gL^{-1}d^{-1}$, with a 95% recovery rate post-separation. This HC fuel was approved by ASTM in 2014 and can be blended up to 10% with CJF. As part of the MegaBio project 2014, Amyris is now working on an integrated DSHC with the goal of obtaining 2L^{-1} of farnesene (Ng et al. 2021).

10.4.4.1 LCA

A study was performed by the Institute for International Trade Negotiations on the life-cycle GHG emissions of BJF synthesized from sugarcane sugars on the basis of Amyris procedure features (Nassar et al. 2012). According to the findings, life-cycle emissions of GHG are approximately 82% lower when compared to standard Jet A/A-1 fuels at 15 gCO_2e/MJ (Nassar et al. 2012). Though the previous study did not take into account that sugarcane could lead to land use change. According to the investigation, the emissions of GHG related to sugarcane synthesis and transportation are mainly driven by agricultural inputs and N_2O pollutants from the soil, which generate 32 gCO_2e/MJ and 45 gCO_2e/MJ, respectively (Total 2012). This study advances our understanding of the environmental impact of a sugarcane-based sustainable jet fuel.

10.5 International Initiatives and Policies

Pertinent shareholders in the airline industry have made significant efforts to address climate change-related issues. Air Transport Action Group (ATAG) has established a set of objectives involving 1.5% fuel effectiveness enhancement annually from 2009 to 2020 (this has been exceeded with an average of 2.1% attained) and maintaining emissions of CO_2 via zero carbon growth from 2020 to 50% mitigation in emissions of CO_2 by 2050 on the basis of the 2005 level. By reducing and offsetting pollutants from the airline industry, the Carbon Offsetting and Reduction Scheme (CORSIA) would enable the global airline sector to attain carbon neutrality by the year 2020, as mandated by the International Civil Aviation Organization (ICAO). Over 87.7% of global airline activities are dedicated to attaining considerable reductions in emissions, as shown by the fact that 70 nations intend to actively engage in the international market-based measure (MBM) plan put in place in May 2017 (ICAO 2016).

Airlines are required by the ICAO to purchase emission credits (such as renewable energy) that are qualified for offset requirements from the carbon markets. The initial voluntary shift will occur between 2021 and 2027; however, it will turn out to be official in 2027. This opens up yet another possibility for BJF to develop into an advantageous fuel substitute for commercial aircraft, assisting them in meeting the annual emission allowance. For nations that adopted the Kyoto Protocol, this regulation has binding legal effect. CORSIA is adopting a route-oriented method where all operators on a similar path will have similar compliance compulsions. Although ICAO continues to work on the execution of the proposal for assessing and balancing restrictions, CORSIA has established a timeframe for the implementation of the latest proposal for the mitigation of carbon emissions. The proposed timeframe has three parts, with the voluntary participation phases being phase I in 2021–2023 and phase II in 2024–2027. Between 2028 and 2035, the aviation sector will be mandated to adhere to the carbon offset criteria, and this will be implemented by mandating compensation for any additional carbon emissions produced throughout overseas flights that exceed 2020 levels.

The Paris Agreement unifies countries working to address climatic changes inside the United Nations Framework Convention on Climate Change (UNFCCC), which was prepared in December 2015, approved by 195 countries globally in 2016, and went into force in 2016. Emissions Trading Schemes (ETSs) for emissions of GHG are implemented in a number of nations and areas (Talberg and Swoboda 2013). The European Union Emissions Trading Scheme (EU ETS) is now a widely used procedure. EU ETS launched Phase I (2005–2007) in January 2005 and became one of many choices that enabled the assessment and encouraged the decrease of GHG pollutants. Additionally, the Environmental Protection Agency (EPA) has decided that emissions from jets are associated with climatic changes, and it is anticipated that they will eventually move through with limits that are at least as strict as the ICAO's norms (US Environmental Protection Agency 2016).

10.6 Sustainability of BJF

The economic analysis will be crucial for determining the process and incentive as well as for comprehending the implications of using BJF. Alternatively, subsidies are types of economic sustenance provided by the government for activities that are thought to be eco-friendly. A subsidy encourages a polluter to reduce emissions rather than penalizing them (Noh et al. 2016). Although familiarity with BJF is seen as necessary for making wise decisions in the airline sector, the aviation industry and authorities need to refocus their efforts in order to advance this new alternative energy and foster eco-friendly development. The value necessary to progressively connect with different airline divisions, the organizational strategy, structure, and systems, as well as the outcomes and feedback (including any increase or decrease in cost). It is significant to be aware of the positive influences that using bioenergy has on engine operation and procedures in order to support the global aviation industry. In fact, it is acknowledged that a successful application of BJF around the world requires the assistance of the policy and regulatory environment. Future studies should examine the how, when, and by whom of initiatives to increase the significance of BJF for the sustainability of the aviation sector.

10.7 Conclusion

The development of BJF needs to be accelerated urgently to achieve the carbon-neutral emission goal in the airline industry. Airline fuel made from bio-resources has the ability to replace traditional fossil-based aviation fuel, meeting the needs of the rapidly expanding aviation market while lowering GHG emissions. This chapter discusses several facets of BJF development, involving the inspiration of substituting fossil-to-biomass-derived aviation fuel, possible bio-renewable raw materials, an outline of approved pathways in BJF production and their life-cycle assessment, international policies, and worldwide initiatives. The choice of raw materials and techniques for BJF synthesis must be rationalized on the basis of cost of production and impact on the environment, whereas avoiding competition with the present marketplace for road transportation of biofuels. To increase the cost-effectiveness of BJF production, additional study and improvement must concentrate on optimizing, incorporating, and increasing BJF technology. To accelerate the adoption of BJF, multistakeholder cooperation must be encouraged alongside government action through policy assistance.

References

Ağbulut Ü, Sirohi R, Lichtfouse E, Chen WH, Len C, Show PL, Le AT, Nguyen XP, Hoang AT (2023) Microalgae bio-oil production through pyrolysis and hydrothermal liquefaction: mechanism and characteristics. Bioresour Technol 376:128860

Agusdinata DB, Zhao F, Ileleji K, DeLaurentis D (2011) Life cycle assessment of potential biojet fuel production in the United States. Environ Sci Technol 45(21):9133–9143

Ail SS, Dasappa S (2016) Biomass to liquid transportation fuel via Fischer Tropsch synthesis–technology review and current scenario. Renew Sust Energ Rev 58:267–286

Alalwan HA, Alminshid AH, Aljaafari HA (2019) Promising evolution of biofuel generations. Subject review. Renew Energy Focus 28:127–139

Alcazar-Ruiz A, Ortiz ML, Dorado F, Sanchez-Silva L (2022) Gasification versus fast pyrolysis bio-oil production: a life cycle assessment. J Clean Prod 336:130373

Alenezi R, Baig M, Wang J, Santos R, Leeke GA (2010) Continuous flow hydrolysis of sunflower oil for biodiesel. Energy Sources Part A Recov Utilizat Env Effects 32(5):460–468

Ali MH, Zailani S, Iranmanesh M, Foroughi B (2019) Impacts of environmental factors on waste, energy, and resource management and sustainable performance. Sustainability 11(8):2443

Ashok B, Nanthagopal K, Saravanan B, Azad K, Patel D, Sudarshan B, Ramasamy RA (2019) Study on isobutanol and Calophyllum inophyllum biodiesel as a partial replacement in CI engine applications. Fuel 235:984–994

ASTM (2019) ASTM D1655-19a, Standard Specification for Aviation Turbine Fuels. Pennsylvania, USA, American Society for Testing and Materials (ASTM)

ATAG (2017) Beginner's guide to sustainable aviation fuel. Geneva: air transport action group (ATAG), 1–24. Available online at https://aviationbenefits.org/media/166152/beginners-guide-to-saf_web.pdf. Accessed 23 July 2019

AviationPros (2015) Gevo's jet fuel to be used in first ever test flight flown on fuel derived from wood waste. Available online at https://www.aviationpros.com/contact-us. Accessed 29 July 2019

Ayompe LM, Schaafsma M, Egoh BN (2021) Towards sustainable palm oil production: the positive and negative impacts on ecosystem services and human wellbeing. J Clean Prod 278:123914

Bailis RE, Baka JE (2010) Greenhouse gas emissions and land use change from Jatropha curcas-based jet fuel in Brazil. Environ Sci Technol 44(22):8684–8691

Baliban RC, Elia JA, Floudas CA, Xiao X, Zhang Z, Li J et al (2013) Thermochemical conversion of duckweed biomass to gasoline, diesel, and jet fuel: process synthesis and global optimization. Ind Eng Chem Res 52(33):11436–11450

Bernstein L, Bosch P, Canziani O, Chen Z, Christ R, Riahi K (2008) IPCC, 2007: climate change 2007: synthesis report

Bwapwa JK, Anandraj A, Trois C (2018) Conceptual process design and simulation of microalgae oil conversion to aviation fuel. Biofuels Bioprod Biorefin 12(6):935–948

Caicedo M, Barros J, Ordás B (2016) Redefining agricultural residues as bioenergy feedstocks. Materials 9(8):635

Cavelius P, Engelhart-Straub S, Mehlmer N, Lercher J, Awad D, Brück T (2023) The potential of biofuels from first to fourth generation. PLoS Biol 21(3):e3002063

Cheng F, Brewer CE (2017) Producing jet fuel from biomass lignin: potential pathways to alkyl-benzenes and cycloalkanes. Renew Sust Energ Rev 72:673–722

De Corato U, De Bari I, Viola E, Pugliese M (2018) Assessing the main opportunities of integrated biorefining from agro-bioenergy co/by-products and agroindustrial residues into high-value added products associated to some emerging markets: a review. Renew Sust Energ Rev 88:326–346

Devi A, Bajar S, Kour H, Kothari R, Pant D, Singh A (2022) Lignocellulosic biomass valorization for bioethanol production: a circular bioeconomy approach. Bioenergy Res 15(4):1820–1841

Dieterich V, Buttler A, Hanel A, Spliethoff H, Fendt S (2020) Power-to-liquid via synthesis of methanol, DME or Fischer–Tropsch-fuels: a review. Energy Environ Sci 13(10):3207–3252

Doliente SS, Narayan A, Tapia JFD, Samsatli NJ, Zhao Y, Samsatli S (2020) Bio-aviation fuel: a comprehensive review and analysis of the supply chain components. Front Energy Res 8:110

Elgowainy A, Han J, Wang M, Carter N, Stratton R, Hileman J, Malwitz A, Balasubramanian S (2012) Life-cycle analysis of alternative aviation fuels in GREET (No. ANL/ESD/12-8). Argonne National lab. (ANL), Argonne (United States)

Elkelawy M, Bastawissi HAE, Radwan AM, Ismail MT, El-Sheekh M (2022) Biojet fuels production from algae: conversion technologies, characteristics, performance, and process simulation. In: Handbook of algal biofuels. Elsevier: Amsterdam, Netherlands, pp 331–361

Gegg P, Budd L, Ison S (2014) The market development of aviation biofuel: drivers and constraints. J Air Transp Manag 39:34–40

Geleynse S, Brandt K, Garcia-Perez M, Wolcott M, Zhang X (2018) The alcohol-to-jet conversion pathway for drop-in biofuels: techno-economic evaluation. ChemSusChem 11(21):3728–3741

Gogulancea V, Rolfe A, Jaffar M, Brandoni C, Atsonios K, Detsios N, Dieringer P, Huang Y (2023) Technoeconomic and environmental assessment of biomass chemical looping gasification for advanced biofuel production. Int J Energy Res 1–17

Goh BHH, Ong HC, Cheah MY, Chen WH, Yu KL, Mahlia TMI (2019) Sustainability of direct biodiesel synthesis from microalgae biomass: a critical review. Renew Sust Energ Rev 107:59–74

Gutiérrez-Antonio C, Gómez-Castro FI, de Lira-Flores JA, Hernández S (2017) A review on the production processes of renewable jet fuel. Renew Sust Energ Rev 79:709–729

Hari TK, Yaakob Z, Binitha NN (2015) Aviation biofuel from renewable resources: routes, opportunities and challenges. Renew Sust Energ Rev 42:1234–1244

Harvey BG, Quintana RL (2010) Synthesis of renewable jet and diesel fuels from 2-ethyl-1-hexene. Energy Environ Sci 3(3):352–357

Hendricks RC, Bushnell DM, Shouse DT (2011) Aviation fueling: a cleaner, greener approach. Int J Rotating Mach:1–13

Herr A, Braid A, Carter J, McIvor J, Murphy HT, O'Connell D, Raisbeck-Brown N, Poole M (2016) Cut your grass and eat it too–is aviation biofuel production and grazing in the Australian tropics possible? Renew Sust Energ Rev 53:1377–1388

Hileman JI, Ortiz DS, Bartis JT, Wong HM, Donohoo PE, Weiss MA, Waitz IA (2009) Near-term feasibility of alternative jet fuels (No. Partner-COE-2009-001). Rand Corporation

Holmgren K, Hagberg L (2009) Life cycle assessment of climate impact of Fischer-Tropsch diesel based on peat and biomass. IVL Svenska Miljöinstitutet

Huber GW, Iborra S, Corma A (2006) Synthesis of transportation fuels from biomass: chemistry, catalysts, and engineering. Chem Rev 106(9):4044–4098

ICAO (2016) Carbon offsetting and reduction scheme for international aviation (CORSIA)

ICAO (2019) CORSIA supporting document: eligible fuels – life cycle assessment methodology. International Civil Aviation Organization, Montreal

Jiménez-Díaz L, Caballero A, Pérez-Hernández N, Segura A (2017) Microbial alkane production for jet fuel industry: motivation, state of the art and perspectives. Microb Biotechnol 10(1):103–124

de Jong S, Hoefnagels R, Wetterlund E, Pettersson K, Faaij A, Junginger M (2017) Cost optimization of biofuel production–the impact of scale, integration, transport and supply chain configurations. Appl Energy 195:1055–1070

Kargbo H, Harris JS, Phan AN (2021) "Drop-in" fuel production from biomass: critical review on techno-economic feasibility and sustainability. Renew Sust Energ Rev 135:110168

Kargbo HO, Ng KS, Phan AN (2022) Evaluation of the economic feasibility of a two-stage gasification system for hydrogen, liquid fuels and energy production from residues. Energy Convers Manag 253:115126

Karmee SK (2017) Fuel not food—towards sustainable utilization of gutter oil. Biofuels 8(3):339–346

Karthick C, Nanthagopal K (2021) A comprehensive review on ecological approaches of waste to wealth strategies for production of sustainable biobutanol and its suitability in automotive applications. Energy Convers Manag 239:114219

Lan EI, Liao JC (2013) Microbial synthesis of n-butanol, isobutanol, and other higher alcohols from diverse resources. Bioresour Technol 135:339–349

Lee RA, Lavoie JM (2013) From first-to third-generation biofuels: challenges of producing a commodity from a biomass of increasing complexity. Anim Front 3(2):6–11

Li Y, Meng L, Nithyanandan K, Lee TH, Lin Y, Chia-fon FL, Liao S (2016) Combustion, performance and emissions characteristics of a spark-ignition engine fueled with isopropanol-n-butanol-ethanol and gasoline blends. Fuel 184:864–872

Lim JHK, Gan YY, Ong HC, Lau BF, Chen WH, Chong CT, Ling TC, Klemeš JJ (2021) Utilization of microalgae for bio-jet fuel production in the aviation sector: challenges and perspective. Renew Sust Energ Rev 149:111396

Liu Y, Cruz-Morales P, Zargar A, Belcher MS, Pang B, Englund E, Dan Q, Yin K, Keasling JD (2021) Biofuels for a sustainable future. Cell 184(6):1636–1647

Lu HR, El Hanandeh A (2017) Assessment of bioenergy production from mid-rotation thinning of hardwood plantation: life cycle assessment and cost analysis. Clean Techn Environ Policy 19:2021–2040

Marano JJ, Ciferno JP (2001) Life-cycle greenhouse gas emissions inventory for Fischer-Tropsch fuels. US Department of Energy. National Energy Technology Laboratory, June

Martinelli M, Gnanamani MK, LeViness S, Jacobs G, Shafer WD (2020) An overview of Fischer-Tropsch synthesis: XtL processes, catalysts and reactors. Appl Catal A Gen 608:117740

Martinez Hernandez E, Ng KS (2018) Design of biorefinery systems for conversion of corn Stover into biofuels using a biorefinery engineering framework. Clean Techn Environ Policy 20:1501–1514

Martinez-Villarreal S, Breitenstein A, Nimmegeers P, Saura PP, Hai B, Asomaning J, Eslami AA, Billen P, Passel SV, Bressler DC, Debecker DP, Remacle C, Richel A (2022) Drop-in biofuels production from microalgae to hydrocarbons: microalgal cultivation and harvesting, conversion pathways, economics and prospects for aviation. Biomass Bioenergy 165:106555

Mat Aron NS, Khoo KS, Chew KW, Show PL, Chen WH, Nguyen THP (2020) Sustainability of the four generations of biofuels–a review. Int J Energy Res 44(12):9266–9282

Mathioudakis V, Gerbens-Leenes PW, Van der Meer TH, Hoekstra AY (2017) The water footprint of second-generation bioenergy: a comparison of biomass feedstocks and conversion techniques. J Clean Prod 148:571–582

de Medeiros AM, de Sousa Castro K, de Macêdo MLG, de Morais Araújo AM, da Silva DR, Gondim AD (2022) Catalytic pyrolysis of coconut oil with Ni/SBA-15 for the production of bio jet fuel. RSC Adv 12(16):10163–10176

Meijaard E, Brooks TM, Carlson KM, Slade EM, Garcia-Ulloa J, Gaveau DL, Lee JSH, Santika T, Juffe-Bignoli D, Struebig MJ, Wich SA (2020) The environmental impacts of palm oil in context. Nature plants 6(12):1418–1426

Mesfun S, Sanchez DL, Leduc S, Wetterlund E, Lundgren J, Biberacher M, Kraxner F (2017) Power-to-gas and power-to-liquid for managing renewable electricity intermittency in the Alpine region. Renew Energy 107:361–372

Millinger M, Ponitka J, Arendt O, Thrän D (2017) Competitiveness of advanced and conventional biofuels: results from least-cost modelling of biofuel competition in Germany. Energy Policy 107:394–402

Moioli E, Salvati F, Chiesa M, Siecha RT, Manenti F, Laio F, Rulli MC (2018) Analysis of the current world biofuel production under a water–food–energy nexus perspective. Adv Water Resour 121:22–31

Mu D, Seager T, Rao PS, Zhao F (2010) Comparative life cycle assessment of lignocellulosic ethanol production: biochemical versus thermochemical conversion. Environ Manag 46:565–578

Muhammad G, Alam MA, Mofijur M, Jahirul MI, Lv Y, Xiong W, Ong HC, Xu J (2021) Modern developmental aspects in the field of economical harvesting and biodiesel production from microalgae biomass. Renew Sust Energ Rev 135:110209

Murphy HT, O'Connell DA, Raison RJ, Warden AC, Booth TH, Herr A, Braid AL, Crawford DF, Hayward JA, Jovanovic T, McIvor JG (2015) Biomass production for sustainable aviation fuels: a regional case study in Queensland. Renew Sust Energ Rev 44:738–750

Nassar AM, Moura P, Granço G, Harfuch L (2012) Benchmark of cane-derived renewable jet fuel against major sustainability standards. ICONE, July.

Nazari MT, Mazutti J, Basso LG, Colla LM, Brandli L (2021) Biofuels and their connections with the sustainable development goals: a bibliometric and systematic review. Environ Dev Sustain 23(8):11139–11156

Ng KS, Farooq D, Yang A (2021) Global biorenewable development strategies for sustainable aviation fuel production. Renew Sust Energ Rev 150:111502

Ng KS, Sadhukhan J (2011) Process integration and economic analysis of bio-oil platform for the production of methanol and combined heat and power. Biomass Bioenergy 35(3):1153–1169

Noh HM, Benito A, Alonso G (2016) Study of the current incentive rules and mechanisms to promote biofuel use in the EU and their possible application to the civil aviation sector. Transp Res Part D Transp Environ 46:298–316

Okolie JA, Nanda S, Dalai AK, Kozinski JA (2021) Chemistry and specialty industrial applications of lignocellulosic biomass. Waste Biomass Valoriz 12:2145–2169

Pan W, Li K, Teng Y (2018) Rethinking system boundaries of the life cycle carbon emissions of buildings. Renew Sust Energ Rev 90:379–390

Pandiyan K, Singh A, Singh S, Saxena AK, Nain L (2019) Technological interventions for utilization of crop residues and weedy biomass for second generation bio-ethanol production. Renew Energy 132:723–741

Paschalidou A, Tsatiris M, Kitikidou K (2016) Energy crops for biofuel production or for food?-SWOT analysis (case study: Greece). Renew Energy 93:636–647

Pereira LG, Cavalett O, Bonomi A, Zhang Y, Warner E, Chum HL (2019) Comparison of biofuel life-cycle GHG emissions assessment tools: the case studies of ethanol produced from sugarcane, corn, and wheat. Renew Sust Energ Rev 110:1–12

Radich T (2015) The flight paths for biojet fuel. Washington, DC, 20585

Rahmawati Z, Santoso L, McCue A, Jamari NLA, Ninglasari SY, Gunawan T, Fansuri H (2023) Selectivity of reaction pathways for green diesel production towards biojet fuel applications. RSC Adv 13(20):13698–13714

Rao P, Rathod V (2019) Valorization of food and agricultural waste: a step towards greener future. Chem Rec 19(9):1858–1871

Ren X, Ghazani MS, Zhu H, Ao W, Zhang H, Moreside E, Zhu J, Yang P, Zhong N, Bi X (2022) Challenges and opportunities in microwave-assisted catalytic pyrolysis of biomass: a review. Appl Energy 315:118970

Richter S, Braun-Unkhoff M, Naumann C, Riedel U (2018) Paths to alternative fuels for aviation. CEAS Aeronaut J 9(3):389–403

Roth A, Riegel F, Batteiger V (2018) Potentials of biomass and renewable energy: the question of sustainable availability. Biokerosene: Status and prospects, 95–122

dos Santos RG, Alencar AC (2020) Biomass-derived syngas production via gasification process and its catalytic conversion into fuels by Fischer Tropsch synthesis: a review. Int J Hydrog Energy 45(36):18114–18132

Schorling M, Enders C, Voigt CA (2015) Assessing the cultivation potential of the energy crop Miscanthus× giganteus for Germany. GCB Bioenergy 7(4):763–773

Seo MW, Lee SH, Nam H, Lee D, Tokmurzin D, Wang S, Park YK (2022) Recent advances of thermochemical conversion processes for biorefinery. Bioresour Technol 343:126109

Shokravi H, Shokravi Z, Heidarrezaei M, Ong HC, Koloor SSR, Petrů M, Lau WJ, Ismail AF (2021) Fourth generation biofuel from genetically modified algal biomass: challenges and future directions. Chemosphere 285:131535

Soleimani SS, Adiguzel A, Nadaroglu H (2017) Production of bioethanol by facultative anaerobic bacteria. J Inst Brew 123(3):402–406

Staples MD, Malina R, Suresh P, Hileman JI, Barrett SR (2018) Aviation CO2 emissions reductions from the use of alternative jet fuels. Energy Policy 114:342–354

Stratton R, Wong HM, Hileman J (2010) Life cycle greenhouse gas emissions from alternative jet fuels (No. PARTNER-COE-2010-001). Partnership for Air Transportation Noise and Emissions Reduction

Su Y, Zhang P, Su Y (2015) An overview of biofuels policies and industrialization in the major biofuel producing countries. Renew Sust Energ Rev 50:991–1003

Sudhakar MP, Kumar BR, Mathimani T, Arunkumar K (2019) A review on bioenergy and bioactive compounds from microalgae and macroalgae-sustainable energy perspective. J Clean Prod 228:1320–1333

Talberg A, Swoboda K (2013) Emissions trading schemes around the world, Department of Parliamentary Services. Retrieved from http://parlinfo.aph.gov.au/parlInfo/download/library/prspub/2501441/upload_binary/2501441.pdf;fileType=application/pdf

Tao L, Milbrandt A, Zhang Y, Wang WC (2017) Techno-economic and resource analysis of hydroprocessed renewable jet fuel. Biotechnol Biofuels 10:1–16

Tao L, Tan EC, McCormick R, Zhang M, Aden A, He X, Zigler BT (2014) Techno-economic analysis and life-cycle assessment of cellulosic isobutanol and comparison with cellulosic ethanol and n-butanol. Biofuels Bioprod Biorefin 8(1):30–48

Taylor CE, Baltrus JP, Driscoll D (2011) Fischer-Tropsch fuels. National Energy Technology Laboratory, Golden

The Engineering ToolBox (2003) US. Standard atmosphere. Available online at https://www.engineeringtoolbox.com/standard-atmosphere-d_604.html. Accessed 31 March 2019

The global aviation industry, I (2010) The right flightpath to reduce aviation emissions. In: UNFCCC Climate Talks

Tiwari R, Mishra R, Choubey A, Kumar S, Atabani AE, Badruddin IA, Khan TY (2023) Environmental and economic issues for renewable production of bio-jet fuel: a global prospective. Fuel 332:125978

Total A (2012) Breaking the barriers with break through jet fuel solutions. Berlin Airshow. Berlin

US Environmental Protection Agency (2016) EPA determines that aircraft emissions contribute to climate change endangering public health and the environment

Vasquez MC, Silva EE, Castillo EF (2017) Hydrotreatment of vegetable oils: a review of the technologies and its developments for jet biofuel production. Biomass Bioenergy 105:197–206

Vijay V, Pimm SL, Jenkins CN, Smith SJ (2016) The impacts of oil palm on recent deforestation and biodiversity loss. PLoS One 11(7):e0159668

Wang WC (2016) Techno-economic analysis of a bio-refinery process for producing hydro-processed renewable jet fuel from Jatropha. Renew Energy 95:63–73

Wang M, Dewil R, Maniatis K, Wheeldon J, Tan T, Baeyens J, Fang Y (2019) Biomass-derived aviation fuels: challenges and perspective. Prog Energy Combust Sci 74:31–49

Wang WC, Tao L (2016) Bio-jet fuel conversion technologies. Renew Sust Energ Rev 53:801–822

Warshay B, Pan J, Sgouridis S (2011) Aviation industry's quest for a sustainable fuel: considerations of scale and modal opportunity carbon benefit. Biofuels 2(1):33–58

Wilbrand K (2018) Potential of fossil kerosene. Status and Prospects, Biokerosene, pp 43–57

Xie X, Wang M, Han J (2011) Assessment of fuel-cycle energy use and greenhouse gas emissions for Fischer– Tropsch diesel from coal and cellulosic biomass. Environ Sci Technol 45(7):3047–3053

Yang J, Xin Z, Corscadden K, Niu H (2019) An overview on performance characteristics of bio-jet fuels. Fuel 237:916–936

Yu Y, Wu J, Ren X, Lau A, Rezaei H, Takada M, Bi X, Sokhansanj S (2022) Steam explosion of lignocellulosic biomass for multiple advanced bioenergy processes: a review. Renew Sust Energ Rev 154:111871

Chapter 11
Current Technological Status and Future Prospect of Biojet Fuel Production

Preeti Yadav, Yatika Dixit, and Arun Kumar Sharma

Abstract The aviation industry has garnered significant global attention due to its unsustainable development caused by excessive greenhouse gas emissions and reliance on traditional petroleum jet fuel. One of the most viable approaches is the exploration and implementation of initiatives aimed at the development and industrialization of alternative aviation fuels derived from renewable resources, such as biomass. The utilization of renewable biojet fuel has promise in mitigating CO_2 emissions throughout its life cycle, rendering biojet fuels a compelling alternative to conventional aviation fuels. This chapter presents a comprehensive review of feedstock selection, advancements in feedstock cultivation, conversion technologies, refining and upgrading processes, and considerations of sustainability and environmental impacts. The utilization of biomass-derived jet fuel, commonly referred to as biojet fuel, has emerged as a crucial component in the aviation sector's approach to mitigating both financial expenses and ecological consequences. A collaborative effort is underway among researchers from many sectors, including the aviation industry, government agencies, biofuel companies, agricultural organizations, and academia to advance the development of a commercially feasible and environmentally friendly method for producing durable renewable jet fuel. This approach aims to achieve both cost-effectiveness in production and minimal greenhouse gas emissions. This chapter reviews the challenges and potential associated with the production of biojet fuel, as well as highlighting interesting areas of study in this field. The utilization of jet fuel obtained from biomass holds promise in substituting a substantial proportion of conventional jet fuel needed to fulfil commercial aviation requirements. The potential for substantial production of biojet fuels and reduction of CO_2 emissions can be realized through the widespread implementation of biojet fuels, considering the availability of biomass feedstock in the future.

Keywords Feedstock · Conversion · Pyrolysis · Challenges · Emission

P. Yadav · Y. Dixit · A. K. Sharma (✉)
Department of Bioscience and Biotechnology, Banasthali Vidyapith, Newai, Rajasthan, India

A. Kuila (ed.), *Biojet Fuel: Current Technology and Future Prospect*, Clean Energy Production Technologies, https://doi.org/10.1007/978-981-99-8783-2_11

11.1 Introduction

In recent decades, the aviation industry has experienced exponential growth, becoming a vital component of the global transportation system (Wang et al. 2019b). However, this expansion has also brought to the forefront the industry's significant environmental impact, particularly with regard to greenhouse gas emissions (O'Shea et al. 2020). As air travel demand continues to surge, the need to find sustainable and environmentally friendly alternatives to conventional jet fuels has become increasingly urgent (Said et al. 2022). In response to these challenges, biojet fuels have emerged as a promising solution with the potential to revolutionize the aviation sector.

Biojet fuels, also known as sustainable aviation fuels (SAFs), are derived from renewable biomass sources, such as agricultural residues, non-food crops, algae, and waste materials (Dyk and Saddler 2021). By utilizing these renewable resources, biojet fuels offer a sustainable option for powering aircraft. Through this alternative, they significantly reduce the aviation industry's carbon footprint and help mitigate its contribution to climate change (Terrenoire et al. 2019).

One critical aspect of biojet fuel production lies in feedstock selection and its potential impact on food security and land-use change (Pulighe and Pirelli 2023). Addressing these concerns is essential, as the indiscriminate use of agricultural crops for fuel production can create competition with food production, leading to potential food shortages and land-use conflicts. To address these concerns, researchers and stakeholders focus on utilizing non-food feedstocks and adopting sustainable agricultural practices that do not compete with food production. By doing so, biojet fuel production can be made more sustainable and avoid exacerbating existing food security challenges (Stavi et al. 2021).

The development of advanced feedstocks, such as algae and lignocellulosic biomass, further enhances the sustainability of biojet fuel production (Why et al. 2019). Algae, for example, can be cultivated on non-arable land, minimizing the use of valuable resources like water and fertilizers (Khan et al. 2023). Lignocellulosic biomass, derived from agricultural and forestry residues, offers an abundant and renewable source of feedstock without competing with food production (Nahak et al. 2022). By utilizing marginal lands and waste resources, biojet fuels can be produced without impacting ecosystems or food supplies.

Beyond the environmental benefits, the adoption of biojet fuels also boons economic opportunities for airlines and fuel producers. Governments and consumers increasingly prioritize sustainability, driving the demand for environmentally friendly products, including bio-based aviation fuels. By adopting biojet fuels, airlines can demonstrate their commitment to corporate social responsibility and gain a competitive edge in a market where environmental consciousness plays an increasingly significant role in consumer decisions (Amicarelli et al. 2021). Moreover, public perception plays a crucial role in driving the adoption of biojet fuels. Engaging with the public through awareness campaigns and educational initiatives can accelerate the transition to sustainable aviation, fostering a positive perception and acceptance of biojet fuels as a viable and responsible alternative. The production of biojet fuels involves several technological processes that contribute to their viability as sustainable aviation fuel. These processes include feedstock selection and optimization, conversion technologies, refining and upgrading processes, and sustainability assessments (Tan et al. 2022). The selection of appropriate feedstocks is crucial to ensuring the overall sustainability and viability of biojet fuel production.

The conversion technologies used to convert biomass into biojet fuels can vary, encompassing biochemical and thermochemical processes (Jayakumar et al. 2023). Biochemical processes involve the use of enzymes or microorganisms to break down the biomass into biofuels, while thermochemical processes use heat and pressure to convert the biomass into fuels. Each method has its strengths and challenges, and ongoing research seeks to optimize these technologies for commercial-scale production (Gnanasekaran et al. 2023). Following the conversion stage, refining and upgrading processes are essential to meet stringent aviation fuel standards. These processes ensure that the produced biojet fuels have the necessary properties and compatibility with existing infrastructure and aircraft engines. By refining and upgrading the fuels, their quality is enhanced, making them suitable replacements for conventional jet fuels (Lahijani et al. 2022).

Sustainability assessments, including life cycle assessments (LCAs), are critical in evaluating the environmental performance of biojet fuel production. LCAs consider the entire life cycle of biojet fuels, from feedstock cultivation to fuel combustion, to determine their overall impact on the environment (Zhu et al. 2022). By identifying areas of improvement and optimizing processes, biojet fuels can continuously evolve to become even more environmentally friendly. While biojet fuels hold immense promise for the aviation industry's sustainable future, there are technological challenges that need to be addressed for widespread adoption. Scaling up biojet fuel production to meet the growing demand remains a significant obstacle. Additionally, economic considerations and policy support are essential in creating an enabling environment for the development and deployment of biojet fuels (Lim et al. 2023).

Nevertheless, ongoing research, coupled with supportive policies and collaborations between governments, industry stakeholders, and researchers, is expected to pave the way for a future where biojet fuels become a mainstream and integral component of the aviation industry. The transition to biojet fuels will contribute to a greener and more sustainable future for air travel, allowing the aviation sector to play its part in global efforts to combat climate change and create a more environmentally conscious world (Dyk and Saddler 2021).

11.2 Biojet Fuels

As the biojet industry grapples with its significant environmental impact, the search for sustainable alternatives to conventional jet fuels has gained paramount importance. In response to this pressing challenge, biojet fuels have emerged as a promising solution. By reducing greenhouse gas emissions and mitigating the industry's contribution to climate change, biojet fuels hold the potential to revolutionize the aviation sector and pave the way for a more sustainable future of air travel (Lim et al. 2023).

11.2.1 The Need for Sustainable Aviation

The aviation industry has experienced unprecedented growth and has become an integral part of the global economy, connecting people and goods worldwide. However, this expansion has also brought significant environmental and climate challenges that demand urgent attention.

One of the primary environmental issues associated with aviation is its substantial contribution to greenhouse gas emissions, particularly carbon dioxide (CO_2) (Ghannouchi et al. 2023). The combustion of conventional jet fuels releases large amounts of CO_2 into the atmosphere, contributing to the greenhouse effect and global warming. According to the International Civil Aviation Organization (ICAO), aviation accounts for approximately 25% of global CO_2 emissions (Keselova et al. 2019). While this percentage might appear relatively small compared to other sectors, the rapid growth of air travel and the lack of scalable alternatives to traditional jet fuels make addressing aviation emissions crucial to overall climate change mitigation efforts (Pilat et al. 2018).

Moreover, aviation emissions have a more significant impact on climate change than just CO_2. Other emissions, such as nitrogen oxides (NO_2), sulfur oxides (SO_2), and particulate matter, can lead to the formation of contrails and cirrus clouds, which have a potent warming effect on the atmosphere (Garde and Zingg 2022). Additionally, biojet emissions at high altitudes have a more significant radiative forcing effect than those emitted at ground level, further exacerbating their impact on the climate. The Intergovernmental Panel on Climate Change (IPCC) and various environmental organizations have consistently emphasized the need to address aviation's environmental impact to limit global warming and its associated adverse effects (Raimi et al. 2021).

Sustainable biojet has thus emerged as a critical imperative for the industry. The concept encompasses various strategies and technologies to mitigate aviation's environmental impact. One of the most promising approaches is developing and adopting sustainable aviation fuels, such as biojet fuels (Ng et al. 2021).

These fuels offer a potential pathway to significantly reduce greenhouse gas emissions from aviation, as they can be produced from renewable sources and have the potential to be nearly carbon-neutral (Zhang and Chen 2022). Furthermore, sustainable aviation encompasses other measures, such as improvements in aircraft design, operational efficiencies, air traffic management, and more efficient engines. By embracing a holistic approach to sustainability, the aviation industry can substantially reduce its environmental impact and contribute to global efforts to combat climate change.

11.2.2 Feedstock Consideration for Biomass-Derived Biojet Fuel

Feedstock consideration is a pivotal aspect of biojet fuel production, involving the meticulous selection of raw materials used to produce renewable aviation fuels. The choice of feedstock profoundly influences the sustainability, environmental impact, economic feasibility, and scalability of biojet fuel production. Various feedstocks are being researched and developed, each offering unique advantages and facing specific challenges (Doliente et al. 2020). Feedstocks can be categorized into first-, second-, third-, and fourth-generation feedstocks. The availability and potential production of agricultural feedstocks are intertwined (Lee and Lavoie 2013). Oil palm has the highest output yield of 19.2 t/ha/year. Microalgae's potential production has been claimed to be substantially higher at 91 t/ha/year for third-generation (3-G) feedstocks, while this number is unknown due to the fact that algae cultivation is often conducted on a lab- or pilot-scale (Alalwan et al. 2019).

11.2.2.1 First-Generation Feedstock

Edible food crops, which involves oil palm, corn, sugarcane, sugar beets, and wheat, collapse under a categorization of first-generation (1-G) feedstocks (Lee and Lavoie 2013). The production of 1-G feedstocks faces a significant challenge posed by competition for land, water, and energy resources with food production (Moioli et al. 2018). This issue is particularly evident in the cultivation of oil palm, which is simultaneously a popular food crop and an intriguing feedstock for biofuel production (Vijay et al. 2016).

Oil palm: The hydroprocessed esters and fatty acids (HEFA) technology stands as the sole commercially accessible renewable jet fuel technology as of currently (Roth et al. 2017). The food sector is the primary driver behind the global demand for palm oil, with Malaysia and Indonesia now supplying over 80% of this need. The production of biodiesel, which exhibits a greater energy yield per unit of energy input compared to other edible oils (Pirker et al. 2016).

Although edible crop oils have been extensively utilized as feedstock for the production of biofuels, there has been increasing scrutiny surrounding these first-generation biofuels. This scrutiny stems from various concerns, such as their limited ability to effectively reduce greenhouse gas emissions and the controversy surrounding the diversion of food crops for fuel production. Consequently, there has been a notable focus on the production of biojet fuel using second-generation technology (Doliente et al. 2020).

11.2.2.2 Second-Generation Feedstock

In contrary to first-generation (1-G) feedstocks, which necessitate a choice between supplying food or energy, second-generation (2-G) biomass resources have the capability to serve both functions (Alalwan et al. 2019). The two primary kinds in this context are energy crops and waste biomass. The sugars included in 2-G feedstocks are enclosed within the resilient and resistant lignocellulosic structure of plant cell walls. Consequently, prior to their conversion into biofuel, these sugars necessitate pretreatment by the utilization of enzymes/microorganisms or thermochemical processes (Boichenko et al. 2013). However, the primary difficulties with the utilization of 2-G feedstocks are their relatively high availability and low level of usage (Alalwan et al. 2019).

The primary concerns are on the technical challenges and significant expenses associated with these conversion methods. Lignocellulosic second-generation (2-G) feedstocks present a viable alternative to first-generation (1-G) crops owing to their abundant availability and less rivalry for utilization (Correa et al. 2019). The use of waste biomass offers a wide array of advantages, including waste management and environmental conservation, as well as the promotion of circular economies (Ahorsu et al. 2018).

Renewable energy crops: Jatropha (*Jatropha curcas*) and castor bean (*Ricinus communis*) are considered as renewable energy crops, mostly cultivated for their oil-seeds. These oil-seeds, however, are deemed unsuitable for human (Shahare et al. 2017). The extraction of BF (bioavailable organic fertilizer) from castor bean oil may be achieved using several methods such as transesterification, catalytic cracking (pyrolysis), or hydroprocessing (Molefe et al. 2019). In an innovative endeavor conducted in India, a commercial aircraft flown by SpiceJet effectively employed a biojet fuel mixture derived from Jatropha seeds. The flight showcased the pragmatic utilization of non-edible plant oils as a feasible raw material for the production of environmentally friendly aviation fuels. The program demonstrated the feasibility of producing biojet fuel using locally accessible, non-edible feedstock, so establishing a foundation for further study and investment (Doliente et al. 2020).

Numerous approaches, such as thermochemical and biochemical treatments, have been recommended for the production of bioavailable organic fertilizer (BOF) utilizing a range of grass and wood energy crops (Kandaramath Hari et al. 2015). Grass energy crops have considerable potential as a viable alternative for biofuel production. This is mostly due to their substantial lignocellulose content and the use of advanced harvesting equipment, which enhances their feasibility as a sustainable energy source (Crawford et al. 2016). Hydrocarbons derived from poplar biomass via pyrolysis and fermentation processes have the potential to be transformed into jet fuel through the process of hydrogenation. The use of rapidly expanding eucalyptus trees (*Eucalyptus* spp.) as a raw material for biomass-to-liquid (BOF) synthesis is observed in Brazil (Zhang et al. 2016).

Waste biomass: Since waste biomass is co-produced from agro-forestry, residential, commercial, and industrial activities, it requires no additional land and has lower water footprints than cultivated crops. No aviation fuel from municipal solid waste (MSW) test flights have been documented as of yet. The main difficulties in using waste biomass as BOF feedstock stem from its logistical complexity and its erratic supply (Mawhood et al. 2016).

11.2.2.3 Third-Generation Feedstocks

In the past few years, there has been a significant focus on microalgae due to its ability to attain high yields with low land utilization. Growth of algae in contaminated water or water that is not suited for agricultural purposes can lead to a reduction in operational expenses and provide advantages in wastewater treatment. Algae need a lower volume of water compared to the majority of 1-G feedstocks, such as canola (5500 L) and soybeans (15,000 L), in order to produce an equivalent quantity of biodiesel (1 L) (Alalwan et al. 2019).

There is a growing body of research focused on the simplification and diversification of production methods through the use of thermochemical processes, including pyrolysis and hydrothermal liquefaction technologies. The industrial-scale generation of biofuels is a highly regarded use of microalgae (Chiaramonti and Horta Nogueira 2017).

To achieve the most favorable economic and environmental outcomes in the context of microalgae supply chain for BOF (biochemical and biofuel) supplies, it is essential to take into account the geographical and temporal aspects of microalgae farming (Doliente et al. 2020). The implementation of algae-based solutions has demonstrated the potential to significantly mitigate carbon dioxide (CO_2) emissions by national airlines, with reductions of up to 85% projected by the year 2050. The examination of supply chains is necessary for the development of microalgae-based biofuels (Behrendt et al. 2017). It is important to note, however, that the economic feasibility of present algal technologies is not anticipated to be realized until around ten years from now.

11.2.2.4 Fourth-Generation Feedstocks

The potential of non-biological resources and genetically modified organisms, together known as fourth-generation (4-G) feedstocks. Several genetically modified organisms, such as cyanobacteria, fungus, and yeast, have been engineered to exhibit enhanced oil and/or sugar production, as well as negative carbon capabilities (Alalwan et al. 2019). The introduction of these organisms into the global supply chain poses potential risks to both human health and the environment, necessitating further investigation and the development of containment and mitigation strategies (Abdullah et al. 2019). The field of bioaviation refers to the study and application of biological principles and technologies in fuel derived from 4-G feedstocks exhibits the potential to achieve zero carbon emissions and facilitate the integration of electricity, heating, and aviation sectors. Consequently, when these technologies attain commercial maturity, they may be considered the most sustainable option (Richter et al. 2018).

11.3 Advancement in Feedstock Selection for Biojet Fuel Production

11.3.1 Feedstock Selection

The selection of suitable feedstocks is a critical factor in the success of biojet fuel production, as it determines the sustainability, economic viability, and environmental impact of the fuel (Tiwari et al. 2023). Various feedstock options are available, each with its own advantages and challenges. Among the most promising options are non-food energy crops, such as jatropha, camelina, and Pongamia, which have high oil content and can grow on marginal lands, minimizing competition for arable land (Mofijur et al. 2023). Algae also offer an intriguing feedstock option, with the potential to yield significant amounts of oil per unit area and the ability to be cultivated in diverse environments, including non-arable land and wastewater. Algae's capacity to sequester carbon dioxide during cultivation further enhances their environmental appeal, but scaling up cultivation and developing cost-effective harvesting methods remain areas of ongoing research (Khan et al. 2023).

Various initiatives and successful tests have demonstrated the potential of these feedstocks for biojet fuel production. The U.S. Navy, for instance, tested flights using a blend of camelina-based biojet and conventional jet fuel (Dangol et al. 2020), while the European Commission's ITAKA project utilized camelina-based biojet fuel for commercial flights. The Jatropha Global Biofuel Alliance (JGBA) and other initiatives have also made significant progress in increasing jatropha oil yields and developing sustainable cultivation practices.

Companies like Neste and Fulcrum BioEnergy have successfully produced biojet fuel from various waste and residues, showcasing the feasibility of this approach (Ng et al. 2021; Dyk and Saddler 2021). These promising developments indicate that a diverse range of feedstocks, including algae, camelina, jatropha, waste oils, residues, and municipal solid waste, hold immense potential in shaping the future of biojet fuel production.

Selecting appropriate feedstocks is crucial for the advancement of biojet fuel production. These feedstock options offer sustainable and environmentally friendly alternatives to conventional jet fuel, contributing to a more sustainable aviation sector. Continued research, technological advancements, and supportive policies will be instrumental in driving the commercialization of biojet fuel and achieving a greener future for the aviation industry.

11.3.2 Advances in Feedstock Cultivation and Harvesting Techniques

Advances in feedstock cultivation and harvesting techniques have played a pivotal role in shaping the current technological status of biojet fuel production (Chopra et al. 2022). For algae-based feedstock, closed photobioreactors have emerged as a promising solution, allowing better control over environmental factors to optimize algae growth and lipid accumulation, resulting in higher oil yields (Sarwer et al. 2022). Additionally, raceway ponds have been refined to efficiently mix and expose algae to light, further enhancing productivity (Khan et al. 2022). Genetic engineering has also been explored to modify algae strains for higher lipid content and resilience (Shahid et al. 2020).

In the case of cellulosic biomass, the cultivation process has seen significant improvements. Energy crops like switchgrass and miscanthus have been identified for their high cellulose content and ability to grow on marginal lands without affecting food crops (Qaseem and Wu 2021). Sustainable agricultural practices, such as reduced tillage and crop rotation, have been adopted to maintain soil health and minimize environmental impacts. Researchers are also investigating genetically engineered energy crops with improved enzymatic digestibility to enhance cellulose conversion into fermentable sugars (Sirangelo et al. 2023). Precision farming technologies, including remote sensing through satellites and drones, provide real-time data on crop health and resource requirements, enabling optimal harvesting times and increased feedstock quality (Mirkouei 2020). Variable rate technology (VRT) ensures the precise application of inputs based on specific field requirements, promoting uniform crop growth and more efficient harvesting (Ahmad and Sharma 2023).

Robotic harvesters with cameras and sensors can accurately identify and harvest energy crops, minimizing manual labor (Rehman et al. 2022). Modified combine harvesters have been adapted for energy crop harvesting on a large scale. Additionally, specialized biomass choppers facilitate the accessible collection and transport of biomass for biojet fuel production. Overall, these advancements in feedstock cultivation and harvesting techniques hold great promise for the future of biojet fuel production. Continued research and development in these areas will likely lead to further improvements in the economic viability and sustainability of biojet fuels, offering a greener and more environmentally friendly alternative for the aviation industry (Khan et al. 2023).

11.3.3 Genetic Engineering and Breeding for Improved Feedstock Traits

Biojet fuel, derived from renewable biomass sources, has emerged as a promising alternative to traditional fossil fuels for aviation, addressing the increasing concerns about greenhouse gas emissions and the need for sustainable energy solutions (Wang et al. 2019a). Researchers have actively explored innovative technologies, including genetic engineering and breeding techniques, to enhance biojet fuel production by improving feedstock traits.

One area of focus in genetic engineering is algae, known for its rapid growth, high lipid content, and adaptability to diverse environments (Khoo et al. 2023).

Researchers have successfully manipulated the genetic material of algae to increase lipid productivity. Key genes involved in lipid biosynthesis, such as acetyl-CoA carboxylase (ACCase) and diacylglycerol acyltransferase (DGAT), have been overexpressed (Shahid et al. 2020). Furthermore, CRISPR-Cas9 technology has been utilized to precisely edit genes responsible for lipid metabolism, resulting in genetically modified algae strains with improved oil yields (Muthukrishnan 2022). For example, in 2020, researchers from different countries, used CRISPR-Cas9 to engineer a high-lipid variant of *Nannochloropsis oceanica*, a species of algae, resulting in a 50% increase in lipid content, making it a more efficient feedstock for biojet fuel production (Harada et al. 2020; Khan and Fu 2020).

Another candidate for genetic engineering is *Jatropha curcas*, a non-edible plant known for its high oil content. Genetic engineering has enhanced Jatropha's agronomic traits, such as reducing toxic components and improving stress tolerance (Fu et al. 2019). Patel et al. (2022) downregulated the expression of phorbol ester biosynthetic genes in Jatropha through RNA interference (RNAi), making Jatropha seeds safer for biojet fuel production.

Traditional breeding techniques, combined with modern biotechnology tools, have also played a significant role in developing superior biofuel feedstock varieties. Sugarcane, a widely cultivated crop for biofuel production, has been a target for traditional breeding methods to enhance sugar yield, increase biomass production, and improve resistance to pests and diseases (Budeguer et al. 2021).

Camelina sativa, also known as false flax, is a drought-resistant plant with potential as a biojet fuel feedstock (Dangol et al. 2020). Breeders have been working on developing camelina varieties with improved oil content, fatty acid profiles, and overall agronomic performance (Pozzo et al. 2022). The US Department of Agriculture (USDA) has been involved in a breeding program to enhance camelina as an oilseed crop for renewable jet fuel production (Ghidoli et al. 2023). By selecting and crossbreeding high-performing camelina varieties, they have developed strains with higher oil yields and improved adaptability to different climates.

These advancements in genetic engineering and breeding techniques hold immense promise for the future of biojet fuel production. Genetically modified algae with increased lipid content and selectively bred sugarcane with improved sugar yields are just a few examples showcasing the potential of these technologies to revolutionize the biofuel sector (Budeguer et al. 2021; Harada et al. 2020; Khan and Fu 2020). As research continues, biojet fuel production is expected to become more efficient and sustainable, driven by further advancements in genetic engineering and breeding practices.

11.4 Current Conversion Technologies for Biojet Fuel Production

11.4.1 Thermochemical Conversion Processes

The aviation industry is one of the major contributors to global carbon dioxide emissions, and finding sustainable alternatives to conventional jet fuel is crucial for mitigating the impact of aviation on the environment (Terrenoire et al. 2019). Biojet fuel, derived from renewable biomass sources, offers a promising solution to reduce greenhouse gas emissions while ensuring energy security. Among the various pathways for biojet fuel production, thermochemical conversion processes have gained significant attention due to their potential to convert a wide range of biomass feedstocks into liquid hydrocarbons suitable for aviation use (Jha et al. 2022). This section provides an overview of thermochemical conversion processes and their significance in biojet fuel production.

11.4.1.1 Pyrolysis

Pyrolysis is a thermal decomposition process that breaks down biomass into its constituent components in the absence of oxygen. The absence of oxygen prevents combustion and allows the biomass to undergo complex chemical reactions, leading to the formation of three main products: bio-oil, biochar, and syngas (Prasad Reddy Kannapu et al. 2022).

The bio-oil produced from pyrolysis is a dark, viscous liquid with a wide range of oxygenated hydrocarbons (Onwudili and Scaldaferri 2023). Its properties depend on the feedstock used and the pyrolysis conditions. The bio-oil contains both valuable compounds, such as sugars, aldehydes, and phenols, as well as impurities, such as water and acids (Chan et al. 2020). As such, the bio-oil requires further upgrading through processes like hydrotreatment or hydrodeoxygenation to improve its stability, reduce oxygen content, and increase its energy density (Dimitriadis et al. 2021). Once upgraded, the bio-oil can be used as a renewable replacement for conventional jet fuel. While, the solid residue left after pyrolysis is called biochar. It consists mainly of carbon-rich material and retains a significant portion of the original carbon content of the biomass (Aup-Ngoen and Noipitak 2020). Biochar is a stable carbon material and has applications in agriculture as a soil amendment. It enhances soil fertility, water retention, and microbial activity, leading to increased crop productivity and improved soil carbon sequestration (Elkhlifi et al. 2023).

However, pyrolysis produces a mixture of gases known as syngas (Zhou et al. 2020). This syngas is mainly composed of hydrogen (H_2), carbon monoxide (CO), methane (CH_4), and other light hydrocarbons. The composition of the syngas varies depending on the feedstock and pyrolysis conditions (Cerone et al. 2020). The syngas can be utilized for power generation, as it has a high energy content, or further processed through Fischer-Tropsch synthesis to produce liquid fuels, including biojet fuel.

Solena Fuels and Red Rock Biofuels are companies that use pyrolysis technology to produce biojet fuel (Dyk and Saddler 2021). Solena Fuels' GreenSky California project aims to convert municipal solid waste into biojet fuel using pyrolysis (Shahabuddin et al. 2020), while Red Rock Biofuels uses pyrolysis to convert forest residues into renewable jet fuel (Björnsson and Ericsson 2022).

11.4.1.2 Gasification

Gasification is a thermochemical process that converts biomass into a gaseous mixture known as syngas (Halba et al. 2023). Unlike pyrolysis, gasification occurs with a controlled amount of oxygen or steam. The process occurs at high temperatures (700 °C to 1000 °C) and under pressure. During gasification, the biomass undergoes several stages (Qi et al. 2023): (a) Drying: Initially, the biomass is dried to remove moisture and increase its energy content (Perazzini et al. 2021). (b) Pyrolysis: At higher temperatures, the biomass undergoes pyrolysis, producing volatile compounds that form the basis of the syngas (Zeng et al. 2020). (c) Gasification: The volatile compounds react with oxygen or steam to produce syngas, which mainly consist of hydrogen (H_2), carbon monoxide (CO), carbon dioxide (CO_2), and methane (CH_4) (Qing et al. 2022).

The AltAir Fuels facility in California uses gasification technology to convert agricultural residues and waste feedstocks into biojet fuel through subsequent Fischer-Tropsch synthesis (Porwal et al. 2020).

11.4.1.3 Fischer-Tropsch Synthesis

Fischer-Tropsch synthesis is a catalytic process that converts syngas (a mixture of hydrogen and carbon monoxide) into hydrocarbons (Davlatova, 2023). The process was initially developed in the 1920s by German scientists Franz Fischer and Hans Tropsch as a way to produce liquid fuels from coal (Dinç and Gürbüz 2022).

The Fischer-Tropsch process involves several steps:

(a) Syngas Preparation: Syngas, obtained from gasification or other sources like pyrolysis, is purified to remove impurities such as sulfur and trace metals (Teimouri et al. 2022).
(b) Fischer-Tropsch Reaction: The purified syngas is fed into a reactor containing a catalyst (usually based on iron or cobalt). The catalyst facilitates a series of polymerization and hydrogenation reactions, leading to the formation of long-chain hydrocarbons. These hydrocarbons are primarily paraffin (alkanes) with a wide range of chain lengths (Martinelli et al. 2020).
(c) Product Upgrading: The hydrocarbons produced through Fischer-Tropsch synthesis have different boiling points and properties (Klerk et al. 2022). The prod-

ucts need to be further processed to obtain a mixture of hydrocarbons with properties similar to conventional jet fuel. The resulting hydrocarbons, often referred to as synthetic fuel or syncrude, are chemically similar to conventional jet fuel. They can be used directly as a drop-in replacement for jet fuel or blended with conventional jet fuel to reduce the overall carbon footprint of aviation (Petersen et al. 2021).

Thermochemical conversion processes, including pyrolysis, gasification, and Fischer-Tropsch synthesis, are crucial in the production of biojet fuel from renewable biomass sources (Jha et al. 2022). Each process has advantages and challenges, and their combination offers a promising pathway to a more sustainable aviation industry. Continued research and development efforts are necessary to optimize these processes, improve efficiency, and make them economically viable on a larger scale for widespread biojet fuel production and the reduction of greenhouse gas emissions.

11.4.2 Catalytic Conversion Processes

11.4.2.1 Hydroprocessing

Hydroprocessing is a catalytic conversion process utilized in biojet fuel production to upgrade biomass-derived feedstocks (Lahijani et al. 2022). The process involves the use of hydrogen gas and catalysts to transform the raw biomass into high-quality biojet fuel. The first step in hydroprocessing is the pretreatment of the biomass feedstock, where solid impurities and contaminants are removed to ensure smooth processing (Haider et al. 2021). Subsequently, hydrogenation takes place, during which the biomass is exposed to high temperatures and pressures in the presence of hydrogen. This leads to a series of hydrogenation and dehydrogenation reactions that break down triglycerides and fatty acids present in the biomass, converting them into smaller hydrocarbons (Du et al. 2023).

The choice of catalyst is a critical aspect of hydroprocessing, as it influences the efficiency and selectivity of the reactions. Typical catalysts include metal sulfides or supported metal catalysts, carefully chosen based on the specific feedstock and desired product composition (Ahmad et al. 2021). After hydrogenation, the product mixture is fractionated to separate the various components based on their boiling points. The desired biojet fuel components are collected in this step (Matuszewska et al. 2021). Finally, the biojet fuel product may undergo additional treatment to refine its properties further, such as increasing its energy density and improving its cold-flow properties (Lahijani et al. 2022).

Hydroprocessing has shown great promise in producing biojet fuel, but it is challenging. Catalyst deactivation due to impurities present in the feedstock is one such obstacle. Additionally, the process requires a considerable amount of hydrogen, raising concerns about sustainable hydrogen production. To address these issues, researchers are actively improving catalyst stability, developing more efficient hydrogen production methods, and exploring using non-conventional feedstocks to make hydroprocessing a more economically and environmentally viable option for biojet fuel production.

11.4.2.2 Hydrotreating

Hydrotreating is a specific catalytic conversion process within the realm of hydroprocessing, focusing on the treatment of bio-oils obtained through pyrolysis or liquefaction of biomass (Alekseeva et al. 2020). The primary goal of hydrotreating is to reduce the oxygen content in bio-oils and remove heteroatoms, such as sulfur, nitrogen, and metals, to make them suitable for use in conventional jet engines (Han et al. 2021).

The hydrotreating process involves several steps. First, the bio-oil is heated to the desired temperature to facilitate the subsequent reactions (Lahijani et al. 2022). Then, hydrogen gas is introduced into the system, and the bio-oil undergoes hydrogenation reactions in the presence of a suitable catalyst. These reactions lead to the breaking of oxygen-carbon bonds and the removal of oxygen from the bio-oil (Zhang et al. 2021b). The choice of catalyst is crucial in hydrotreating, as it determines the selectivity of the reactions. Typical catalysts include supported nickel or cobalt catalysts (Saidi and Moradi 2021). After hydrotreating, the product mixture is fractionated to separate the desirable biojet fuel components from other by-products (Cavalcanti et al. 2022). Additional treatments may be applied to improve the stability, energy content, and cold-flow properties of the final biojet fuel product. Hydrotreating has shown promising results in producing biojet fuel with properties suitable for use in existing jet engines.

Despite its potential, hydrotreating faces challenges related to the complexity of bio-oil compositions, catalyst deactivation, and the formation of unwanted by-products (Zhang et al. 2021a). Researchers are working on developing tailored catalysts that are more selective and stable in the presence of impurities found in bio-oils. Additionally, reactor design and process optimization advancements aim to improve the overall efficiency of hydrotreating for biojet fuel production.

11.4.2.3 Deoxygenation

Deoxygenation is a crucial step in the conversion of biomass into biojet fuel (Zhou et al. 2023). The presence of oxygen in biomass-derived feedstocks reduces the energy density of the final fuel product and can lead to instability during storage. Catalytic deoxygenation processes are designed to remove oxygen atoms from biomass components, such as carbohydrates and fatty acids, to produce hydrocarbon-rich biojet fuel components (Zamri et al. 2023).

Different deoxygenation pathways include decarboxylation, decarbonylation, and hydrodeoxygenation (HDO) (Silva and Andrade 2023). In decarboxylation, carboxylic acid groups in biomass components are removed as carbon dioxide, forming hydrocarbons (Zhao et al. 2021). Decarbonylation involves the removal of carbonyl groups, leading to the generation of hydrocarbons and water (Lu et al. 2021). HDO, the most common deoxygenation route, combines hydrogenation with oxygen removal. The oxygen in the biomass is replaced with hydrogen, producing water as a by-product (Akmach et al. 2023). The deoxygenation process can be influenced by temperature, pressure, hydrogen availability, and the type of catalyst used (Di Vito Nolfi et al. 2021). Catalysts with specific functionalities are chosen to achieve the desired deoxygenation pathway while minimizing undesired side reactions.

Deoxygenation processes are challenging due to the diverse range of oxygen-containing functional groups in biomass, leading to a complex mixture of products. Developing highly efficient and selective catalysts is crucial for improving deoxygenation yields and minimizing energy-intensive separation processes (Jung et al. 2021). Researchers are exploring advanced catalyst materials and innovative reactor configurations to enhance deoxygenation processes for biojet fuel production. Successful deoxygenation processes promise to create high-quality biojet fuel with properties comparable to traditional jet fuel.

11.4.3 Biochemical Conversion

11.4.3.1 Alcohol-to-Jet (ATJ)

This method involves subjecting alcohol molecules generated from sugar, starch, or lignocellulosic materials to dehydration and oligomerization. The Alcohol-to-Jet (ATJ) method involves converting alcohols into a blend stock for alternative jet fuel. This process relies on catalytic stages that have been traditionally used in the petroleum refining and petrochemical sector (Geleynse et al. 2018). This particular pathway offers a viable method for generating a sustainable alternative jet fuel (SAJF) using a diverse range of resources. It presents a promising opportunity in the short term for alcohol producers to enter the SAJF market, while also addressing the increasing demand for SAJF within the aviation industry. The fuel blend stock generated via the ATJ (Alcohol-to-Jet) method is commonly referred to as ATJ-SPK (synthetic paraffinic kerosene) and has obtained approval from ASTM D75 (Lim et al. 2023).

11.4.3.2 Lignin-to-Jet (LTJ)

In the conventional jet fuel process, lignin is obtained from the residuals after cellulose and hemicellulose hydrolysis (Ruan et al. 2019). However, obtaining pure lignin is challenging, and it requires extraction and purification through processes like organosolv and ionic liquid extraction. Lignin-derived bio-oils are then collected through depolymerization using methods like fast pyrolysis, hydrolysis, and hydrogenolysis (Pappa et al. 2022). To upgrade the lignin-derived bio-oils, the focus is on producing aromatic hydrocarbons and cycloalkanes (Hu et al. 2021). The main method used for upgrading is hydrodeoxygenation. This makes lignin a potential source for primary and cyclic components in alternative jet fuel production, eliminating the need for blending.

11.4.3.3 Direct Sugar to Hydrocarbons (DSHC)

It is a process that uses genetically modified microbes to convert sugar into hydrocarbons or lipids. In one such scenario, yeasts produce isoprenoids like farnesene, which are then hydrogenated to produce farnesane, a fuel with favorable properties. The modified bacteria feed on extracted sugars. Microorganisms ferment carbohydrates into metabolic intermediates like fatty acids (Crawford et al. 2016). The cells transform microorganism-produced fatty acids into hydrocarbons through enzymatic processes. The resulting hydrocarbons may have impurities or different chain lengths. Hydrocracking and isomerization can be used to refine jet fuel hydrocarbons. Quality control tests guarantee refined hydrocarbons fulfil aviation fuel requirements. Biojet fuel can be combined with conventional jet fuel or other additives to meet performance specifications (Doliente et al. 2020).

11.4.4 Electrofuels

Electrofuels, also known as e-fuels, offer a promising solution for producing sustainable biojet fuels, contributing to a greener aviation industry (Brynolf et al. 2022). This innovative technology converts renewable electricity, carbon dioxide (CO_2), and water (H_2O) into synthetic hydrocarbons through an electrochemical process (Hussain et al. 2023). The key principle involves using renewable electricity from sources like solar and wind to drive water electrolysis and capture CO_2 from the atmosphere or industrial emissions (Sankaran 2023). The resulting hydrogen gas (H_2) is then combined with CO_2 to create syngas, which is further transformed into hydrocarbons through catalytic processes (Shi et al. 2020).

In recent years, significant progress has been made in developing Electrofuels. Researchers have focused on improving the efficiency of electrolysis, exploring novel electrode materials and advanced electrolyte compositions to increase hydrogen gas yields while reducing energy consumption (Burton et al. 2021). Advanced catalysts have also been developed to efficiently convert syngas into hydrocarbons, ensuring the desired properties of the biojet fuels. Additionally, integrated systems that optimize scalability and commercial production have been explored. Despite its promise, Electrofuels face some challenges that need to be addressed for widespread implementation (Carvalho et al. 2021). Energy efficiency during electrolysis is a primary concern, especially when using fluctuating renewable energy sources.

Efforts are underway to improve electrolysis efficiency and find ways to store excess renewable energy for continuous production. Catalyst deactivation during the conversion process is another challenge, requiring the development of durable catalysts for consistent fuel production (Galadima and Muraza 2019).

Cost-effectiveness remains a significant hurdle for Electrofuels, but ongoing research aims to reduce production expenses and make it competitive with traditional fossil fuels (Chattopadhyay and Srivastava 2021). These fuels can be blended with conventional jet fuels, enabling a gradual transition to sustainable aviation practices without major infrastructure modifications. Moreover, Electrofuels have applications beyond aviation, serving as a drop-in replacement for other transportation sectors and offering energy storage solutions (Gray et al. 2021).

Electrofuels represent a promising frontier in sustainable biojet fuel production (Grahn et al. 2022). By harnessing renewable electricity and CO_2 utilization, this technology has the potential to significantly reduce the aviation industry's carbon footprint (Lai et al. 2022). Ongoing research and development are essential to address challenges, improve efficiency, and make Electrofuels a transformative force in achieving a carbon-neutral future.

11.4.5 Microbial Conversion Processes

The aviation sector's heavy reliance on petroleum-based jet fuels has contributed to rising carbon dioxide emissions and climate change concerns (Yusaf et al. 2022). To combat this pressing issue, the development of alternative jet fuels with lower carbon footprints has become essential. Among these alternatives, biojet fuels offer a promising solution as they are derived from renewable biomass feedstocks, significantly reducing net carbon emissions over their life cycle. Notably, microbial conversion processes, such as fermentation and algal biofuels, have emerged as sustainable and economically viable pathways for biojet fuel production (Fu et al. 2022).

Fermentation, an established process for producing bioethanol and biodiesel, has shown immense potential in biojet fuel production. Microorganisms like bacteria and yeasts can ferment biomass-derived sugars into valuable bio-based jet fuel precursors, including fatty acids and alcohols (Shanmugam et al., 2023). Advancements in metabolic engineering have led to optimized microbial strains, increasing yield, enhancing selectivity, and improving tolerance to inhibitory compounds in lignocellulosic feedstocks (Joshi et al. 2022). By adopting biorefinery concepts, multiple value-added products can be generated from a single biomass feedstock, enhancing the economic viability of fermentation-based biojet fuel production.

Another promising feedstock for biojet fuel production is algae, microscopic photosynthetic organisms. Algae offer exceptional advantages due to their high lipid content and rapid growth rates (Khan et al. 2023). They can be cultivated in various systems, from open ponds to closed photobioreactors, depending on environmental factors and desired productivity. Algae cultivation does not compete for arable land, as they can thrive on non-arable or saline land, thus minimizing the impact on food production. Certain algae species can even be engineered to accumulate higher lipid content, making them highly desirable for sustainable aviation fuel production (Correa et al. 2020). After harvesting, lipids can be extracted from the algae biomass and processed through various downstream techniques to obtain high-quality bio-jet fuel.

Significant progress in microbial conversion technologies is evident, with pilot and demonstration-scale facilities established globally (Ewing et al. 2022). Successful test flights by commercial airlines and military organizations using bio-jet fuels derived from microbial conversion processes underscore the commercial potential of these technologies (Why et al. 2019). Ongoing research focuses on developing robust and adaptable microbial strains, exploring novel feedstock sources, and optimizing bioprocess engineering to improve overall biojet fuel yields and reduce production costs.

Microbial conversion processes, particularly fermentation and algal biofuels, offer a compelling solution to the aviation industry's quest for sustainable jet fuels. These technologies provide numerous advantages, including the use of renewable feedstocks, reduction of greenhouse gas emissions, and decreased dependence on fossil fuels (Maliha and Abu-Hijleh 2022). As the aviation sector increasingly embraces environmentally responsible practices, investment in developing and deploying microbial conversion technologies is crucial to achieve a greener and more sustainable future for aviation fuel production.

11.4.6 Hydrodeoxygenation (HDO) Technique

Hydrodeoxygenation is a crucial step in producing biojet fuels, as it addresses one of the primary challenges of biomass-derived feedstocks: their high oxygen content (Lahijani et al. 2022). Biomass, such as plant oils, animal fats, and waste oils, contains significant amounts of oxygenated compounds like alcohol, aldehydes, and carboxylic acids (Okolie et al. 2021). These oxygen-containing functional groups contribute to the bio-oils lower energy density and inferior stability, making them unsuitable for direct use as jet fuel. The HDO process involves the removal of oxygen atoms from these oxygenated compounds, thereby transforming them into hydrocarbons. This conversion increases the energy density of the bio-oils and improves their combustion characteristics, enabling them to meet the stringent performance requirements of jet engines (Attia et al. 2020).

The HDO reaction transforms oxygenated compounds using hydrogen gas and a catalyst under high temperature and pressure. Hydrogen reacts with the compounds, converting them into hydrocarbons and water (Vutolkina et al. 2022). The catalyst enhances reaction speed and selectivity, while challenges include catalyst choice and controlling conditions like temperature, pressure, and hydrogen-to-biomass ratio for optimal results. Various catalyst types are explored, and precise control is needed to balance efficient oxygen removal without excessive hydrogen use or side reactions.

Another challenge lies in the diversity of biomass feedstocks, each with its unique composition and oxygen-containing compounds. Tailoring the HDO process to accommodate different feedstocks and achieve consistent product quality is an ongoing area of research. Additionally, scaling up the HDO process from laboratory-scale to commercial production is a complex task that requires careful consideration of engineering aspects and economic viability (Lynd et al. 2022).

11.4.7 Hydroisomerization Technique

Hydroisomerization is another critical refining technique utilized in the production of biojet fuels (Misra et al. 2023). This process aims to improve the biofuel's cold-flow properties and low-temperature performance by converting straight-chain hydrocarbons into branched isomers (Chen et al. 2020). Branched isomers have lower melting points and improved fluidity at low temperatures, making them less prone to wax formation and enhancing the fuel's ability to flow even in cold climates (Adu-Mensah et al. 2019). During hydroisomerization, the feedstock undergoes molecular rearrangement in the presence of hydrogen gas and specialized catalysts. The process requires high temperature and pressure conditions to promote the breaking and rearrangement of carbon-carbon bonds, forming branched isomers (Ibrahim et al. 2020).

The choice of catalyst plays a critical role in determining the efficiency and selectivity of the hydroisomerization process (Verma et al. 2023). Solid acid catalysts, such as zeolites and modified zeolites, have shown promise in catalyzing the isomerization reactions effectively. Additionally, researchers are exploring new catalytic materials and improving catalyst design to enhance process efficiency and stability. The hydroisomerization process also faces feedstock variability and catalyst deactivation challenges (Tan et al. 2021).

The future prospects of hydrodeoxygenation and hydroisomerization techniques in biojet fuel production are promising. Researchers continuously explore novel catalysts and improve reaction conditions to enhance conversion efficiency and selectivity. Integrating these refining processes into existing petroleum refining infrastructure is a step forward in commercial-scale biojet fuel production. Moreover, ongoing research in process intensification, reactor design, and catalytic material advancements will lead to more cost-effective and sustainable biojet fuel production methods. As the aviation industry intensifies its focus on reducing carbon emissions and transitioning to renewable fuels, these refining and upgrading techniques will play a vital role in shaping the future of biojet fuel production.

11.5 Refining and Upgrading of Biojet Fuels

11.5.1 Integrated Biorefineries

The research investigated presents a holistic approach to the development of an advanced hybrid biorefinery capable of processing diverse biomass feedstocks, such as energy crops (e.g., Jatropha energy crop), dry biomass (e.g., municipal solid waste), and wet biomass (e.g., livestock manure) (Malode et al. 2021). The hybrid system included many advanced processes such as hydroprocessing, Fischer-Tropsch, gasification, dry-reforming, and hydrothermal liquefaction. A prediction model was employed to evaluate the most suitable insertion streams for biomass (Osman et al. 2021). Furthermore, there were comprehensive efforts made to include various materials, heat, water, and electricity in order to optimize the production of JBF, while simultaneously addressing its environmental consequences and managing expenses. The system generated a combined volume of 328 million liters of JBF, 94 million liters of petrol, and 44 million liters of diesel (Alherbawi et al. 2023).

The analysis of characterization indicated that the produced JBF (Jet Biofuel) satisfied or surpassed all relevant international standards. At the highest allowable concentration of jet biofuel mix, which is 50%, the resulting JBF has the potential to substitute 15.3% of Qatar's jet fuel use, so enabling it to fuel about one third of its aircraft fleet (Doliente et al. 2020). Based on the model put out, the minimum attainable selling price for JBF in the year 2019 was determined to be $0.43 per kilogram, representing a reduction of 22% compared to the prevailing market price of standard Jet-A fuel. Based on the environmental study conducted on the model, JBF demonstrates a reduction of 41% in greenhouse gas emissions when compared to Jet-A fuel during the entirety of its lifespan (Alherbawi et al. 2021).

Recent researches introduced a biorefinery system that utilizes pyrolysis as the primary process for the production of jet biofuel. The proposed system employs hydroprocessing as a means to transform bio-oil into a liquid fuel suitable for transportation purposes. Hydrogen is created through the processes of steam reforming and pressure swing adsorption, while power generation takes place on the premises (Sadhukhan and Sen 2021). The previous approach employed by a refinery with a feed capacity of 1.6 million tons per year resulted in a minimum selling price (MSP) range of JBF between $0.60 and $1.40 per kilogram. The Brazilian sugarcane biorefinery achieved the lowest minimum selling price (MSP) of 1.69 $/kg by employing the ATJ route for sugar conversion, and by utilizing rapid pyrolysis and upgrading processes to convert bagasse into Jet biofuel (Alherbawi et al. 2023).

Recent research suggested that the integration of the Hydroprocessed Esters and Fatty Acids (HEFA) and Fischer-Tropsch Synthesis with Solid Acid Catalyst (FT-SPK) pathways has the potential to enhance the maximum selling price (MSP) of jet biofuel from $0.45 to $0.99 per kilogram (Starck et al. 2016). Furthermore, integration of an alcohol-to-jet (ATJ) process into existing palm oil biorefineries as a means of producing Jet biofuel from second-generation biomass. The estimated minimum selling price (MSP) for this method was $0.58 per kilogram (Geleynse et al. 2020).

Nevertheless, the existence of a JBF biorefinery design has not been documented. Such a design would need a significant level of integration and involvement of many technologies to effectively process diverse feedstocks derived from various biomasses. The integration of biorefineries can enhance refinery economics by valorizing process waste through the use of by-products (Tanzil et al. 2022). The design of a hybrid biorefinery is challenging due to the variety of feedstock options, their heterogeneity, and the impact of seasonal variations. The design of an integrated biorefinery that is effective requires the use of conversion pathways that are efficient, the optimization of the supply chain for biomass, the growth of the base of feedstock, and the trade of wastes and by-products (Alherbawi et al. 2023).

11.5.2 Catalysts Used in Biojet Fuel Production for Quality Enhancement

Catalysts play a crucial role in refining, influencing the efficiency and selectivity of reactions. Researchers are exploring various catalyst formulations, support materials, and reaction conditions to optimize the conversion of biomass-derived molecules into aviation-grade hydrocarbons. Heterogeneous catalysts, with easy separation and recyclability, have gained attention for this purpose (Zhao et al. 2017). Metal-based catalysts supported on materials like alumina, silica, or zeolites have shown promise in converting bio-oils and fatty acids into hydrocarbons with improved properties. Research focuses on catalyst structure-activity relationships, metal dispersion enhancement, and novel catalytic materials to improve refining process efficiency and selectivity. Hydrocracking, complementing hydrotreatment, breaks down larger molecules into smaller, more valuable hydrocarbons. This process is beneficial for improving biojet fuel's cold-flow properties, crucial for high-altitude and long flights in colder climates (Babu et al. 2022).

Iron (Fe) and cobalt (Co) are the predominant catalysts employed in contemporary commercial Fischer-Tropsch (FT) processes for the generation of biojet fuel. The current ASTM definition does not encompass FTS biojet generation with other catalysts. The efficiency of FTS is contingent upon the catalyst employed. The CO hydrogenation process is widely acknowledged to benefit from the use of group VIII transition metal oxides due to their exceptional catalytic properties in terms of lifetime, activity, and selectivity (Alherbawi et al. 2023).

The degree of purity shown by the syngas plays a crucial role in determining the lifespan of the catalyst. The metal catalysts that exhibit the highest activity for Fischer-Tropsch synthesis (FTS) are ranked as follows: ruthenium (Ru) demonstrates the highest activity, followed by iron (Fe), nickel (Ni), and cobalt (Co). It has been asserted that syngas catalysts of superior quality have the potential to remain functional for a period ranging from three to five years. While ruthenium catalysts have superior effectiveness, they also exhibit a much higher cost compared to their iron, nickel, and cobalt equivalents, with a price that is almost 100 times more (Jahangiri et al. 2014).

The use of nickel as a catalyst for methanation in Fischer-Tropsch synthesis (FTS) is somewhat restricted when compared to alternative catalysts employed in FTS. Iron demonstrates Water-Gas-Shift (WGS) reactivity; yet, it functions as an acidic catalyst, hence promoting carbon deposition and coking phenomena. Consequently, these undesirable effects result in diminished product yields and reduced catalytic lifespans. Although it is almost 200 times more expensive than Fe, the greater yields and longer catalyst lifetimes that can be achieved with cobalt are well worth the additional cost (Sarkari et al. 2014).

Iron (Fe) serves as a cocatalyst in the manufacture of biojet fuel. The concept of limited temperature refers to a certain range of temperatures within which a system or process operates. This range is defined by upper fast reactors are a type of nuclear reactor that utilize high-energy neutrons to sustain a self-sustaining chain reaction. These reactors catalyst deactivation and an increase in methane (CH_4) selectivity are seen within the temperature range of 200–350° C (Zhao et al. 2017).

11.5.3 Blending and Compatibility Considerations for Biojet Fuels

As the aviation sector continues to expand, concerns about its contribution to climate change intensify, given its significant carbon dioxide emissions. Biojet fuels, known as "drop-in" fuels due to their compatibility with existing engines and infrastructure, offer a promising solution to decarbonize aviation (Shahriar and Khanal 2022). This review paper explores key aspects of biojet fuel production, focusing on blending strategies and compatibility considerations critical for their successful integration into the aviation fuel supply chain.

Blending biojet fuels with conventional jet fuels is a pivotal strategy to introduce renewable alternatives without costly infrastructure modifications. Balancing factors such as greenhouse gas emissions reduction, fuel stability, energy density, and combustion characteristics is essential to achieve an optimal blend. Extensive research has been conducted to understand the impact of various biojet fuel blends on engine performance, emissions, and combustion efficiency (Lim et al. 2023). These studies contribute to identifying suitable blend ratios that meet environmental standards and aircraft engine requirements.

Introducing biojet fuels requires comprehensive assessments of their compatibility with existing aviation infrastructure, aircraft systems, and materials. Differences in chemical compositions and physical properties can affect fuel storage, distribution systems, and engine components. Compatibility studies evaluate potential risks and implement necessary modifications to ensure safe integration (Doliente et al. 2020). This includes evaluating interactions with fuel system seals, lubricants, pipelines, and addressing issues related to freezing points and thermal stability. Optimizing conversion processes and refining techniques to improve production efficiency while maintaining fuel quality is also a technical challenge. Successful commercial deployment relies on robust supply chains, infrastructural investments, and supportive policies. Transitioning to biojet fuels offers the aviation industry an unprecedented opportunity to significantly reduce its carbon footprint and contribute to global climate goals (Lim et al. 2023).

This review paper highlights blending strategies and compatibility considerations to facilitate their successful integration. By understanding challenges and opportunities associated with biojet fuel production, stakeholders can collaboratively address technical barriers, accelerate research, and foster supportive policies. With a holistic approach, biojet fuels can transform the aviation sector into an environmentally responsible and resilient industry, paving the way for a cleaner and greener future for air travel worldwide (Maliha and Abu-Hijleh 2022).

11.6 Sustainability and Environmental Impact

11.6.1 Life Cycle Assessment of Biojet Fuel Production

Biofuels must be compatible with conventional petroleum (petro) fuels before they can be used in modern automobile or machine engines. Various international bodies have issued compatibility standards; for example, the United States has issued the American Society for Testing and Materials (ASTM) standard, and the European Union has issued the European Nation 14,214 standard (Wood 2022). Blends of biodiesel (made from microalgae like *Streptomyces platensis*) and gasoline (a petroleum by-product) are currently available and are used in engines if algal biofuels do not meet these standards.

Analyzing the efficacy of third-generation biofuels necessitates the use of life cycle assessment (LCA), a powerful approach for examining many environmental elements of a given system. The life cycle assessment (LCA) method has been utilized extensively in the past to evaluate the ecological effects of biomass-related systems (Maliha and Abu-Hijleh 2022; Sandmann et al. 2021).

LCA is a standardized, encompassing and internationally compiled methodology, there is no one universal approach to functioning it. LCA is a growing necessity that offers a sustainable foundation for the selection process and expanding customer perspectives on products like biojet fuel, which has a significant environmental impact due to the amount of resources used and the pollution produced throughout its production and distribution (Capaz et al. 2021) (Table 11.1).

The emissions of greenhouse gases (GHGs) originating from the combustion of jet fuel, accounting for approximately 2% of total GHG emissions, have experienced a substantial increase in recent years as a result of the rapid expansion of the aviation sector. There exists optimism over the potential of biojet fuel to mitigate greenhouse gas emissions throughout its entire life cycle. The first life cycle assessment (LCA) was purportedly conducted by the Midwest Research Institute in 1969, as indicated by the research sources (Sun et al. 2016). The main types of technologies for generating biojet fuel are lipid conversion, thermochemical conversion, and biochemical conversion, which encompass processes such as alcohol-to-jet and direct sugar-to-hydrocarbon conversion (Mat Aron et al. 2020). In recent years, there has been a notable decrease in the quantity of life cycle assessment (LCA) studies pertaining to biojet fuels, but a considerable number of such research still exist. Due to the phenomenon of climate change and the aviation sector's ambitious objectives, the predominant focus of research has been directed towards mitigating the release of greenhouse gas emissions (Capaz et al. 2021).

The estimations of HRJ's greenhouse gas (GHG) emissions resulting from microalgae cultivation exhibit a significant degree of variability, spanning a broad spectrum (14.1476 g CO_2/MJ). This large range is mostly attributable to the diverse growth conditions employed and the methodologies employed for allocation. In a research conducted by Pandey et al. (2013), the assessment of biojet fuel generation from microalgae in China was conducted using energy allocation, resulting in a calculated emission of 160 g CO_2eq/MJ. The system expansion approach was employed to examine the impact of CO_2 emissions from power plants on algae growth. The researchers determined that in a high yield and algal lipid content scenario, the emissions were estimated to be 14.1 g CO_2eq/MJ. In contrast, a low yield

Table 11.1 Breakdown of jet biofuel's life cycle carbon, land, water, and energy footprints (Alherbawi et al. 2023)

Stage	Substage	Value (m^3/lifespan)	Value (m^2/lifespan)	Value (ton CO_2^{-e}/lifespan)	Value (MJ/lifespan)
Cultivation	Land setup		454×10^6	3.82×10^3	4428×10^6
	Fertilizers			259.2×10^3	$20,635 \times 10^6$
	Irrigation			1665.2×10^3	$19,305 \times 10^6$
	Machineries			1222×10^3	$14,170 \times 10^6$
	Growing	$10,074 \times 10^6$		$\times 10^3$	
Production	Refinery	0.76×10^6	28.2×10^6	16.9×10^3	25.4×10^6
	Landfill diversion	$\times 10^6$	-197×10^6	−653	
	Raw materials	21.4×10^6		6.9×10^3	8437×10^6
	Processing emissions	-22.12×10^6		409×10^3	-9210×10^6
	Electricity substitution			-1058×10^3	
End use	Fuel combustion			2.2×10^7	
Total		$10,053 \times 10^6$	453.8×10^6	$16,675 \times 10^3$	
Energy footprint		10.4 (cm^2/MJ jet biofuel)	0.023 (m^3/MJ jet biofuel)	53(gCO_2^{-e}/MJ biojet fuel)	0.13 (MJ/MJ jet biofuel)

and algal lipid content scenario resulted in emissions of 193.2 g CO_2eq/MJ (Björnsson and Ericsson 2022).

The major emphasis of the F-T study was on agricultural wastes such as maize stover and non-food energy crops like switchgrass. The life cycle emissions of jet fuels derived from maize stover, forest waste, and switchgrass were found to range from 5 to 15 g CO_2/MJ (Clippinger and Davis 2019). The carbon dioxide emissions associated with F-T fuel derived from switchgrass would experience an increase from 17.7 g CO_2/MJ to 22.0 g CO_2/MJ. According to the International Civil Aviation Organization (ICAO), it is projected that biojet will account for around 2% of the overall use of jet fuel worldwide by the year 2025. However, attaining this level of production would necessitate substantial levels of investment and regulatory assistance that have not been witnessed before (Mat Aron et al. 2020).

The environmental consequences associated with biomass-related systems have frequently been assessed using life cycle assessment (LCA) methodology (Sandmann et al. 2021). This encompasses the environmental impacts associated with the production of ethanol, bioenergy, and other by-products. Numerous research studies and international organizations have conducted investigations on the life cycle assessment (LCA) of algal biofuel production. Hence, the objective of this study is to present a comprehensive analysis of the sustainability, energy consumption, and cost efficiency of third-generation biofuel production. This research encompasses key aspects pertaining to the manufacturing process (Maliha and Abu-Hijleh 2022).

11.6.2 Carbon Footprint Reduction Strategies

To aid in the promotion of sustainable aviation and the altercation against climate change, it is essential to significantly reduce the carbon footprint of biojet fuel production. Feedstock selection is one of the best strategies to reduce carbon footprints (Doliente et al. 2020). The best feedstocks for producing biojet fuel are those that have a low carbon intensity and don't cut into agricultural output. Biofuel feedstock production should be encouraged to employ sustainable farming practices. Emissions from land use and agriculture can be reduced by the adoption of techniques like no-till farming, crop rotation, and precision agriculture (Mathur et al. 2022). Conversion technologies that efficiently transform feedstocks into biojet fuel are a priority. Use alternative techniques of manufacturing that produce less carbon dioxide (such as pyrolysis, gasification, or algae-based production) can significantly reduce the carbon footprints. Biojet fuel-producing facilities should have carbon capture systems (Jayakumar et al. 2023). As a result, the amount of carbon dioxide released into the atmosphere during production can be mitigated by capturing and storing the gas.

Carbon emissions from production of biojet fuel by renewable energy integration sources such as solar, wind, or hydroelectricity process can drastically reduce by decreasing reliance on fossil fuels as a source of energy. Co-products utilization from biojet fuel production to boost profits and cut down on waste. Using by-product biomass for charcoal synthesis or as a nutrient-rich animal feed, for instance, can improve the process's overall sustainability (Doliente et al. 2020) (Table 11.1).

To find the most likely sources of carbon emissions during the production of biojet fuel, a thorough life cycle assessment (LCA) should be conducted. Using this approach, you can pinpoint exactly what needs fixing. The carbon emissions from producing biojet fuel can be mitigated by funding reforestation and land restoration programs (Maliha and Abu-Hijleh 2022). These initiatives have the potential to improve the global carbon balance through carbon sequestration.

Sustainability certification programs and standards should be supported and followed in the manufacturing of biojet fuel. Environmentally sound procedures can be guaranteed by obtaining certification from organizations like the Roundtable on Sustainable Biomaterials (RSB) or the International Sustainability and Carbon Certification (ISCC) (Sukamto 2023). Biojet fuel production can be made more efficiently and sustainably if more money is put into research and development of cutting-edge biofuel technologies. By combining these measures, the carbon footprint of producing biojet fuel can be greatly reduced, making it a competitive and sustainable option to conventional aviation fuels made from fossil fuels (Lim et al. 2023).

11.7 Technological Challenges of Biojet Fuel Production

11.7.1 Low Oil Prices and Competition with Traditional Jet Fuel

Analysts have acknowledged that the alignment with conventional jet fuel is a significant obstacle, given the limited availability of crude oil, escalating expenses, fluctuating prices, and the imperative for ensuring energy stability. O'Connell et al. (2019) have highlighted the significance of the cost of traditional jet fuel in influencing the viability of the alternative aviation fuel industry. In recent decades, there has been a fluctuation in crude oil prices, which therefore affects the pricing of conventional jet fuel, with periods of both increase and decrease. This assertion has particular validity subsequent to the occurrence of price surges in the years 2008 and 2014 (Martinez-Hernandez et al. 2019). As an illustration, it may be observed that in the year 2015, the price of a barrel of Brent crude oil saw a decline to around $40 USD, as compared to its previous value of over $100 USD in 2011 (Olcay et al. 2018). Due to the prevailing cheap cost of oil, the price of biojet fuel is significantly higher compared to that of conventional jet fuel, rendering it economically unfeasible (Carter et al. 2011). In addition, the military's usage of biojet fuel is primarily driven by worries over energy security, which are currently being alleviated by the prevailing low oil prices.

Biojet will engage in competition not only with other biofuels, but also with alternative fuel products derived from coal and natural gas, known as synfuel, for the purpose of obtaining feedstock (Doliente et al. 2020). Aviation synthetic fuel, also referred to as synfuel or synjet, presents a somewhat lower level of environmental advantages compared to biojet fuel, however it currently boasts a more cost-effective production process. In contrast to the limited adoption of alternative jet fuel in the United States, Sasol has successfully implemented significant volumes of synfuel, including synjet, on a commercial scale in South Africa (Lim et al. 2023).

11.7.2 High Production Costs

The primary challenge frequently mentioned in the biojet sector was the significant expense associated with production. In recent interviews it is revealed that high expenses emerged as the second most significant challenge, following the primary concern of obtaining adequate funding (Bittner et al. 2015). According to several sources, the cost of biojet fuel is stated to be between two to four times more than that of conventional jet fuel (Mawhood et al. 2016). However, other writers have claimed even higher price differentials, with estimates reaching as high as seven or eight times more expensive. Furthermore, certain researchers have demonstrated that the documented expenses associated with current biojet operations, as well as the prices paid by consumers such as the US Department of Defense, exceed the projections of prevailing models.

This indicates that numerous studies may significantly underestimate the actual manufacturing costs of biojet fuel (Do and Lim 2016). Between the years 2007 and 2012, the United States Department of Defense allocated an average expenditure of US $10.11 per liter towards the procurement of high-energy ethanol-fuel-air (HEFA) biojet. Additionally, an average of $15.59 per liter was allocated towards alcohol-to-jet (ATJ) biojet, while direct sugar-to-hydrocarbons (DSHC) biojet had an average allocation of $6.80 per liter (Capaz et al. 2021). Notwithstanding the range of perspectives on this matter, a substantial body of research overwhelmingly suggests that the present expenses associated with biojet production are prohibitively expensive from an economic standpoint, hence rendering its manufacturing unjustifiable (McGarvey and Tyner 2018). The primary elements contributing to the cost problem are feedstock availability, capital requirements, economies of scale, suboptimal manufacturing processes, immature technology, accreditation expenses, and many other incidental costs. However, it is important to note that cost estimations are heavily influenced by these factors as well as other relevant considerations (Lim et al. 2023).

11.7.3 Infrastructure Barriers

A further obstacle to the widespread use of biojet fuel is the insufficiency of necessary infrastructure, coupled with restrictions on the utilization of current pipelines and blending facilities (Bond et al. 2014). Researchers cited perceived logistic and infrastructural problems as the second most significant obstacle to the implementation of biojet (Bardell and Ashton 2020). European Union (EU) has implemented regulations that prohibit the transportation of alternative fuels over established fossil fuel pipelines, and also restrict the blending of alternative fuels at the majority of mixing stations. Consequently, the use of substantial quantities of biojet fuel will require additional infrastructure for transport, storage, blending, and fuel testing specifically designed for airports (Kandaramath Hari et al. 2015). In the absence of adequate infrastructure, the production of biojet fuel would necessitate on-site blending and transportation to airports, resulting in increased costs compared to conventional jet fuel. One additional concern is to the limited availability of transportation infrastructure for feedstock, particularly in developing nations (Lee and Mo 2011). Nevertheless, the majority of writers considered the necessary modifications for the utilization of biojet fuel to be a relatively insignificant barrier. Specifically ranked supply chain logistics and infrastructure as the sixth and fifth least significant limitations for biojet production, respectively. Biojet fuel manufacturing was hindered by infrastructural restrictions (Bond et al. 2014).

11.7.4 Strict Fuel Standards

In conjunction with ecological considerations, biojet manufacturers have a noteworthy obstacle in complying with stringent composition and performance criteria imposed on all jet fuel. In contrast to terrestrial transportation vehicles, aircrafts are required to navigate through challenging environments and frequently undergo refueling operations across various global locations (Lim et al. 2023). Consequently, it is imperative for aviation fuel to possess excellent reliability and compatibility with diverse current fleets. Consequently, the production of biojet necessitates more costly and time-consuming procedures compared to other types of biofuels (Connelly et al. 2015). Additionally, prior to utilization, biojet must be blended with conventional jet fuel. Prior to 2010, it was concluded by experts that bio-feedstocks were unable to meet the necessary specifications for aviation fuel (Dominguez-García et al. 2017). Numerous articles acknowledge the difficulties associated with meeting jet fuel standards. However, a consensus has been reached among the bulk of these publications that the technological feasibility obstacle has been overcome in the past decade. This was evidenced by the acceptance of over five different biojet pathways by ASTM (Neuling and Kaltschmitt 2018). Nevertheless, although technological limitations are no longer a hindrance, the expenses associated with the necessary technologies to meet the fuel requirement continue to be a matter of worry.

11.8 Future Prospective

11.8.1 Promising Research Area

Biojet fuel production offers promising prospects for a greener aviation sector, addressing environmental and economic concerns. However, balancing land use and food security is a crucial consideration, necessitating the selection of sustainable feedstocks to avoid competition with food production and mitigate adverse impacts like deforestation and food shortages (Zhang et al. 2020).

Lignocellulosic biomass, including agricultural residues, forestry waste, and dedicated energy crops, holds great promise as a biojet fuel feedstock. Abundant availability and non-competition with food crops make them environmentally sustainable options, exemplified by the Red Rock Biofuels plant in Oregon, USA, utilizing forestry residues for large-scale sustainable aviation fuel production (Ng et al. 2021). Microalgae strains offer significant potential for biojet fuel production, surpassing conventional crop yields and efficiency. Their closed-loop cultivation systems, utilizing sunlight and CO_2, make them efficient biofuel producers, supporting carbon capture efforts (Lim et al. 2021).

Innovations in genetic engineering have led to enhanced lipid or oil content in crops like high-oil maize and canola, promising higher biofuel yields. However, careful regulation is crucial to address environmental and public acceptance concerns. Overcoming challenges related to feedstock variability, costly cultivation, and advancing conversion technologies is essential for the economic viability of large-scale biojet fuel production (Kargbo et al. 2021). By investing in sustainable feedstock options and refining technologies, the biojet fuel industry can lead towards a more environmentally friendly aviation sector with reliable fuel quality.

The future of microbial conversion technologies for biojet fuel production looks promising, with continuous research aimed at addressing existing challenges and uncovering new opportunities. Advancements in metabolic engineering and synthetic biology are expected to lead to tailor-made microorganisms capable of efficiently converting diverse feedstocks into biojet fuels (Keasling et al. 2021). Additionally, integrating renewable energy sources, such as solar and wind power, into algal cultivation systems can enhance algal biofuel production's sustainability and overall energy balance. Process intensification and optimization innovations will likely reduce production costs, making biojet fuels more economically competitive in the aviation market.

Biojet fuel production has emerged as a promising solution to address aviation emissions' environmental challenges. Derived from renewable biomass feedstocks like vegetable oils, animal fats, algae, and waste materials, biojet fuels offer a more sustainable alternative to traditional fossil-based jet fuels (Doliente et al. 2020). To make biojet fuel a viable and widely adopted solution in the aviation industry, enhancing its quality, performance, and cost-effectiveness through advanced refining processes and specialized catalysts is crucial. Waste biomass, being a by-product of agro-forestry, residential, commercial, and industrial operations, possesses the advantage of not necessitating more land and exhibiting reduced water footprints compared to cultivated crops (Mawhood et al. 2016).

Hydrotreatment, a pivotal step in refining, involves using hydrogen and catalysts to remove impurities and stabilize the fuel. With catalysts typically based on metals like nickel, cobalt, or palladium, hydrotreatment facilitates chemical reactions that reduce oxygen, sulfur, and nitrogen content in the feedstock (Lahijani et al. 2022). As a result, the energy density, thermal stability, and overall quality of the biojet fuel are significantly improved.

To meet stringent aviation specifications, biojet fuel undergoes fractional distillation and blending with conventional jet fuel. This fine-tunes the properties of the final product, ensuring compliance with aviation standards set by organizations like ASTM International and IATA (Yildiz 2022).

Despite significant progress, challenges remain in scaling up these technologies for large-scale biojet fuel production while maintaining cost-effectiveness. Biomass feedstock availability and sustainability also influence economic viability. Continued research and innovation are necessary to optimize refining processes, reduce energy consumption, and minimize waste generation, making biojet fuel economically competitive with conventional jet fuel (Goh et al. 2020).

11.8.2 Opportunities for Biojet Fuel Production

11.8.2.1 Increasing Emissions/Demand for Jet Fuel

The primary factor often cited as a catalyst for biojet manufacturers is the increasing cost of air travel and the subsequent emissions it generates. Recent study discovered comparable findings, indicating that the primary opportunity for biojet generation lies in the reduction of greenhouse gas (GHG) emissions (Barbosa 2017). There are two potential advancements that possess the capacity to augment the demand for biojet fuel. There are two factors that contribute to the growth of the sector. Firstly, it is projected that the industry will see an annual expansion of 4–5%. Secondly, there is an improvement in fuel economy, which historically has shown a more modest annual rise of 1.5% (Deane and Pye 2018).

According to the International Civil Aviation Organization (ICAO) (Diederichs et al. 2016), it is projected that global jet fuel consumption would see a substantial increase of 2.8–3.9 times its 2010 levels by the year 2040, mostly as a result of the aforementioned two trends. The key catalyst for increasing demand, notably in China, India, and the Middle East, is projected to be the swift economic growth observed in developing nations (Dietrich et al. 2018). The aforementioned regions encompass Asia, Africa, and South America (Do and Lim 2016). In a study conducted by Dodd et al. (2018), it is said that the immediate consequences of the COVID-19 pandemic, such as the significant decline in jet fuel demand over the years 2020 and 2021, are improbable to have an impact on the long-term trend of rising air traffic.

11.8.2.2 Demand for Diversification of Jet Fuel Supply

The demand for diversification of jet fuel supply has emerged as a pressing concern in the aviation industry due to several factors. Heavy reliance on conventional petroleum-based jet fuels poses various challenges, including price volatility, supply chain disruptions, and geopolitical risks. Fluctuations in global oil prices can significantly impact airlines' operating costs, leading to financial instability and unpredictability in the industry (Barbosa 2017). Moreover, geopolitical tensions and conflicts in major oil-producing regions can disrupt the supply and availability of traditional jet fuels, further accentuating the need for a more resilient and diversified fuel supply (Bond et al. 2014).

To address these challenges, there is a growing recognition of the importance of reducing the aviation industry's dependence on fossil fuels and transitioning towards more sustainable and environmentally friendly alternatives. The concept of energy security, which focuses on ensuring a stable and reliable energy supply, has gained traction in the aviation sector, driving the exploration of alternative fuel sources like biojet fuels. Biojet fuels offer a compelling solution to the demand for diversification (Matuszewska et al. 2021). Produced from renewable feedstocks such as plant oils, algae, agricultural residues, and waste materials, biojet fuels present a promising opportunity to reduce carbon emissions and promote a more sustainable aviation sector. These fuels can be blended with conventional jet fuels or used as drop-in replacements, ensuring compatibility with existing aircraft and infrastructure (Connelly et al. 2015). Additionally, the diversification of jet fuel supply aligns with global efforts to address climate change and reduce greenhouse gas emissions. As countries and international organizations implement policies to limit carbon emissions and encourage sustainable practices, airlines are increasingly motivated to adopt biojet fuels to meet environmental regulations and enhance environmental performance (Lim et al. 2023).

11.8.2.3 Potential Profitability and Positive Public Perception of Biofuels

The potential profitability and positive public perception of biofuels, particularly biojet fuels, are critical to widespread adoption in the aviation industry. As the world seeks to transition towards more sustainable practices, biofuels offer a unique opportunity to reduce greenhouse gas emissions and mitigate the environmental impact of air travel (Cantarella et al. 2015). Airlines and aviation companies increasingly recognize the economic benefits of investing in sustainable initiatives. While biofuel production and distribution costs may be higher than conventional jet fuels, technological advancements and economies of scale steadily drive down costs, making biofuels more economically competitive (Martinez-Hernandez et al. 2019).

Moreover, many governments and regulatory bodies are implementing policies and incentives to encourage the use of biojet fuels. Biofuel mandates, tax incentives, and carbon pricing mechanisms are being introduced to support sustainable aviation practices (Lu 2018). This, in turn, can create a favorable market for biojet fuels, attracting investments and promoting profitability in the sector. Public perception plays a crucial role as the aviation industry transitions to more sustainable practices (Maniatis 2013). Consumers are becoming increasingly environmentally conscious and are seeking eco-friendly travel options. Furthermore, the positive public perception of biofuels is not limited to passengers but extends to investors and stakeholders (Liu et al. 2013). As environmental concerns become more prominent, investors increasingly seek companies that prioritize sustainability and demonstrate responsible corporate citizenship (Bond et al. 2014).

11.8.2.4 Supportive Government Policy

The manufacturers of biojet fuel have identified government policy as the third most favorable opportunity. Favorable government regulation was identified as the third and fourth most potential chances for biojet, respectively. Biofuels have many opportunities within the ambit of governmental regulations implemented globally (Bardell and Ashton 2020). The most often cited policy supports in the literature study were carbon pricing systems, namely the European Union's Emission Trading Scheme (ETS) and the sector-specific Carbon Offsetting and Reduction Scheme for International Aviation (CORSIA), with a total of 63 mentions (Filimonau et al. 2016).

Biofuels, such as biojet, possess a carbon neutral status, enabling users to either diminish their overall carbon footprint or generate credits that may be traded with other entities contributing to pollution (Bond et al. 2014). Nevertheless, a substantial number of countries, amounting to at least 80, have made a commitment to adopt the CORSIA carbon pricing system once it is operational. The European Union promotes the utilization of biofuels through the Renewable Energy Directives, namely RED-I and RED-II. These directives offer credits to biojet users, which may be utilized to fulfil the renewable energy consumption targets (Connelly et al. 2015).

The emergence of general carbon and biofuel restrictions has given rise to several opportunities in the field of biojet. However, it is worth noting that specialized rules pertaining to biojet are relatively few. Only a limited number of countries have implemented such forms of assistance, with Norway (since 2015) (Do and Lim 2016) and Indonesia (starting in 2018) (Choi and Ritchie 2014) being two notable examples. Indonesia and France have expressed intentions to increase their utilization of biojet fuel in the forthcoming years. Conversely, Spain is contemplating the establishment of a target to achieve a 2% biojet fuel consumption rate by the year 2025 (Lim et al. 2023).

11.9 Conclusion

The utilization of renewable bioresources for the generation of jet fuels holds significant promise for the aviation sector in its efforts to decrease reliance on fossil fuels and attain carbon emission reduction objectives. The feedstock utilized in the creation of biojet fuel is diverse, with distinct production paths necessitating specific feedstock sources. The primary obstacles to the commercialization of jet fuel pertain to the availability of feedstock, economic considerations, and sustainability concerns. The rationale for choosing feedstock and technologies for the manufacture of SAF should be supported by a thorough analysis of production costs and environmental impact. It is important to ensure that these choices do not create competition with the current biofuel market for road transportation. Additional investigation and advancement should be directed on enhancing, incorporating, and expanding SAF technology in order to enhance the economic efficiency of SAF manufacturing. The acceleration of SAF adoption necessitates the implementation of policy support that promotes collaboration among several stakeholders, in addition to governmental action.

References

Abdullah B, Syed Muhammad SAF, Shokravi Z, Ismail S, Kassim KA, Mahmood AN, Aziz MMA (2019) Fourth generation biofuel: a review on risks and mitigation strategies. Renew Sust Energ Rev 107. https://doi.org/10.1016/j.rser.2019.02.018

Adu-Mensah D, Mei D, Zuo L, Zhang Q, Wang J (2019) A review on partial hydrogenation of biodiesel and its influence on fuel properties. Fuel 251. https://doi.org/10.1016/j.fuel.2019.04.036

Ahmad U, Sharma L (2023) A review of best management practices for potato crop using precision agricultural technologies. Smart Agric Technol 4. https://doi.org/10.1016/j.atech.2023.100220

Ahmad MS, Cheng CK, Bhuyar P, Atabani AE, Pugazhendhi A, Chi NTL, Witoon T, Lim JW, Juan JC (2021) Effect of reaction conditions on the lifetime of SAPO-34 catalysts in methanol to olefins process – a review. Fuel 283. https://doi.org/10.1016/j.fuel.2020.118851

Ahorsu R, Medina F, Constantí M (2018) Significance and challenges of biomass as a suitable feedstock for bioenergy and biochemical production: a review. Energies 11(12). https://doi.org/10.3390/en11123366

Akmach D, Bathla S, Tran CC, Kaliaguine S, Mushrif SH (2023) Transition metal carbide catalysts for upgrading lignocellulosic biomass-derived oxygenates: a review of the experimental and computational investigations into structure-property relationships. Catal Today 423. https://doi.org/10.1016/j.cattod.2023.114285

Alalwan HA, Alminshid AH, Aljaafari HAS (2019) Promising evolution of biofuel generations. Subject review. Renew Energ Focus 28. https://doi.org/10.1016/j.ref.2018.12.006

Alekseeva MV, Bulavchenko OA, Saraev AA, Kremneva AM, Shashkov MV, Zaikina OO, Gulyaeva YK, Grachev AN, Kikhtyanin O, Kubička D, Yakovlev VA (2020) Effect of temperature on the hydrotreatment of sewage sludge-derived pyrolysis oil and behavior of Ni-based catalyst. Catalysts 10(11). https://doi.org/10.3390/catal10111273

Alherbawi M, McKay G, Mackey HR, Al-Ansari T (2021) A novel integrated pathway for jet biofuel production from whole energy crops: a Jatropha curcas case study. Energy Convers Manag 229. https://doi.org/10.1016/j.enconman.2020.113662

Alherbawi M, McKay G, Al-Ansari T (2023) Development of a hybrid biorefinery for jet biofuel production. Energy Convers Manag 276. https://doi.org/10.1016/j.enconman.2022.116569

Amicarelli V, Lagioia G, Patruno A, Grosu RM, Bux C (2021) Enhancing the sustainability of the aviation industry: airlines' commitment to "green" practices. Amfiteatru Economic 23(15). https://doi.org/10.24818/EA/2021/S15/934

Attia M, Farag S, Chaouki J (2020) Upgrading of oils from biomass and waste: catalytic hydrodeoxygenation. Catalysts 10(12). https://doi.org/10.3390/catal10121381

Aup-Ngoen K, Noipitak M (2020) Effect of carbon-rich biochar on mechanical properties of PLA-biochar composites. Sustain Chem Pharm 15. https://doi.org/10.1016/j.scp.2019.100204

Babu S, Singh Rathore S, Singh R, Kumar S, Singh VK, Yadav SK, Yadav V, Raj R, Yadav D, Shekhawat K, Ali Wani O (2022) Exploring agricultural waste biomass for energy, food and feed production and pollution mitigation: a review. Bioresour Technol 360. https://doi.org/10.1016/j.biortech.2022.127566

Barbosa FC (2017) Biojet fuel – a tool for a sustainable aviation industry – a technical assessment. SAE technical papers, 2017-Novem (November). https://doi.org/10.4271/2017-36-0142

Bardell NS, Ashton MJ (2020) Some issues affecting potential stakeholder uptake of sustainable aviation fuel within Australia: a case study conducted at Darwin international airport. Aust J Mech Eng 18(sup1). https://doi.org/10.1080/14484846.2018.1478700

Behrendt D, Schreiber C, Pfaff C, Müller A, Grobbelaar J, Nedbal L (2017) Algae as a potential source of biokerosene and diesel – opportunities and challenges. In: Biokerosene: status and Prospects. https://doi.org/10.1007/978-3-662-53065-8_14

Bittner A, Tyner WE, Zhao X (2015) Field to flight: a techno-economic analysis of the corn Stover to aviation biofuels supply chain. Biofuels Bioprod Biorefin 9(2). https://doi.org/10.1002/bbb.1536

Björnsson L, Ericsson K (2022) Emerging technologies for the production of biojet fuels from wood—can greenhouse gas emission reductions meet policy requirements? Biomass Convers Biorefin. https://doi.org/10.1007/s13399-022-02916-0

Boichenko S, Vovk O, Iakovlieva A (2013) Overview of innovative technologies for aviation fuels production. Chem Chem Technol 7(3). https://doi.org/10.23939/chcht07.03.305

Bond JQ, Upadhye AA, Olcay H, Tompsett GA, Jae J, Xing R, Alonso DM, Wang D, Zhang T, Kumar R, Foster A, Sen SM, Maravelias CT, Malina R, Barrett SRH, Lobo R, Wyman CE, Dumesic JA, Huber GW (2014) Production of renewable jet fuel range alkanes and commodity chemicals from integrated catalytic processing of biomass. Energy Environ Sci 7(4). https://doi.org/10.1039/c3ee43846e

Brynolf S, Hansson J, Anderson JE, Skov IR, Wallington TJ, Grahn M, Korberg AD, Malmgren E, Taljegård M (2022) Review of electrofuel feasibility – prospects for road, ocean, and air transport. Prog Energy 4(4). https://doi.org/10.1088/2516-1083/ac8097

Budeguer F, Enrique R, Perera MF, Racedo J, Castagnaro AP, Noguera AS, Welin B (2021) Genetic transformation of sugarcane, current status and future prospects. Front Plant Sci 12. https://doi.org/10.3389/fpls.2021.768609

Burton NA, Padilla RV, Rose A, Habibullah H (2021) Increasing the efficiency of hydrogen production from solar powered water electrolysis. Renew Sust Energ Rev 135. https://doi.org/10.1016/j.rser.2020.110255

Cantarella H, Nassar AM, Cortez LAB, Baldassin R (2015) Potential feedstock for renewable aviation fuel in Brazil. Environ Dev 15. https://doi.org/10.1016/j.envdev.2015.05.004

Capaz RS, Guida E, Seabra JEA, Osseweijer P, Posada JA (2021) Mitigating carbon emissions through sustainable aviation fuels: costs and potential. Biofuels Bioprod Biorefin 15(2). https://doi.org/10.1002/bbb.2168

Carter NA, Stratton RW, Bredehoeft MK, Hileman JI (2011) Energy and environmental viability of select alternative jet fuel pathways. 47th AIAA/ASME/SAE/ASEE joint propulsion conference and exhibit 2011. https://doi.org/10.2514/6.2011-5968

Carvalho F, Müller-Casseres E, Poggio M, Nogueira T, Fonte C, Wei HK, Portugal-Pereira J, Rochedo PRR, Szklo A, Schaeffer R (2021) Prospects for carbon-neutral maritime fuels production in Brazil. J Clean Prod 326. https://doi.org/10.1016/j.jclepro.2021.129385

Cavalcanti CJS, Ravagnani MASS, Stragevitch L, Carvalho FR, Pimentel MF (2022) Simulation of the soybean oil hydrotreating process for green diesel production. Cleaner Chem Eng 1. https://doi.org/10.1016/j.clce.2022.100004

Cerone N, Zimbardi F, Contuzzi L, Baleta J, Cerinski D, Skvorčinskienė R (2020) Experimental investigation of syngas composition variation along updraft fixed bed gasifier. Energy Convers Manag 221. https://doi.org/10.1016/j.enconman.2020.113116

Chan YH, Loh SK, Chin BLF, Yiin CL, How BS, Cheah KW, Wong MK, Loy ACM, Gwee YL, Lo SLY, Yusup S, Lam SS (2020) Fractionation and extraction of bio-oil for production of greener fuel and value-added chemicals: recent advances and future prospects. Chem Eng J 397. https://doi.org/10.1016/j.cej.2020.125406

Chattopadhyay J, Srivastava R (2021) Metal hollow spheres as promising electrocatalysts in electrochemical conversion of CO2 to fuels. In: Nano-Catalysts for Energy Applications, pp 190–197. https://doi.org/10.1201/9781003082729-11

Chen YK, Hsieh CH, Wang WC (2020) The production of renewable aviation fuel from waste cooking oil. Part II: catalytic hydro-cracking/isomerization of hydro-processed alkanes into jet fuel range products. Renew Energy 157. https://doi.org/10.1016/j.renene.2020.04.154

Chiaramonti D, Horta Nogueira LA (2017) Aviation biofuels: processes, opportunities, constraints, and perspectives. In: Biofuels production and processing technology. https://doi.org/10.1201/9781315155067

Choi AS, Ritchie BW (2014) Willingness to pay for flying carbon neutral in Australia: an exploratory study of offsetter profiles. J Sustain Tour 22(8). https://doi.org/10.1080/09669582.2014.894518

Chopra J, Rangarajan V, Sen R (2022) Recent developments in oleaginous yeast feedstock based biorefinery for production and life cycle assessment of biofuels and value-added products. Sustainable Energy Technol Assess 53. https://doi.org/10.1016/j.seta.2022.102621

Clippinger J, Davis R (2019) Techno-economic analysis for the production of algal biomass via closed Photobioreactors: future cost potential evaluated across a range of cultivation system designs. National Renew Energy Lab (issue September)

Connelly EB, Colosi LM, Clarens AF, Lambert JH (2015) Risk analysis of biofuels industry for aviation with scenario-based expert elicitation. Syst Eng 18(2). https://doi.org/10.1002/sys.21298

Correa DF, Beyer HL, Fargione JE, Hill JD, Possingham HP, Thomas-Hall SR, Schenk PM (2019) Towards the implementation of sustainable biofuel production systems. Renew Sust Energ Rev 107. https://doi.org/10.1016/j.rser.2019.03.005

Correa DF, Beyer HL, Possingham HP, García-Ulloa J, Ghazoul J, Schenk PM (2020) Freeing land from biofuel production through microalgal cultivation in the Neotropical region. Environ Res Lett 15(9). https://doi.org/10.1088/1748-9326/ab8d7f

Crawford JT, Shan CW, Budsberg E, Morgan H, Bura R, Gustafson R (2016) Hydrocarbon bio-jet fuel from bioconversion of poplar biomass: techno-economic assessment. Biotechnol Biofuels 9(1). https://doi.org/10.1186/s13068-016-0545-7

Dangol N, Shrestha DS, Duffield JA (2020) Life-cycle energy, GHG and cost comparison of camelina-based biodiesel and biojet fuel. Biofuels 11(4). https://doi.org/10.1080/17597269.2017.1369632

Davlatova M (2023) Study of the process of obtaining hydrocarbons on the basis of synthesis gas and the fischer-tropsch synthesis reaction. E3S Web of Conferences, 390. https://doi.org/10.1051/e3sconf/202339005033

de Klerk A, Chauhan G, Halmenschlager C, Link F, Montoya Sánchez N, Gartley B, El-Sayed HEM, Sehdev R, Lehoux R (2022) Sustainable aviation fuel: Pathways to fully formulated synthetic jet fuel via Fischer–Tropsch synthesis. Energy Sci Eng. https://doi.org/10.1002/ese3.1379

Deane JP, Pye S (2018) Europe's ambition for biofuels in aviation – A strategic review of challenges and opportunities. Energ Strat Rev 20. https://doi.org/10.1016/j.esr.2017.12.008

Di Vito Nolfi G, Gallucci K, Rossi L (2021) Green diesel production by catalytic hydrodeoxygenation of vegetables oils. Int J Environ Res Public Health 18(24). https://doi.org/10.3390/ijerph182413041

Diederichs GW, Ali Mandegari M, Farzad S, Görgens JF (2016) Techno-economic comparison of biojet fuel production from lignocellulose, vegetable oil and sugar cane juice. Bioresour Technol 216. https://doi.org/10.1016/j.biortech.2016.05.090

Dietrich RU, Albrecht FG, Maier S, König DH, Estelmann S, Adelung S, Bealu Z, Seitz A (2018) Cost calculations for three different approaches of biofuel production using biomass, electricity and CO2. Biomass Bioenergy 111. https://doi.org/10.1016/j.biombioe.2017.07.006

Dimitriadis A, Chrysikou LP, Meletidis G, Terzis G, Auersvald M, Kubička D, Bezergianni S (2021) Bio-based refinery intermediate production via hydrodeoxygenation of fast pyrolysis bio-oil. Renew Energy 168. https://doi.org/10.1016/j.renene.2020.12.047

Dinç Ö, Gürbüz GB (2022) The benefits of using indirect liquefaction including Fischer-Tropsch process for the automotive industry. Int J High School Res 4(4):77–82. https://doi.org/10.36838/v4i4.15

Do TX, Lim YI (2016) Techno-economic comparison of three energy conversion pathways from empty fruit bunches. Renew Energy 90. https://doi.org/10.1016/j.renene.2016.01.030

Dodd T, Orlitzky M, Nelson T (2018) What stalls a renewable energy industry? Industry outlook of the aviation biofuels industry in Australia, Germany, and the USA. Energy Policy 123. https://doi.org/10.1016/j.enpol.2018.08.048

Doliente SS, Narayan A, Tapia JFD, Samsatli NJ, Zhao Y, Samsatli S (2020) Bio-aviation fuel: a comprehensive review and analysis of the supply chain components. Front Energy Res 8. https://doi.org/10.3389/fenrg.2020.00110

Domínguez-García S, Gutiérrez-Antonio C, De Lira-Flores JA, Ponce-Ortega JM (2017) Optimal planning for the supply chain of biofuels for aviation in Mexico. Clean Techn Environ Policy 19(5). https://doi.org/10.1007/s10098-017-1337-x

Du J, Shen T, Hu J, Zhang F, Yang S, Liu H, Wang H (2023) Study on thermochemical conversion of triglyceride biomass catalyzed by biochar catalyst. Energy 277. https://doi.org/10.1016/j.energy.2023.127733

Elkhlifi Z, Iftikhar J, Sarraf M, Ali B, Saleem MH, Ibranshahib I, Bispo MD, Meili L, Ercisli S, Torun Kayabasi E, Alemzadeh Ansari N, Hegedűsová A, Chen Z (2023) Potential role of biochar on capturing soil nutrients, carbon sequestration and managing environmental challenges: a review. Sustainability (Switzerland) 15(3). https://doi.org/10.3390/su15032527

Ewing TA, Nouse N, van Lint M, van Haveren J, Hugenholtz J, van Es DS (2022) Fermentation for the production of biobased chemicals in a circular economy: a perspective for the period 2022-2050. Green Chem 24(17). https://doi.org/10.1039/d1gc04758b

Filimonau V, Mika M, Pawlusiński R (2016) Public attitudes to biofuel use in aviation: evidence from an emerging tourist market. J Clean Prod 172. https://doi.org/10.1016/j.jclepro.2017.11.101

Fu Q, Tao YB, Xu ZF (2019) Genetic transformation and transgenics of jatropha curcas, a biofuel plant. In: Jatropha, Challenges for a New Energy Crop: Volume 3: A Sustainable Multipurpose Crop. https://doi.org/10.1007/978-981-13-3104-6_5

Fu R, Kang L, Zhang C, Fei Q (2022) Application and progress of techno-economic analysis and life cycle assessment in biomanufacturing of fuels and chemicals. Green Chem Eng. https://doi.org/10.1016/j.gce.2022.09.002

Galadima A, Muraza O (2019) Catalytic thermal conversion of CO2 into fuels: perspective and challenges. Renew Sust Energ Rev 115. https://doi.org/10.1016/j.rser.2019.109333

Garde M, Zingg D (2022) Climate change mitigation in the aviation industry: an analysis of the climate impact attributed to aircraft in flight to enable climate optimization

Geleynse S, Brandt K, Garcia-Perez M, Wolcott M, Zhang X (2018) The alcohol-to-jet conversion pathway for drop-in biofuels: techno-economic evaluation. ChemSusChem 11(21). https://doi.org/10.1002/cssc.201801690

Geleynse S, Jiang Z, Brandt K, Garcia-Perez M, Wolcott M, Zhang X (2020) Pulp mill integration with alcohol-to-jet conversion technology. Fuel Process Technol 201. https://doi.org/10.1016/j.fuproc.2020.106338

Ghannouchi I, Ouni F, Aloulou F (2023) Investigating the impact of transportation system and economic growth on carbon emissions: application of GMM system for 33 european countries. Environ Sci Pollut Res. https://doi.org/10.1007/s11356-023-28595-6

Ghidoli M, Ponzoni E, Araniti F, Miglio D, Pilu R (2023) Genetic improvement of Camelina sativa (L.) Crantz: opportunities and challenges. Plants 12(3). https://doi.org/10.3390/plants12030570

Gnanasekaran L, Priya AK, Thanigaivel S, Hoang TKA, Soto-Moscoso M (2023) The conversion of biomass to fuels via cutting-edge technologies: explorations from natural utilization systems. Fuel 331. https://doi.org/10.1016/j.fuel.2022.125668

Goh BHH, Chong CT, Ge Y, Ong HC, Ng JH, Tian B, Ashokkumar V, Lim S, Seljak T, Józsa V (2020) Progress in utilisation of waste cooking oil for sustainable biodiesel and biojet fuel production. Energy Convers Manage 223. https://doi.org/10.1016/j.enconman.2020.113296

Grahn M, Malmgren E, Korberg AD, Taljegard M, Anderson JE, Brynolf S, Hansson J, Skov IR, Wallington TJ (2022) Review of electrofuel feasibility – cost and environmental impact. Progr Energy 4(3). https://doi.org/10.1088/2516-1083/ac7937

Gray N, McDonagh S, O'Shea R, Smyth B, Murphy JD (2021) Decarbonising ships, planes and trucks: an analysis of suitable low-carbon fuels for the maritime, aviation and haulage sectors. Adv Appl Energy 1. https://doi.org/10.1016/j.adapen.2021.100008

Haider MS, Castello D, Rosendahl LA (2021) The art of smooth continuous Hydroprocessing of biocrudes obtained from hydrothermal liquefaction: Hydrodemetallization and propensity for coke formation. Energy Fuels 35(13). https://doi.org/10.1021/acs.energyfuels.1c01228

Halba A, Thengane SK, Arora P (2023) A critical outlook on Lignocellulosic biomass and plastics co-gasification: a mini-review. Energy and fuels 37(1). https://doi.org/10.1021/acs.energyfuels.2c02907

Han X, Wang H, Zeng Y, Liu J (2021) Advancing the application of bio-oils by co-processing with petroleum intermediates: a review. Energy Convers Manage X 10. https://doi.org/10.1016/j.ecmx.2020.100069

Harada R, Nomura T, Yamada K, Mochida K, Suzuki K (2020) Genetic engineering strategies for Euglena gracilis and its industrial contribution to sustainable development goals: a review. Front Bioeng Biotechnol 8. https://doi.org/10.3389/fbioe.2020.00790

Hu L, Wei XY, Kang YH, Guo XH, Xu ML, Zong ZM (2021) Mordenite-supported ruthenium catalyst for selective hydrodeoxygenation of lignin model compounds and lignin-derived bio-oil to produce cycloalkanes. J Energy Inst 96. https://doi.org/10.1016/j.joei.2021.03.017

Hussain I, Alasiri H, Ullah Khan W, Alhooshani K (2023) Advanced electrocatalytic technologies for conversion of carbon dioxide into methanol by electrochemical reduction: recent progress and future perspectives. Coord Chem Rev 482. https://doi.org/10.1016/j.ccr.2023.215081

Ibrahim M, Jalil AA, Khusnun NF, Fatah NAA, Hamid MYS, Gambo Y, Abdulrasheed AA, Hassan NS (2020) Enhanced n-hexane hydroisomerization over bicontinuous lamellar silica mordenite supported platinum (Pt/HM@KCC-1) catalyst. Int J Hydrog Energy 45(36). https://doi.org/10.1016/j.ijhydene.2019.04.067

Jahangiri H, Bennett J, Mahjoubi P, Wilson K, Gu S (2014) A review of advanced catalyst development for Fischer-Tropsch synthesis of hydrocarbons from biomass derived syn-gas. Cat Sci Technol 4(8). https://doi.org/10.1039/c4cy00327f

Jayakumar M, Bizuneh Gebeyehu K, Deso Abo L, Wondimu Tadesse A, Vivekanandan B, Prabhu Sundramurthy V, Bacha W, Ashokkumar V, Baskar G (2023) A comprehensive outlook on topical processing methods for biofuel production and its thermal applications: current advances, sustainability and challenges. Fuel 349. https://doi.org/10.1016/j.fuel.2023.128690

Jha S, Nanda S, Acharya B, Dalai AK (2022) A review of thermochemical conversion of waste biomass to biofuels. Energies 15(17). https://doi.org/10.3390/en15176352

Joshi A, Verma KK, Rajput VD, Minkina T, Arora J (2022) Recent advances in metabolic engineering of microorganisms for advancing lignocellulose-derived biofuels. Bioengineered 13(4). https://doi.org/10.1080/21655979.2022.2051856

Jung S, Kwon EE, Park YK (2021) Catalytic hydrodeoxygenation for upgrading of lignin-derived bio-oils. Biomass Biofuels Biochemicals Lignin Biorefinery. https://doi.org/10.1016/B978-0-12-820294-4.00010-7

Kandaramath Hari T, Yaakob Z, Binitha NN (2015) Aviation biofuel from renewable resources: routes, opportunities and challenges. Renew Sust Energ Rev 42. https://doi.org/10.1016/j.rser.2014.10.095

Kargbo H, Harris JS, Phan AN (2021) "Drop-in" fuel production from biomass: critical review on techno-economic feasibility and sustainability. Renew Sust Energ Rev 135. https://doi.org/10.1016/j.rser.2020.110168

Keasling J, Garcia Martin H, Lee TS, Mukhopadhyay A, Singer SW, Sundstrom E (2021) Microbial production of advanced biofuels. Nat Rev Microbiol 19(11). https://doi.org/10.1038/s41579-021-00577-w

Keselova M, Liptakova D, Galanda J, Koscak P (2019) The current impact of aviation performance on carbon dioxide production. MOSATT 2019 – modern safety technologies in transportation international scientific conference, proceedings. https://doi.org/10.1109/MOSATT48908.2019.8944124

Khan S, Fu P (2020) Biotechnological perspectives on algae: a viable option for next generation biofuels. Curr Opin Biotechnol 62. https://doi.org/10.1016/j.copbio.2019.09.020

Khan S, Naushad M, Iqbal J, Bathula C, Sharma G (2022) Production and harvesting of microalgae and an efficient operational approach to biofuel production for a sustainable environment. Fuel 311. https://doi.org/10.1016/j.fuel.2021.122543

Khan S, Das P, Abdul Quadir M, Thaher MI, Mahata C, Sayadi S, Al-Jabri H (2023) Microalgal feedstock for biofuel production: recent advances, challenges, and future perspective. Fermentation 9(3). https://doi.org/10.3390/fermentation9030281

Khoo KS, Ahmad I, Chew KW, Iwamoto K, Bhatnagar A, Show PL (2023) Enhanced microalgal lipid production for biofuel using different strategies including genetic modification of microalgae: a review. Prog Energy Combust Sci 96. https://doi.org/10.1016/j.pecs.2023.101071

Lahijani P, Mohammadi M, Mohamed AR, Ismail F, Lee KT, Amini G (2022) Upgrading biomass-derived pyrolysis bio-oil to bio-jet fuel through catalytic cracking and hydrodeoxygenation: a review of recent progress. Energy Convers Manage 268. https://doi.org/10.1016/j.enconman.2022.115956

Lai YY, Christley E, Kulanovic A, Teng CC, Björklund A, Nordensvärd J, Karakaya E, Urban F (2022) Analysing the opportunities and challenges for mitigating the climate impact of aviation: a narrative review. Renew Sust Energ Rev 156. https://doi.org/10.1016/j.rser.2021.111972

Lee RA, Lavoie JM (2013) From first- to third-generation biofuels: challenges of producing a commodity from a biomass of increasing complexity. Anim Front 3(2). https://doi.org/10.2527/af.2013-0010

Lee J, Mo J (2011) Analysis of technological innovation and environmental performance improvement in aviation sector. Int J Environ Res Public Health 8(9). https://doi.org/10.3390/ijerph8093777

Lim JHK, Gan YY, Ong HC, Lau BF, Chen WH, Chong CT, Ling TC, Klemeš JJ (2021) Utilization of microalgae for bio-jet fuel production in the aviation sector: challenges and perspective. Renew Sust Energ Rev 149. https://doi.org/10.1016/j.rser.2021.111396

Lim M, Luckert MK (Marty), Qiu F (2023) Economic opportunities and challenges in biojet production: A literature review and analysis. Biomass Bioenergy 170. https://doi.org/10.1016/j.biombioe.2023.106727

Liu G, Yan B, Chen G (2013) Technical review on jet fuel production. Renew Sust Energ Rev 25. https://doi.org/10.1016/j.rser.2013.03.025

Lu C (2018) When will biofuels be economically feasible for commercial flights? Considering the difference between environmental benefits and fuel purchase costs. J Clean Prod 181. https://doi.org/10.1016/j.jclepro.2018.01.227

Lu H, Yu TY, Xu PF, Wei H (2021) Selective decarbonylation via transition-metal-catalyzed carbon-carbon Bond cleavage. Chem Rev 121(1). https://doi.org/10.1021/acs.chemrev.0c00153

Lynd LR, Beckham GT, Guss AM, Jayakody LN, Karp EM, Maranas C, McCormick RL, Amador-Noguez D, Bomble YJ, Davison BH, Foster C, Himmel ME, Holwerda EK, Laser MS, Ng CY, Olson DG, Román-Leshkov Y, Trinh CT, Tuskan GA et al (2022) Toward low-cost biological and hybrid biological/catalytic conversion of cellulosic biomass to fuels†. Energy Environ Sci 15(3). https://doi.org/10.1039/d1ee02540f

Maliha A, Abu-Hijleh B (2022) A review on the current status and post-pandemic prospects of third-generation biofuels. Energy Syst. https://doi.org/10.1007/s12667-022-00514-7

Malode SJ, Prabhu KK, Mascarenhas RJ, Shetti NP, Aminabhavi TM (2021) Recent advances and viability in biofuel production. Energy Conver Manage X 10. https://doi.org/10.1016/j.ecmx.2020.100070

Maniatis K (2013) A performing biofuels supply chain for EU aviation. European Commission (Issue August)

Martinelli M, Gnanamani MK, LeViness S, Jacobs G, Shafer WD (2020) An overview of Fischer-Tropsch synthesis: XtL processes, catalysts and reactors. Appl Catal A Gen 608. https://doi.org/10.1016/j.apcata.2020.117740

Martinez-Hernandez E, Ramírez-Verduzco LF, Amezcua-Allieri MA, Aburto J (2019) Process simulation and techno-economic analysis of bio-jet fuel and green diesel production—minimum selling prices. Chem Eng Res Des 146. https://doi.org/10.1016/j.cherd.2019.03.042

Mat Aron NS, Khoo KS, Chew KW, Show PL, Chen WH, Nguyen THP (2020) Sustainability of the four generations of biofuels – A review. Int J Energy Res 44(12). https://doi.org/10.1002/er.5557

Mathur S, Waswani H, Singh D, Ranjan R (2022) Alternative fuels for agriculture sustainability: carbon footprint and economic feasibility. AgriEngineering 4(4). https://doi.org/10.3390/agriengineering4040063

Matuszewska A, Hańderek A, Paczuski M, Biernat K (2021) Hydrocarbon fractions from thermolysis of waste plastics as components of engine fuels. Energies 14(21). https://doi.org/10.3390/en14217245

Mawhood R, Gazis E, de Jong S, Hoefnagels R, Slade R (2016) Production pathways for renewable jet fuel: a review of commercialization status and future prospects. Biofuels Bioprod Biorefin 10(4). https://doi.org/10.1002/bbb.1644

McGarvey E, Tyner WE (2018) A stochastic techno-economic analysis of the catalytic hydrothermolysis aviation biofuel technology. Biofuels Bioprod Biorefin 12(3). https://doi.org/10.1002/bbb.1863

Mirkouei A (2020) A cyber-physical analyzer system for precision agriculture. Environ Sci Current Res 3(1):1–8. https://doi.org/10.24966/escr-5020/100016

Misra P, Alvarez-Majmutov A, Chen J (2023) Isomerization catalysts and technologies for biorefining: opportunities for producing sustainable aviation fuels. Fuel 351. https://doi.org/10.1016/j.fuel.2023.128994

Mofijur M, Ahmed SF, Rony ZI, Khoo KS, Chowdhury AA, Kalam MA, Le VG, Badruddin IA, Khan TMY (2023) Screening of non-edible (second-generation) feedstocks for the production of sustainable aviation fuel. Fuel 331. https://doi.org/10.1016/j.fuel.2022.125879

Moioli E, Salvati F, Chiesa M, Siecha RT, Manenti F, Laio F, Rulli MC (2018) Analysis of the current world biofuel production under a water–food–energy nexus perspective. Adv Water Resour 121:22–31. https://doi.org/10.1016/j.advwatres.2018.07.007

Molefe M, Nkazi D, Mukaya HE (2019) Method selection for biojet and biogasoline fuel production from Castor oil: a review. Energy Fuels 33(7). https://doi.org/10.1021/acs.energyfuels.9b00384

Muthukrishnan L (2022) Bio-engineering of microalgae: challenges and future prospects toward industrial and environmental applications. J Basic Microbiol 62(3–4). https://doi.org/10.1002/jobm.202100417

Nahak BK, Preetam S, Sharma D, Shukla SK, Syväjärvi M, Toncu DC, Tiwari A (2022) Advancements in net-zero pertinency of lignocellulosic biomass for climate neutral energy production. Renew Sust Energ Rev 161. https://doi.org/10.1016/j.rser.2022.112393

Neuling U, Kaltschmitt M (2018) Techno-economic and environmental analysis of aviation biofuels. Fuel Process Technol 171. https://doi.org/10.1016/j.fuproc.2017.09.022

Ng KS, Farooq D, Yang A (2021) Global biorenewable development strategies for sustainable aviation fuel production. Renew Sustain Energy Rev 150. https://doi.org/10.1016/j.rser.2021.111502

O'Connell A, Kousoulidou M, Lonza L, Weindorf W (2019) Considerations on GHG emissions and energy balances of promising aviation biofuel pathways. Renew Sust Energ Rev 101. https://doi.org/10.1016/j.rser.2018.11.033

O'Shea R, Lin R, Wall DM, Browne JD, Murphy JD (2020) Using biogas to reduce natural gas consumption and greenhouse gas emissions at a large distillery. Appl Energy 279. https://doi.org/10.1016/j.apenergy.2020.115812

Okolie JA, Mukherjee A, Nanda S, Dalai AK, Kozinski JA (2021) Next-generation biofuels and platform biochemicals from lignocellulosic biomass. Int J Energy Res 45(10). https://doi.org/10.1002/er.6697

Olcay H, Malina R, Upadhye AA, Hileman JI, Huber GW, Barrett SRH (2018) Techno-economic and environmental evaluation of producing chemicals and drop-in aviation biofuels via aqueous phase processing. Energy Environ Sci 11(8). https://doi.org/10.1039/c7ee03557h

Onwudili JA, Scaldaferri CA (2023) Catalytic upgrading of intermediate pyrolysis bio-oil to hydrocarbon-rich liquid biofuel via a novel two-stage solvent-assisted process. Fuel 352. https://doi.org/10.1016/j.fuel.2023.129015

Osman AI, Mehta N, Elgarahy AM, Al-Hinai A, Al-Muhtaseb AH, Rooney DW (2021) Conversion of biomass to biofuels and life cycle assessment: a review. Environ Chem Lett 19(6). https://doi.org/10.1007/s10311-021-01273-0

Pandey A, Lee DJ, Chisti Y, Soccol CR (2013) Biofuels from algae. https://doi.org/10.1016/C2012-0-00170-6

Pappa C, Feghali E, Vanbroekhoven K, Triantafyllidis KS (2022) Recent advances in epoxy resins and composites derived from lignin and related bio-oils. Curr Opin Green Sustainable Chem 38. https://doi.org/10.1016/j.cogsc.2022.100687

Patel, H. K., Dobariya, J. B., & Patel, P. S. (2022). Genetic engineering: an optimism for sustainable biofuel production https://doi.org/10.1007/978-981-16-8090-8_6

Perazzini H, Perazzini MTB, Freire FB, Freire FB, Freire JT (2021) Modeling and cost analysis of drying of citrus residues as biomass in rotary dryer for bioenergy. Renew Energy 175. https://doi.org/10.1016/j.renene.2021.04.144

Petersen AM, Chireshe F, Okoro O, Gorgens J, Van Dyk J (2021) Evaluating refinery configurations for deriving sustainable aviation fuel from ethanol or syncrude. Fuel Process Technol 219. https://doi.org/10.1016/j.fuproc.2021.106879

Pilat M, Mako S, Szabo S, Tobisova A, Hanak P, Socha L, Rozenberg R (2018) Wizz air cancels the base in Košice despite a minimum of 80% occupancy of flights to four destinations. NTAD 2018 – 13th international scientific conference – new trends in aviation development, proceedings. https://doi.org/10.1109/NTAD.2018.8551727

Pirker J, Mosnier A, Kraxner F, Havlík P, Obersteiner M (2016) What are the limits to oil palm expansion? Glob Environ Chang 40:73–81. https://doi.org/10.1016/j.gloenvcha.2016.06.007

Porwal J, Porwal SK, Singh R, Singh K (2020) Synthetic biofuels and greenhouse gas mitigation. In: Climate change, photosynthesis and advanced biofuels: the role of biotechnology in the production of value-added plant bio-products. https://doi.org/10.1007/978-981-15-5228-1_10

Pozzo S, Piergiovanni AR, Ponzoni E, Brambilla IM, Galasso I (2022) Evaluation of nutritional and antinutritional compounds in a collection of Camelina sativa varieties. J Crop Improv. https://doi.org/10.1080/15427528.2022.2163017

Prasad Reddy Kannapu H, Pyo S, Shiung Lam S, Jae J, Hoon Rhee G, Ali Khan M, Jeon BH, Park YK (2022) MgO-modified activated biochar for biojet fuels from pyrolysis of sawdust on a simple tandem micro-pyrolyzer. Bioresour Technol 359. https://doi.org/10.1016/j.biortech.2022.127500

Pulighe G, Pirelli T (2023) Assessing the sustainability of bioenergy pathways through a land-water-energy nexus approach. Renew Sust Energ Rev 184. https://doi.org/10.1016/j.rser.2023.113539

Qaseem MF, Wu AM (2021) Marginal lands for bioenergy in China; an outlook in status, potential and management. GCB Bioenergy 13(1). https://doi.org/10.1111/gcbb.12770

Qi J, Wang Y, Hu M, Xu P, Yuan H, Chen Y (2023) A reactor network of biomass gasification process in an updraft gasifier based on the fully kinetic model. Energy 268. https://doi.org/10.1016/j.energy.2023.126642

Qing NKC, Samiran NA, Rashid RA (2022) CFD simulation analysis of sub-component in municipal solid waste gasification using plasma downdraft technique. CFD Lett 14(8). https://doi.org/10.37934/cfdl.14.8.6370

Raimi MO, Odubo TV, Omidiji AO (2021) Creating the healthiest nation: climate change and environmental health impacts in Nigeria: a narrative review. Sustainability Environ 6(1). https://doi.org/10.22158/se.v6n1p61

Rehman A, Saba T, Kashif M, Fati SM, Bahaj SA, Chaudhry H (2022) A revisit of internet of things technologies for monitoring and control strategies in smart agriculture. Agronomy 12(1). https://doi.org/10.3390/agronomy12010127

Richter S, Braun-Unkhoff M, Naumann C, Riedel U (2018) Paths to alternative fuels for aviation. CEAS Aeronaut J 9(3). https://doi.org/10.1007/s13272-018-0296-1

Roth A, Riegel F, Batteiger V (2017) Potentials of biomass and renewable energy: the question of sustainable availability. In: Biokerosene Status Prospects. https://doi.org/10.1007/978-3-662-53065-8_6

Ruan H, Qin Y, Heyne J, Gieleciak R, Feng M, Yang B (2019) Chemical compositions and properties of lignin-based jet fuel range hydrocarbons. Fuel 256. https://doi.org/10.1016/j.fuel.2019.115947

Sadhukhan J, Sen S (2021) A novel mathematical modelling platform for evaluation of a novel biorefinery design with green hydrogen recovery to produce renewable aviation fuel. Chem Eng Res Des 175. https://doi.org/10.1016/j.cherd.2021.09.014

Said Z, Nguyen TH, Sharma P, Li C, Ali HM, Nguyen VN, Pham VV, Ahmed SF, Van DN, Truong TH (2022) Multi-attribute optimization of sustainable aviation fuel production-process from microalgae source. Fuel 324. https://doi.org/10.1016/j.fuel.2022.124759

Saidi M, Moradi P (2021) Catalytic hydrotreatment of lignin-derived pyrolysis bio-oils using cu/γ--Al2O3 catalyst: reaction network development and kinetic study of anisole upgrading. Int J Energy Res 45(6). https://doi.org/10.1002/er.6642

Sandmann M, Smetana S, Heinz V, Rohn S (2021) Comparative life cycle assessment of a mesh ultra-thin layer photobioreactor and a tubular glass photobioreactor for the production of bioactive algae extracts. Bioresour Technol 340. https://doi.org/10.1016/j.biortech.2021.125657

Sankaran K (2023) Turning black to green: circular economy of industrial carbon emissions. Energy Sustain Dev 74. https://doi.org/10.1016/j.esd.2023.05.003

Sarkari M, Fazlollahi F, Ajamein H, Atashi H, Hecker WC, Baxter LL (2014) Catalytic performance of an iron-based catalyst in Fischer-Tropsch synthesis. Fuel Process Technol 127. https://doi.org/10.1016/j.fuproc.2014.05.003

Sarwer A, Hamed SM, Osman AI, Jamil F, Al-Muhtaseb AH, Alhajeri NS, Rooney DW (2022) Algal biomass valorization for biofuel production and carbon sequestration: a review. Environ Chem Lett 20(5). https://doi.org/10.1007/s10311-022-01458-1

Shahabuddin M, Alam MT, Krishna BB, Bhaskar T, Perkins G (2020) A review on the production of renewable aviation fuels from the gasification of biomass and residual wastes. Bioresour Technol 312. https://doi.org/10.1016/j.biortech.2020.123596

Shahare VV, Kumar B, Singh P (2017) Biofuels for sustainable development: a global perspective. In: Green Technologies and Environmental Sustainability. https://doi.org/10.1007/978-3-319-50654-8_3

Shahid A, Rehman A u, Usman M, Ashraf MUF, Javed MR, Khan AZ, Gill SS, Mehmood MA (2020) Engineering the metabolic pathways of lipid biosynthesis to develop robust microalgal strains for biodiesel production. Biotechnol Appl Biochem 67(1). https://doi.org/10.1002/bab.1812

Shahriar MF, Khanal A (2022) The current techno-economic, environmental, policy status and perspectives of sustainable aviation fuel (SAF). Fuel 325. https://doi.org/10.1016/j.fuel.2022.124905

Shanmugam S, Hari A, Pugazhendhi A, Kikas T (2023) Integrated catalytic upgrading of biomass-derived alcohols for advanced biofuel production. Energies 16(13). https://doi.org/10.3390/en16134998

Shi K, Yan J, Menéndez JA, Luo X, Yang G, Chen Y, Lester E, Wu T (2020) Production of H2-rich syngas from Lignocellulosic biomass using microwave-assisted pyrolysis coupled with activated carbon enabled reforming. Front Chem 8. https://doi.org/10.3389/fchem.2020.00003

Silva GCR, de Andrade MHC (2023) Simulation of deoxygenation of vegetable oils for diesel-like fuel production in continuous reactor. Biomass Convers Biorefin 13(3). https://doi.org/10.1007/s13399-021-01441-w

Sirangelo TM, Ludlow RA, Chenet T, Pasti L, Spadafora ND (2023) Multi-omics and genome editing studies on plant cell walls to improve biomass quality. Agriculture (Switzerland) 13(4). https://doi.org/10.3390/agriculture13040752

Starck L, Pidol L, Jeuland N, Chapus T, Bogers P, Bauldreay J (2016) Production of Hydroprocessed Esters and Fatty Acids (HEFA) – optimisation of process yield. Oil Gas Sci Technol Revue d'IFP Energies Nouvelles 71(1). https://doi.org/10.2516/ogst/2014007

Stavi I, Paschalidou A, Kyriazopoulos AP, Halbac-Cotoara-zamfir R, Siad SM, Suska-Malawska M, Savic D, de Pinho JR, Thalheimer L, Williams DS, Hashimshony-Yaffe N, van der Geest K, Cordovil CMDS, Ficko A (2021) Multidimensional food security nexus in drylands under the slow onset effects of climate change. Land 10(12). https://doi.org/10.3390/land10121350

Sukamto S (2023) Certification of people's oil palm plantations as a model of sustainable plantation management. Baltic J Law Politics 16(3):1870–1882. https://doi.org/10.2478/bjlp-2023-00000153

Sun Z, Liu J, Zhou ZG (2016) Algae for biofuels: an emerging feedstock. An emerging feedstock. In: Handbook of biofuels production: processes and technologies: Second edition. https://doi.org/10.1016/B978-0-08-100455-5.00022-9

Tan Y, Hu W, Du Y, Li J (2021) Species and impacts of metal sites over bifunctional catalyst on long chain n-alkane hydroisomerization: a review. Appl Catal A General 611. https://doi.org/10.1016/j.apcata.2020.117916

Tan ECD, Harris K, Tifft SM, Steward D, Kinchin C, Thompson TN (2022) Adoption of biofuels for marine shipping decarbonization: a long-term price and scalability assessment. Biofuels Bioprod Biorefin 16(4). https://doi.org/10.1002/bbb.2350

Tanzil AH, Brandt K, Zhang X, Wolcott M, Silva Lora EE, Stockle C, Garcia-Perez M (2022) Evaluation of bio-refinery alternatives to produce sustainable aviation fuels in a sugarcane mill. Fuel 321. https://doi.org/10.1016/j.fuel.2022.123992

Teimouri Z, Borugadda VB, Dalai AK, Abatzoglou N (2022) Application of computational fluid dynamics for modeling of Fischer-Tropsch synthesis as a sustainable energy resource in different reactor configurations: a review. Renew Sust Energ Rev 160. https://doi.org/10.1016/j.rser.2022.112287

Terrenoire E, Hauglustaine DA, Gasser T, Penanhoat O (2019) The contribution of carbon dioxide emissions from the aviation sector to future climate change. Environ Res Lett 14(8). https://doi.org/10.1088/1748-9326/ab3086

Tiwari R, Mishra R, Choubey A, Kumar S, Atabani AE, Badruddin IA, Khan TMY (2023) Environmental and economic issues for renewable production of bio-jet fuel: a global prospective. Fuel 332. https://doi.org/10.1016/j.fuel.2022.125978

van Dyk S, Saddler J (2021) Progress in commercialization of biojet/sustainable aviation fuels (SAF): technologies, potential and challenges. Task, IEA Bioenergy (Issue May)

Verma V, Mishra A, Anand M, Farooqui SA, Sinha AK (2023) Catalytic hydroprocessing of waste cooking oil for the production of drop-in aviation fuel and optimization for improving jet biofuel quality in a fixed bed reactor. Fuel 333. https://doi.org/10.1016/j.fuel.2022.126348

Vijay V, Pimm SL, Jenkins CN, Smith SJ (2016) The impacts of oil palm on recent deforestation and biodiversity loss. PLoS One 11(7). https://doi.org/10.1371/journal.pone.0159668

Vutolkina AV, Baigildin IG, Glotov AP, Pimerzin AA, Akopyan AV, Maximov AL, Karakhanov EA (2022) Hydrodeoxygenation of guaiacol via in situ H2 generated through a water gas shift

reaction over dispersed NiMoS catalysts from oil-soluble precursors: tuning the selectivity towards cyclohexene. Appl Catal B Environ 312. https://doi.org/10.1016/j.apcatb.2022.121403
Wang M, Dewil R, Maniatis K, Wheeldon J, Tan T, Baeyens J, Fang Y (2019a) Biomass-derived aviation fuels: challenges and perspective. Prog Energy Combust Sci 74. https://doi.org/10.1016/j.pecs.2019.04.004
Wang J, Yang H, Wang H (2019b) The evolution of China's international aviation markets from a policy perspective on air passenger flows. Sustainability (Switzerland) 11(13). https://doi.org/10.3390/su11133566
Why ESK, Ong HC, Lee HV, Gan YY, Chen WH, Chong CT (2019) Renewable aviation fuel by advanced hydroprocessing of biomass: challenges and perspective. Energy Convers Manage 199. https://doi.org/10.1016/j.enconman.2019.112015
Wood DA (2022) Biodiesel from microalgae. 3rd generation biofuels: disruptive technologies to enable commercial production, 417–438. https://doi.org/10.1016/B978-0-323-90971-6.00039-5
Yildiz M (2022) Initial airworthiness requirements for aircraft electric propulsion. Aircr Eng Aerosp Technol 94(8). https://doi.org/10.1108/AEAT-08-2021-0238
Yusaf T, Fernandes L, Talib ARA, Altarazi YSM, Alrefae W, Kadirgama K, Ramasamy D, Jayasuriya A, Brown G, Mamat R, Dhahad H, Al Benedict F, Laimon M (2022) Sustainable aviation—hydrogen is the future. Sustainability (Switzerland) 14(1). https://doi.org/10.3390/su14010548
Zamri MFMA, Shamsuddin AH, Ali S, Bahru R, Milano J, Tiong SK, Fattah IMR, Raja Shahruzzaman RMH (2023) Recent advances of triglyceride catalytic pyrolysis via Heterogenous dolomite catalyst for upgrading biofuel quality: a review. Nano 13(13). https://doi.org/10.3390/nano13131947
Zeng K, Li J, Xie Y, Yang H, Yang X, Zhong D, Zhen W, Flamant G, Chen H (2020) Molten salt pyrolysis of biomass: the mechanism of volatile reforming and pyrolysis. Energy 213. https://doi.org/10.1016/j.energy.2020.118801
Zhang S, Chen W (2022) China's energy transition pathway in a carbon neutral vision. Engineering 14. https://doi.org/10.1016/j.eng.2021.09.004
Zhang X, Lei H, Zhu L, Qian M, Chan JC, Zhu X, Liu Y, Yadavalli G, Yan D, Wang L, Bu Q, Wei Y, Wu J, Chen S (2016) Development of a catalytically green route from diverse lignocellulosic biomasses to high-density cycloalkanes for jet fuels. Catal Sci Technol 6(12). https://doi.org/10.1039/c5cy01623a
Zhang L, Butler TL, Yang B (2020) Recent trends, opportunities and challenges of sustainable aviation fuel. In: Green energy to sustainability: strategies for global industries. https://doi.org/10.1002/9781119152057.ch5
Zhang L, Hu X, Li C, Zhang S, Wang Y, Esmaeili V, Gholizadeh M (2021a) Fates of heavy organics of bio-oil in hydrotreatment: the key challenge in the way from biomass to biofuel. Sci Total Environ 778. https://doi.org/10.1016/j.scitotenv.2021.146321
Zhang M, Hu Y, Wang H, Li H, Han X, Zeng Y, Xu CC (2021b) A review of bio-oil upgrading by catalytic hydrotreatment: advances, challenges, and prospects. Mol Catal 504. https://doi.org/10.1016/j.mcat.2021.111438
Zhao X, Wei L, Cheng S, Julson J (2017) Review of heterogeneous catalysts for catalytically upgrading vegetable oils into hydrocarbon biofuels. Catalysts 7(3). https://doi.org/10.3390/catal7030083
Zhao Z, Abdo SMA, Wang X, Li H, Li X, Gao X (2021) Process intensification on co-pyrolysis of polyethylene terephthalate wastes and biomass via microwave energy: synergetic effect and roles of microwave susceptor. J Anal Appl Pyrolysis 158. https://doi.org/10.1016/j.jaap.2021.105239
Zhou N, Zhou J, Dai L, Guo F, Wang Y, Li H, Deng W, Lei H, Chen P, Liu Y, Ruan R (2020) Syngas production from biomass pyrolysis in a continuous microwave assisted pyrolysis system. Bioresour Technol 314. https://doi.org/10.1016/j.biortech.2020.123756

Zhou Y, Remón J, Jiang Z, Matharu AS, Hu C (2023) Tuning the selectivity of natural oils and fatty acids/esters deoxygenation to biofuels and fatty alcohols: a review. Green Energy Environ 8(3). https://doi.org/10.1016/j.gee.2022.03.001

Zhu X, Xiao J, Wang C, Zhu L, Wang S (2022) Global warming potential analysis of bio-jet fuel based on life cycle assessment. Carbon Neutrality 1(1). https://doi.org/10.1007/s43979-022-00026-4

Chapter 12
Life Cycle Assessment of Bio-Jet Fuel

Samakshi Verma, Y. Lalitha Kameswari, and Sonu Kumar

Abstract The investigation of environmentally friendly substitutes for conventional jet fuels has been motivated by the aviation industry's negative environmental effects, particularly those caused by greenhouse gas emissions. As a potential answer, bio-jet fuels (BJFs) made from sustainable feedstocks have gained popularity. In order to compare the environmental performance of bio-jet fuel to that of traditional fossil-based jet fuel, this study does a thorough life cycle assessment (LCA). The LCA framework takes into account every stage of the manufacture of bio-jet fuel, from feedstock cultivation through final application in aircraft engines. To estimate the environmental impacts throughout multiple life cycle stages, data from numerous sources are combined, including literature, industry reports, and databases. The possibility for global warming, energy use, water use, eutrophication, and land use change are among the environmental factors evaluated. The LCA's first findings show that producing bio-jet fuel typically results in fewer net carbon dioxide emissions than producing conventional jet fuel. The production of biofuel feedstocks and their processing has an impact on things like land use change and agricultural inputs. However, the ability of feedstock crops to store carbon and the adoption of more environmentally friendly agricultural methods frequently mitigate these effects. A lower overall energy consumption during the life cycle of bio-

S. Verma (✉)
Department of Biotechnology, Faculty of Engineering and Technology, Rama University, Kanpur, Uttar Pradesh, India

Y. L. Kameswari
Department of Electronics and Communication Engineering, Koneru Lakshmaiah Education Foundation (Deemed to be University), Vaddeswaram, Guntur, Andhra Pradesh, India

Department of Electrical and Electronics Engineering, MLR Institute of Technology, Hyderabad, Telangana, India

S. Kumar
Department of Electronics and Communication Engineering, K.S.R.M. College of Engineering, Kadapa, Andhra Pradesh, India

Manipur International University, Imphal, Manipur, India

A. Kuila (ed.), *Biojet Fuel: Current Technology and Future Prospect*, Clean Energy Production Technologies, https://doi.org/10.1007/978-981-99-8783-2_12

jet fuel is also a result of the reduced reliance on fossil fuels. Water utilization is a crucial factor to take into account, with some biofuel manufacturing methods using more water than traditional jet fuel refining. The study assesses methods to improve the overall sustainability of the manufacturing of bio-jet fuel and lessen its effects on the environment. This LCA seeks to provide light on the environmental trade-offs related to the adoption of bio-jet fuel through a comparative examination of several bio-jet fuel production pathways and feedstock sources. The findings will help stakeholders in the aviation industry make well-informed judgments about the viability and sustainability of integrating bio-jet fuels. Life cycle assessment carried out in this chapter emphasizes the potential of bio-jet fuel to lessen the environmental impact of the aviation sector. The practicality of bio-jet fuels as a crucial part of a more sustainable aviation future is anticipated to be further enhanced by continuous research and technology improvements, even though challenges regarding feedstock farming practices and water utilization still exist.

Keywords Life cycle assessment (LCA) · Fossil-based jet fuel · Bio-jet fuels

12.1 Introduction

The aviation sector is crucial to international trade and transportation because it links people and commodities everywhere. However, there are substantial environmental issues due to the industry's reliance on fossil fuels, notably jet fuels made from crude oil (Tien et al. 2019). Aircraft engines' increasing emissions of greenhouse gases, particularly carbon dioxide (CO_2), are a factor in climate change and deteriorating air quality. The aviation industry is under pressure to lessen its environmental impact as global efforts to combat climate change grow. The creation and use of bio-jet fuels is one promising strategy for reducing the environmental impact of the aviation sector. Aviation biofuels, commonly referred to as bio-jet fuels, are produced from renewable feedstocks such plant oils, agricultural waste, and algae (Wang and Tao 2016). Bio-jet fuels have the potential to dramatically lower net carbon emissions and dependency on fossil fuels than conventional jet fuels, which are generally carbon-intensive and non-renewable. The idea of bio-jet fuels, is not entirely new; numerous research groups and pilot projects have investigated various production methods for these fuels (Wei et al. 2019). The availability of feedstock, the capacity to scale up production, and the viability of bio-jet fuels all present obstacles to their widespread use (Hari et al. 2015). Therefore, policymakers, industry stakeholders, and researchers must have a thorough grasp of the environmental effects of bio-jet fuel generation in order to make informed decisions. In order to assess the overall environmental impact of bio-jet fuel over the course of its full life cycle, from feedstock production and processing to distribution and end-use combustion in aircraft engines, a strong technique is the life cycle assessment (LCA) (Michaga et al. 2022). An LCA offers a comprehensive picture of the sustainability of bio-jet fuel in comparison to conventional jet fuels by taking into account a variety of environmental variables, including greenhouse gas emissions, energy consumption, water usage, and land use (Fortier et al. 2014). In order to compare the environmental performance of bio-jet fuel generation to that of conventional jet fuels, this study seeks to undertake an exhaustive and meticulous life cycle assessment (LCA). Numerous feedstock sources, production strategies, and related environmental trade-offs will all be taken into account in the LCA (McKechnie et al. 2011). This study aims to advance the ongoing conversation about environmentally friendly aviation and provide ideas for lowering the carbon footprint of the sector by assessing the possible advantages and difficulties of the adoption of bio-jet fuel. The technique used in the LCA will be described in depth in the parts that follow, including the data sources, presumptions, and boundary considerations (Joensuu et al. 2022). The assessment's findings will be presented in the following sections, together with their consequences and some ideas for improving the sustainability of bio-jet fuel production (Michaga et al. 2022). In the end, this study intends to give stakeholders useful information to help them decide how bio-jet fuels might help the aviation industry become more ecologically conscious.

12.2 Historical Aspects and Definitions

Historical Considerations**:** Life cycle assessment (LCA) was developed as a methodological framework for analyzing how items and processes affect the environment throughout the course of their full life cycles. The 1960s and 1970s saw a rise in concerns about pollution, resource depletion, and environmental degradation, which can be linked to the origins of LCA (McManus and Taylor 2015). The need for a systematic method to evaluate the environmental effects of various products and processes emerged as businesses and governments started to address these challenges. In instance, the introduction of ideas like "cradle-to-grave" analysis helped to formalize the idea of analyzing a product's life cycle in the 1960s (Rebitzer et al. 2004). In the late 1970s and early 1980s, the phrase "Life Cycle Assessment" as a whole began to gain popularity. As organizations and researchers started to create frameworks and procedures for conducting LCA, the ISO 14000 series of standards, including ISO 14040 and ISO 14044, which are especially devoted to LCA methodology and principles, were established (Patón-Romero et al. 2019).

Life Cycle Assessment (LCA): In order to evaluate the environmental effects of a product, process, or activity over the course of its full life cycle, from the extraction of raw materials to the end-of-life disposal or recycling, life cycle assessment (LCA) is a systematic methodology (Lucchetti et al. 2019). LCA aims to provide a thorough understanding of the environmental costs and potential advantages of various options, assisting in the decision-making process towards more sustainable practices (Thabrew et al. 2009).

12.2.1 Important LCA Definitions and Concepts

(a) A product's life cycle is the series of steps it takes from the extraction of the raw materials to production, use, and eventual disposal (Braungart et al. 2007).
(b) *Cradle-to-Grave*: Cradle-to-grave refers to the entire life cycle of a product, from the point of production (cradle) to the point of disposal (grave) (Braungart et al. 2007).
(c) *Functional Unit*: A quantifiable reference unit that captures the functionality of the under investigation process or product. It enables insightful comparisons between several options (Böckin et al. 2022).
(d) *System Boundaries*: The boundaries established for the study that specify which life cycle stages are included and which are excluded. Boundaries might be extensive, encompassing the entire life cycle, or narrow, such as "gate-to-gate" (just one specific stage) (Motalebi et al. 2023).
(e) Inventory analysis is the gathering and measurement of information on inputs, outputs, and environmental effects at each stage of the life cycle (Saavedra-Rubio et al. 2022).
(f) *Impact Assessment*: The process of converting inventory data into relevant categories of environmental impact, such as eutrophication, acidification, and global warming potential (Gaurav et al. 2023).
(g) *Interpretation*: The stage in which the impact assessment's findings are examined in light of potential consequences, uncertainties, and constraints. Making conclusions and wise decisions are aided by this phase (Ghoroghi et al. 2022).
(h) *Sensitivity Analysis*: A method for examining how changes in the data or underlying premises affect the final LCA results (Teng et al. 2023).

When several goods or functions use the same process, environmental loads are divided through a process called allocation. There are numerous allocation strategies, including energy, mass, and economic allocation. Cut-off based on their potential to have an impact on the final outcomes, criteria are used to decide whether processes or inputs are substantial enough to be included in the study.

Life cycle assessment (LCA) is a methodical methodology that assesses the environmental effects of items and activities from conception to disposal. It has developed into a useful resource for comprehending how human activity affects the environment and for assisting in making sustainable decisions.

12.3 Biomass to BJF Conversion

The process of transforming sustainable biomass feedstocks into a kind of aviation fuel known as bio-jet fuel or aviation biofuel is known as biomass-to-bio-jet fuel (BJF). The conversion of biomass, which can include different organic materials like plant oils, agricultural residues, algae, and waste materials, into a fuel that can be used as an alternative to conventional fossil-based jet fuels in aircraft engines involves a number of chemical and technological steps (Sharno and Hiloidhari 2022). The production of bio-jet fuel aims to replace or combine with traditional jet fuels, which are predominantly sourced from crude oil, in order to lessen the carbon footprint and environmental impact of flying. The issue of how the aviation sector contributes to greenhouse gas emissions and climate change has drawn attention to bio-jet fuels. A complicated set of procedures is needed to convert renewable feedstock materials from biomass into bio-jet fuel (BJF), a fuel that can be used in aviation applications (Ravindran et al. 2022). Depending on the technologies used and the type of biomass feedstock, many conversion paths may be possible. Biomass is often transformed into bio-jet fuel through a number of important stages, including:

(a) *Selection and Preparation of the Feedstock*:

The choice of biomass feedstocks is made taking into account elements like accessibility, sustainability, and compatibility with conversion methods (Makepa et al. 2023). Feedstocks are made acceptable for further processing by cleaning, drying, and perhaps pretreating. Agricultural waste, wood, algae, plant oils, and other substances can all be used as biomass feedstock (Makepa et al. 2023). For further processing, feedstock is gathered, harvested, and prepared to produce a constant particle size and moisture content. Various methods including cleaning, drying and grinding are employed during the preparation phase (Zein and Antony 2022).

(b) *Conversion Methods*:

There are several ways to turn biomass into bio-jet fuel, including:

1. Gasification and pyrolysis are examples of thermochemical processes that entail heating biomass in the absence of oxygen to create bio-oil or syngas (Inayat et al. 2022).
2. Bio-oil is produced from biomass by hydrothermal methods like hydrothermal liquefaction, which require high pressure and temperature in the presence of water (Eswary Devi et al. 2022).
3. Fermentation is a biochemical process where bacteria transform biomass sugars into bio-based hydrocarbons.
4. Biomass thermal decomposition without oxygen to produce bio-oil, gas, and char by pyrolysis.
5. A crude bio-oil is created when biomass is converted at high temperatures and pressures while being in the presence of water.
6. Fischer-Tropsch synthesis: Creating liquid hydrocarbons from syngas, a mixture of carbon monoxide and hydrogen obtained from biomass (Sharew et al. 2022).
7. Using hydrocracking or hydrotreating to remove contaminants and enhance the fuel characteristics of bio-oil is known as hydroprocessing (Sharew et al. 2022).

(c) *Upgrading and Refining*:

To achieve the criteria for aviation fuel, the intermediate products produced by the conversion operations frequently need to be refined further. To enhance fuel characteristics and get rid of contaminants, refining techniques including hydrocracking, hydrotreating, and distillation can be used (Sarkar et al. 2023). In order to increase the efficiency of the conversion process, pretreatment seeks to disassemble the intricate biomass structures. Mechanical grinding, chemical processing, and thermal procedures including pyrolysis and hydrothermal liquefaction are typical pretreatment techniques (Ramos et al. 2022). To meet the requirements of aviation fuels, intermediate products resulting from the conversion stage frequently need to be refined further (Misra et al. 2023). Hydrocracking, hydrotreating, and distillation are upgrading procedures used to clean up materials and modify the lengths of the carbon chains.

(d) *Testing and Blending*:

To meet the requirements for usage in aircraft engines, the refined bio-jet fuel is mixed with regular jet fuel. The bio-jet fuel is put through quality control and testing to make sure it satisfies performance and legal requirements (Okolie et al. 2023).

(e) *Combining and Quality Assurance*:

To meet the requirements for usage in aviation, the refined bio-jet fuel is mixed with regular jet fuel. Regulatory and performance standards are met by the final bio-jet fuel, thanks to quality control processes (Lahijani et al. 2022).

(f) *Availability and Use*:

The created bio-jet fuel is delivered to airports where it is mixed with regular jet fuel or utilized as a drop-in substitute in aircraft engines. Existing engines can use bio-jet fuel without extensive modifications (Tiwari et al. 2023). Bio-jet fuel has an end-of-life stage, just like any other fuel. It's important to think about disposal and recycling options to reduce your negative environmental effects (Donnelly et al. 2023). There are numerous ways to convert biomass into bio-jet fuel, and the precise technologies and procedures employed can change depending on elements like the type of feedstock, regional availability, technological improvements, and economic considerations (Hussin et al. 2023). To ensure the total environmental benefits of the bio-jet fuel production process, sustainability factors such feedstock procurement, water usage, and land impacts must also be carefully considered (Tiwari et al. 2023). Airports receive the produced bio-jet fuel, which doesn't require major adjustments to be used in aircraft engines. Bio-jet fuel can be used in place of conventional jet fuel or combined with it. Utilizing regenerative feedstocks that may be regenerated over time, biomass-to-bio-jet fuel conversion seeks to lower the carbon emissions linked to aviation (Seber et al. 2022). The efficiency of conversion processes, feedstock procurement procedures, impacts on land use, water usage, and potential rivalry with food and land resources are only a few examples of the aspects that affect how sustainably bio-jet fuel production may be done overall (Tiwari et al. 2023). The viability and environmental advantages of producing biomass-to-bio-jet fuel are being improved via ongoing study and technology development (Ahmed et al. 2023).

12.4 BJF Production Routes

In order to transform biomass feedstocks into a useful aviation fuel, bio-jet fuel (BJF) can be generated using a variety of production pathways, each incorporating a different set of technologies and procedures (Sharno and Hiloidhari 2022). Depending on the kind of biomass used, regional availability, and technological improvements, these paths may change. Here are a few typical BJF production pathways:

(a) *Hydroprocessed Esters and Fatty Acids (HEFA)*:

One of the well-known methods for producing BJF is HEFA.

Plant oils used in cooking as well as soy and camelina oils are used as feedstock. Transesterification is the method used to transform the feedstock into fatty acid methyl esters (FAME). The FAME is then hydroprocessed to clean it up and modify its characteristics for usage in aircraft. Through this process, biomass-derived syngas—a mixture of carbon monoxide and hydrogen—is transformed into liquid hydrocarbons (Emmanouilidou et al. 2023). Biomass-derived syngas is the feedstock. Catalytic conversion of syngas to long-chain hydrocarbons results in the production of aviation fuels.

(b) *ATF (Alcohol-to-Jet)*:

The ATJ process, which involves turning alcohols obtained from biomass into jet fuel, appears promising. Lignocellulosic biomass or agricultural waste can be used as a feedstock to produce alcohols like ethanol (Peters et al. 2023).

Process: To create an appropriate aviation fuel, alcohols are dehydrated and oligomerized to generate longer hydrocarbon chains.

(c) *Hydrothermal Liquefaction Catalyzed (HTL)*:

HTL entails the transformation of wet biomass into bio-oil at high pressure and temperature while water is present. Wet biomass such as algae or agricultural waste is called a feedstock (Mishra et al. 2022).

Process: By exposing biomass to hydrothermal conditions, bio-oil is created, which can then be processed into jet fuel.

(d) *Upgrading and Pyrolysis*:

In order to create bio-oil, which is later improved, biomass is thermally decomposed in the absence of oxygen. Woody biomass or agricultural wastes are used as feedstock (Singh et al. 2023).

Process: To produce aviation-grade fuel, bio-oil is improved and refined using techniques like hydrotreating and hydrocracking.

(e) *Biological Synthesis and Microbial Transformation*:

It is possible to design microorganisms to transform biomass sugars into aerosol-safe hydrocarbons. Sugars generated from lignocellulosic biomass are used as feedstock. Fermentation by microbial engineering results in bio-based hydrocarbons, which can then be processed into BJF.

It's vital to remember that each production route's viability and environmental performance depend on a variety of elements, including the availability of feedstock, energy use, greenhouse gas emissions, water use, and the entire life cycle evaluation (Andooz et al. 2023). Additionally, these production pathways are being improved as a result of technological and scientific advance, which makes the manufacturing of bio-jet fuel more effective and sustainable. A combination of technical, economic, and sustainability factors will determine the production route to take.

12.5 System Boundary and LCA Framework

The system boundary and life cycle assessment (LCA) paradigm for bio-jet fuel (BJF) entails defining the analysis's purview and methodically assessing the environmental effects connected to BJF use and manufacturing. LCA is a thorough process that examines every aspect of a product's life cycle, from the extraction of raw materials to its disposal at the end of its useful life (Barbhuiya and Das 2023). The analysis's scope is determined by the system boundary, which outlines the life cycle stages that will be covered. Producing feedstock entails gathering, and getting the biomass feedstock (such as plant oils, agricultural waste, and algae) ready for further processing. The movement of feedstock from the point of origin to the facility for conversion is taken into account, as well as the accompanying energy use and emissions (Colbertaldo et al. 2023). The system boundary includes all operations required to transform the biomass feedstock into bio-jet fuel. This also applies to the particular conversion technique employed (such as HEFA, FTS, ATJ, or HTL). Energy use and emissions are taken into account while refining and upgrading materials to meet aviation fuel standards (Ahlström et al. 2023). The research takes into account both the distribution of BJF to airports and its blending with regular jet fuel. A crucial component of the system boundary is the combustion of BJF in aircraft engines, including emissions produced while flying (Sharno and Hiloidhari 2022). Taking into account the recycling or disposal of waste and byproducts produced during the manufacture and usage of BJF. The LCA framework consists of the following crucial steps:

(a) *Goals and Purpose*: Define the LCA's objective, the functional unit (for instance, one million passenger-kilometers), and the system's boundaries (Ali et al. 2023).
(b) *Inventory Analysis*: Gather information about each stage of the life cycle's energy use, emissions, resource use, and other environmental aspects (De Wolf et al. 2023).
(c) *Influence Assessment*: Transform the inventory data into categories that reflect how the data may have an influence on the environment, such as the potential

for global warming (expressed in CO_2 equivalents), energy use, water use, land use, etc. (Jennings et al. 2023).

(d) *Normalization and Weighting*: Based on stakeholder preferences, normalize impact data to pertinent reference values (e.g., world averages) and use weighting variables to prioritize various affects (Torkayesh et al. 2022).

(e) *Interpretation*: Examine and interpret the findings while taking uncertainties and trade-offs into account. Find the areas with the biggest environmental consequences. Analyze the impact of important variables and presumptions on the outcomes to determine how robust the conclusions are (Haldar et al. 2023). Based on the identified environmental hotspots, recommendations were made for enhancing the sustainability of BJF production and consumption. LCA offers a thorough analysis of the environmental performance of BJF production, assisting stakeholders in making decisions that will improve bio-jet fuel sustainability overall by streamlining operations, minimizing negative effects, and minimizing environmental impact (Julio et al. 2021). It's important to keep in mind that LCA is an iterative process that may be improved as more precise data become accessible, as technology and practices advance, and as more data becomes available.

12.6 Methods to Deal with Co-Products

Along with the primary fuel product, many co-products and byproducts may be produced during the manufacture of bio-jet fuel (BJF). Remains, waste streams, and other materials produced during the conversion processes can all be considered co-products. Maximizing the sustainability and financial viability of the BJF production process depends on managing and utilizing these co-products well (Le Foll et al. 2023). Here are several strategies for handling BJF co-products:

(a) *Co-Processing and Valorization*:

Many co-products can be utilized or processed further to add value. Co-products, for instance, might be used as raw materials or chemicals in other sectors. In order to benefit from the chemical composition of co-products, co-processing includes incorporating them into already-in place industrial processes (Henchion and Shirsath 2022). For instance, biochar produced from the pyrolysis of biomass can be added to soil or utilized to store carbon.

(b) *Conversion of Waste to Energy*:

Some co-products might be energy-rich and useful as a renewable energy source. Co-products can be used in waste-to-energy processes to produce heat, electricity, or biofuels, such as gasification or incineration (Sharma et al. 2022).

(c) *Production of Biogas or Biomethane*:

Some co-products can be utilized as a feedstock for the production of biogas or anaerobic digestion. In this process, organic materials are broken down by microorganisms to create biogas (methane and carbon dioxide), which can be utilized to create electricity (Manikandan et al. 2023).

(d) *Agriculture and Animal Feed*: Some co-products can contribute to circular agricultural systems by being employed as soil amendments or animal feed, depending on their composition. For instance, waste biomass from the manufacturing of biofuels can be fed to animals (Colombo et al. 2023).
(e) *Resource Recycling*:

Co-products that have the appropriate material characteristics can be recycled or used again. For instance, materials made from co-products or bio-based plastics can be incorporated into industrial procedures (Ribul et al. 2021).

(f) *Utilizing and Capturing Carbon (CCU)*:

Co-products may occasionally have the ability to capture and use carbon. It is possible to catch CO_2-rich streams and use them for things like promoting the growth of algae for biomass production (Eloka-Eboka et al. 2019).

(g) *Applying the Land and Improving the Soil*:

As a result of their high nutritional content, residues from some BJF production paths, such as hydrothermal liquefaction, can be utilized as fertilizers or soil supplements to enhance the quality of the soil and stimulate plant development (Jacob-Lopes et al. 2023).

12.6.1 Practices of the Circular Economy and Recycling

Designing methods to reduce waste, encourage reuse, and improve resource efficiency is a key component in putting circular economy principles into practice. In closed-loop systems, where waste is reduced and materials are retained in use, co-products may be important (Kara et al. 2022). It's vital to remember that the best way to handle co-products depends on a variety of elements, including their makeup, local laws, market demand, practicability, and environmental concerns (Mungodla et al. 2019). Effective co-product management in the BJF production process requires an integrated strategy that takes into account both economic viability and environmental sustainability as shown in Fig. 12.1.

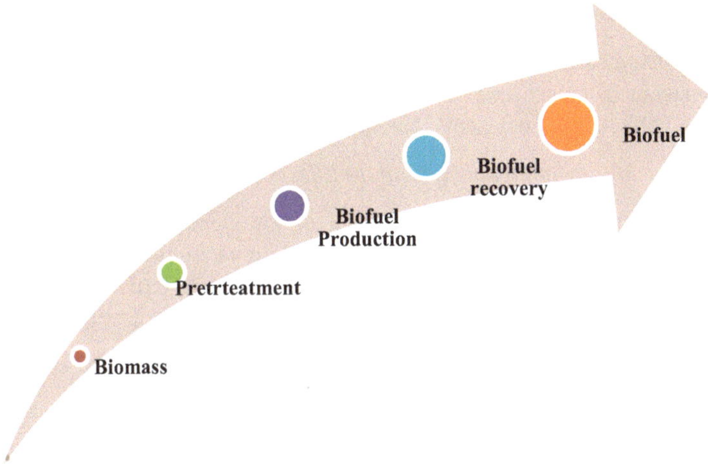

Fig. 12.1 Process in biomass to biofuel conversion

12.7 Land Availability for BJF Production

A key element in assessing the viability and sustainability of manufacturing bio-jet fuel on a wider scale is the availability of land for BJF production. The choice of feedstock, production techniques, and potential effects on ecosystems and food production are all influenced by the availability of land (Ong et al. 2021). Considerations regarding land availability for BJF production include as follows:

(a) *Type of Feedstock and Land Needed*:

The amount of land needed will vary depending on the feedstock. For instance, specialized energy crops like algae or certain oilseed crops may require a different type of land usage than agricultural waste or residues. When determining whether a feedstock is viable for the manufacture of BJF, its yield per unit of land area is a crucial consideration (Cervi et al. 2020).

(b) *Contradictory Land Uses*:

Other significant land uses, such as food production, biodiversity preservation, and ecosystem services, must be balanced with the availability of land for BJF production. Land use conflicts can be reduced by identifying feedstocks that don't directly compete with food crops (Zeng et al. 2022).

(c) *Degraded and Marginal Lands*:

It may be possible to reduce rivalry with food production by making use of marginal or degraded sites that are unsuitable for intensive food crop agriculture. However, it is important to carefully consider the environmental effects and viability of manufacturing bio-jet fuel on such sites (Csikós and Tóth 2023).

(d) *Managing Sustainable Land*:

Agroforestry, rotational cropping, and cover crops are examples of sustainable land management techniques that can help preserve soil fertility and lessen their detrimental effects on ecosystems (Kumar et al. 2023).

(e) *Land Use Modification and Secondary Effects*:

BJF production-related land use change may indirectly affect ecosystems and carbon emissions; particularly if it results in deforestation or the conversion of carbon-rich ecosystems. The overall climatic advantages of BJF production depend heavily on avoiding land use changes that generate large amounts of carbon emissions (Nagy et al. 2022).

(f) *Local Laws and Ordinances*:

The availability of land for the production of BJF can be impacted by laws pertaining to land use, land tenure, and environmental protection. For environmentally friendly industrial techniques to be used, compliance with these rules is crucial (O'Donoghue et al. 2021).

(g) *Efficiency and Productivity in Land Use*:

To make a significant difference in the production of BJF, it is imperative to maximize the productivity and efficiency of land use. Increased feedstock yields per unit of land can be achieved with the use of technological developments, breeding programs, and agronomic techniques (Li et al. 2023).

(h) *Analyzing the Environmental Impact*:

The possible environmental effects of BJF production on land use can be better understood by conducting thorough environmental impact evaluations, including life cycle assessments (LCAs) (Cooreman-Algoed et al. 2023).

(i) *Engaging Local Stakeholders*:

Understanding their viewpoints, issues, and potential advantages of BJF production on local lands requires interaction with indigenous groups, farmers, and local people. Determining the optimal land availability for BJF production ultimately requires a balanced strategy that takes into account environmental, social, and economic factors (Cervi et al. 2021). Establishing a responsible and viable BJF manufacturing pathway requires careful planning, environmentally sound procedures, and collaboration with pertinent stakeholders.

12.8 Environmental Impact Assessment

When producing bio-jet fuel (BJF), from growing the feedstock to using it in aircraft engines, an environmental impact assessment (EIA) is conducted to assess any potential environmental impacts. An EIA aims to inform decision-makers, stakeholders, and the public about the project's possible effects on the environment while also identifying, assessing, and mitigating the environmental impacts of BJF production (Ikiz Kaya et al. 2021). Here is a description of the procedure:

1. Define the scope of the assessment by specifying the stages of BJF production that will be taken into account, the geographic scope of the analysis, and the environmental factors that will be assessed.
2. List the different types of environmental impacts that need to be evaluated. Greenhouse gas emissions, energy use, water use, land use change, air and water pollution, habitat destruction, and other factors may be among them (Rajak 2021).
3. Compile information on the inputs and outputs related to each step of the manufacturing of BJF, including the production of feedstock, processing, blending, distribution, and usage of aircraft. Literature, databases, business reports, and measurements taken specifically at a site can all provide data (Julio et al. 2021).
4. Using impact assessment approaches, calculate the potential environmental effects. For instance, estimate the carbon dioxide equivalents of greenhouse gas emissions to determine the potential for global warming (Rajak 2021).
5. Perform sensitivity studies to comprehend how changes to the data, factors, and assumptions can affect the outcomes. This aids in evaluating how reliable the results are (Cro et al. 2020).
6. Determine potential environmental effects that go above allowable limits and suggest mitigating actions. Among these tactics are resource efficiency improvements, waste reduction, the use of renewable energy sources, and process optimization (Cainelli et al. 2020).
7. Explain the findings in light of the project's importance and possible outcomes. Be straightforward and accessible in your communication of the findings to decision-makers, stakeholders, and the general public (Gibbs et al. 2023).
8. Use the EIA findings to inform your choices about the viability, layout, and execution of BJF manufacturing. This could entail making modifications to the project plans, policies, or production method (Lorenzo 2021).
9. Put in place a monitoring strategy to keep tabs on the real environmental effects of the BJF production process. This enables adaptive management, where decisions can be made based on observations made in the real world (Lorenzo 2021).
10. Write a report on the environmental impact assessment that details every step of the procedure, from data collecting to mitigating measures. The methodology adopted, the presumptions made, and the conclusions reached should all be disclosed in the report (Julio et al. 2021).

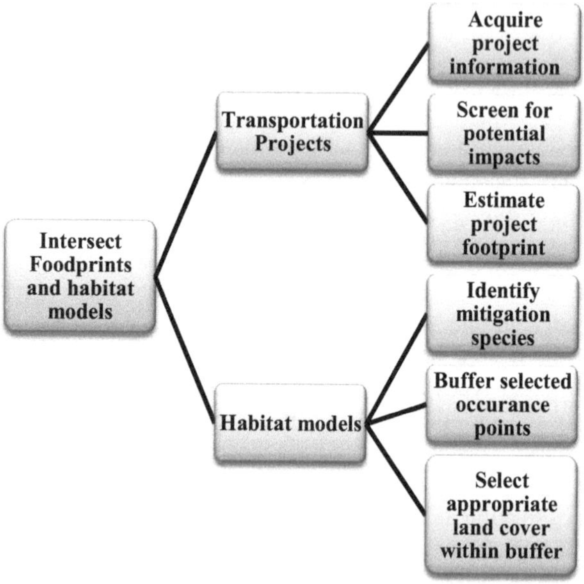

Fig. 12.2 Environmental impact assessment in BJF

EIA is a useful tool for evaluating both the possible advantages and disadvantages of BJF manufacturing on the environment as shown in Fig. 12.2. It ensures that BJF manufacturing supports to environmental protection and lower carbon emissions in aviation, encourages responsible practices, and aids in guiding sustainable decision-making.

12.9 Yield Developments

The term "yield developments" for bio-jet fuel (BJF) describes developments and enhancements to the productivity and efficiency of the BJF production process. With the help of these advancements, more bio-jet fuel will eventually be produced from a given amount of biomass feedstock, improving both the economics and sustainability of BJF production (Julio et al. 2021).

(a) *Feedstock Breeding and Selection*:

In order to enhance the yield of the bio-based feedstock required for BJF manufacturing, research and development activities are concentrated on choosing or breeding feedstock crops with higher oil or carbohydrate content (Pudel and Wiesen 2019).

(b) *Genetic Modification and Engineering*:

Techniques in genetic engineering can be used to increase the yield of crops used as feedstock. This can entail altering plants to yield more oil or other important substances (Khoo et al. 2023).

(c) *Agronomic and Cultivation Techniques*:

Improved cultivation techniques can result in higher biomass yields and better feedstock quality. These practices include optimized planting density, irrigation, and fertilizer management (Osman et al. 2023).

(d) *Efficiency of Harvesting and Collection*:

Technology advancements in harvesting and collection can lower losses and boost the effectiveness of biomass collection, resulting in increased yields (Hiloidhari et al. 2023).

(e) *Technologies for Conversion*:

The production of bio-oil or bio-based intermediates can be increased by improvements in conversion methods, such as more effective pyrolysis, hydrothermal liquefaction, or fermentation processes (Ebhodaghe et al. 2022).

(f) *Process Improvement*:

To increase the amount of useful fuel generated, process optimization focuses on enhancing the effectiveness of each stage of BJF production, from feedstock preprocessing to refining (Cervi et al. 2020).

(g) *Utilization of Co-Products*:

To increase total yield, co-products and byproducts produced during the BJF production process might be used. By adding value to these byproducts, waste is reduced and the economic viability of production is increased (Julio et al. 2021).

(h) *Engineering of Catalysts and Reactions*:

Higher yields of suitable bio-jet fuel constituents may be attained through research into more effective catalysts and reaction conditions for conversion processes like hydroprocessing or pyrolysis (Lahijani et al. 2022).

(i) *Fermentation and Biological Transformation*:

The production of bio-based molecules suited for BJF can be increased by improving biological conversion processes, such as fermentation employing engineered microbes (Rioux et al. 2022).

(j) *Systems for Integrated Production*:

The creation of integrated systems that integrate various processes (such as the cultivation and conversion of feedstocks) might result in synergies that boost overall output and resource efficiency. In order to solve the issues of scalability, feedstock availability, and economic feasibility, yield developments are crucial for BJF production (Cervi et al. 2020). These developments help bio-jet fuel become a more appealing substitute for traditional fossil-based jet fuels as they achieve the twin objectives of lowering greenhouse gas emissions and guaranteeing a sustainable aviation sector.

12.10 Uncertainty and Sensitivity Analysis

The robustness and dependability of data obtained in the context of the manufacture of bio-jet fuel (BJF) must be evaluated, and sensitivity and uncertainty analysis are crucial parts of this process (Cervi et al. 2020). These studies aid in comprehending how changes in the assumptions and input parameters impact the results of the BJF production process.

***Uncertainty Analysis*:** Uncertainty analysis measures how much a BJF production assessment's input parameters are unknown and how that uncertainty affects the assessment's outcomes. It entails locating uncertainty sources, estimating their size, and propagating them throughout the analysis (Cervi et al. 2021). With respect to the variability in the input data and assumptions, the objective is to present a range of alternative outcomes.

***Sensitivity Analysis*:** The BJF production assessment's outputs are affected by changes to certain input parameters or assumptions, which are examined in sensitivity analysis. It enables the identification of the main sources of variability and assists in determining which parameters significantly affect the findings (Sharno and Hiloidhari 2022).

Here is how sensitivity and uncertainty analysis can be used to improve BJF production:

(a) *Uncertainty of a Parameter*:

Decide on important input characteristics, such as feedstock production, efficiency of conversion, amount of energy used, and emissions considerations. Using probability distributions or ranges based on the information at hand and professional judgment; quantify the uncertainty associated with these parameters (Lo et al. 2021).

(b) *Simulation Using Monte Carlo*:

Run the BJF production analysis many times using Monte Carlo simulation, using parameter values from their respective distributions each time (Wood et al. 2021). To comprehend the range of possible outcomes and related uncertainty, analyze the distribution of results.

(c) *Indices of Sensitivity*:

To evaluate the relative significance of each input parameter in causing the variability in the output results, compute sensitivity indices, such as the Sobol indices. Determine "high-impact" parameters that have a big impact on the outcomes and "low-impact" parameters that barely have an impact (Lo et al. 2021).

(d) *Tornado Schematics*:

Make tornado diagrams to illustrate the findings of the sensitivity analysis. These illustrations demonstrate how changes in parameters affect the final outcome (McCabe et al. 2020).

(e) *Scenario Evaluation*:

To comprehend the ramifications on the results of BJF production, run scenario studies by taking extreme circumstances into consideration or changing certain parameters (Presbitero et al. 2021).

(f) *Interpretation*:

Understanding which uncertainties have the most impact on the outcomes of the BJF production assessment will help you interpret the results of uncertainty and sensitivity assessments (Wait 2021).

(g) *Decision-Making and Data Improvement*:

Prioritize data collecting efforts, enhance model accuracy, and make well-informed judgments based on the robustness of the results using the knowledge gained from uncertainty and sensitivity studies (Lo et al. 2021). The potential variances and weaknesses of BJF production assessments are better understood by using both uncertainty and sensitivity analysis. They assist better informed and transparent decision-making in the context of sustainable bio-jet fuel generation and strengthen the credibility of findings.

12.11 Specification for Bio-Jet Fuel

Aviation biofuel, commonly referred to as bio-jet fuel, is a type of renewable jet fuel made from biomass (Balogu et al. 2022). It is intended to be a more environmentally friendly substitute for standard jet fuel, which is predominantly made from fossil fuels. Depending on the feedstock used for production and the precise processing techniques used, bio-jet fuel characteristics can change.

Chemical Make-Up: Hydrocarbon chains, which are the molecules that produce energy through burning, are a component of bio-jet fuel, just like they are in conventional jet fuel (Lahijani et al. 2022).

Saturated and Unsaturated Hydrocarbons: To guarantee optimum combustion characteristics, the fuel should have a well-balanced blend of saturated and unsaturated hydrocarbons (Tiwari et al. 2023).

12.11.1 Physical Characteristics

Density: Bio-jet fuel needs to have a density that works with the infrastructure for storing and transporting aviation fuel (Dahal et al. 2021).

Freezing Point: The fuel must have a low enough freezing point to maintain liquid state under the extremely cold conditions found at high altitudes (Dahal et al. 2021).

Energy Density: To deliver enough energy per unit volume for aeroplane operations, bio-jet fuel should have a high energy density (Dahal et al. 2021).

12.11.2 Chemical Characteristics

Octane Rating: While this is more significant for petrol, diesel and jet fuels also need to have a high cetane rating. Better combustion efficiency and ignition quality are indicated by a higher cetane rating (Labeckas and Slavinskas 2021).

12.11.3 Impurities and Contaminants

Sulfur Concentration: To reduce hazardous emissions, bio-jet fuel should have a low sulfur concentration.

Water Content: To avoid problems with phase separation and microbiological growth, the fuel should have low water content.

Particulate Matter: Particulate matter that could clog fuel filters or harm engines shouldn't be present in bio-jet fuel.

Infrastructure Compatibility: Bio-jet fuel shouldn't require major modifications in order to work with current aircraft engines and fuel distribution infrastructure.

Performance During Combustion: The fuel should have steady combustion properties and emit little pollutants.

12.11.4 Sustainability and Renewability

Bio-jet fuel should be produced from renewable biomass feedstocks such as leftover food from farms, algae, or used cooking oil (Why et al. 2019).

Emissions of Greenhouse Gases: Compared to traditional fossil jet fuel, the production and use of bio-jet fuel should result in fewer net emissions of greenhouse gases (Tiwari et al. 2023).

Certification Requirements: The aviation fuel specification known as ASTM D7566 was developed by ASTM International for synthetic blending ingredients used in the creation of aviation turbine fuels (Rumizen 2021) as shown in the tabular form given in Table 12.1.

It's crucial to remember that regional legislation, technical improvements, and the specific production procedures employed for bio-jet fuel can all have an impact on its specs (Karunanidhi 2015). As the aviation industry continues to prioritize sustainability and works to lower its carbon impact, these criteria can change over time.

12.12 Technology Assessment of BJF Routes

Examining various methods and procedures for transforming biomass feedstocks into aviation fuel is a part of assessing the technological paths for manufacturing bio-jet fuel (BJF).

(a) *Hydroprocessing*:

Triglycerides are subjected to high temperatures and pressures in the presence of hydrogen during the hydrodeoxygenation (HDO) process in order to remove oxygen and transform them into hydrocarbon chains (Melero et al. 2012). A hydrocarbon-rich product is produced as a result, which can then be processed into bio-jet fuel.

Similar to HDO, hydrotreatment involves removing sulfur, nitrogen, and oxygen from oils and fats generated from biomass (Huber and Corma 2007). This is necessary to raise gasoline quality and guarantee compatibility with current infrastructure.

(b) *Synthesis Through Fischer-Tropsch (FT)*:

By using catalytic processes, FT synthesis transforms syngas—a mixture of carbon monoxide and hydrogen—obtained from biomass gasification or other sources into liquid hydrocarbons. The hydrocarbons that are created can be processed to make bio-jet fuel. This method allows for flexible use of a variety of feedstocks, including as garbage, agricultural wastes, and woody biomass (Klankermayer et al. 2016).

Table 12.1 Specification for bio-jet fuel

S. No.	Specifications	Range	Standard values	Measured values
1.	Acidity (total mg KOH/g)	Max	0.1	0.085
2.	Density at 15 °C (kg/m^3)	–	775–840	820
3.	Flashpoint (°C)	Min	38	45
4.	Freezing point (°C)	Max	−40	−38
5.	Net heat of combustion (MJ/kg)	Min	42.8	42.8
6.	Sulfur (total mass %)	Max	0.003	0

(c) *Alcohol-to-Jet Conversion (ATJ)*:

Through catalytic dehydration followed by hydroprocessing, ethanol or other alcohols generated from biomass can be transformed into hydrocarbons suitable for aviation fuel. The ATJ methods offer the benefit of using agricultural waste and cellulosic biomass as feedstocks, and they may be combined with current ethanol production facilities (Geleynse et al. 2018).

(d) *Conversion of Lipids to Hydrocarbons*:

Through procedures such as hydrothermal liquefaction or catalytic conversion, microbial oil produced by algae or other microorganisms can be transformed into hydrocarbons. Production of bio-jet fuel based on algae has the potential to provide high yields and reduce land usage issues related to other biomass feedstocks (Wei et al. 2019).

12.12.1 HTL: Hydrothermal Liquefaction

HTL entails the high temperature and pressure conversion of wet biomass (like algae or sewage sludge) into biocrude oil (Xu et al. 2019). The biocrude can then be transformed further to produce bio-jet fuel. This method has the advantage of being able to process wet or non-food biomass and can handle a range of feedstocks.

(a) *Upgrading and Pyrolysis*:

Biomass is converted during the pyrolysis process into bio-oil, which can then be enhanced using techniques like hydrotreatment to create bio-jet fuel (Lahijani et al. 2022). Pyrolysis has the capacity for effective conversion and can handle a variety of feedstocks.

(b) *Catalytic Cracking*:

Catalytic cracking of biomass-derived feedstocks can yield hydrocarbons suitable for aviation fuel. Catalytic cracking offers adaptability in production and can be adapted to various feedstocks (Corma et al. 2007).

(c) *Mixed-Feedstock Methodologies*:

To manufacture bio-jet fuel, certain techniques blend various feedstocks or utilize waste streams, which increases resource efficiency and sustainability overall. The availability of feedstock, the level of technology maturity, scalability, energy efficiency, environmental impacts, and compatibility with existing infrastructure are all things that should be taken into account while evaluating various technology paths (Rissman et al. 2020). The practicality of a certain technology approach for the manufacture of bio-jet fuel is also significantly influenced by economic reasons, policy backing, and regulatory compliance.

12.13 Comparison of Three Bio-Jet Fuel Paths

The three widely used bio-jet fuel manufacturing processes: hydroprocessing (hydrodeoxygenation), Fischer-Tropsch synthesis, and alcohol-to-jet conversion.

(a) *Hydrooxygenation (Hydrodeprocessing)*:

Wide range of feedstocks, including vegetable oils, animal fats, and other triglyceride-rich materials, can be used as feedstocks.

Process: Produces hydrocarbon-rich byproducts by removing oxygen through high-temperature and high-pressure reactions with hydrogen gas (hydrotreatment or hydrodeoxygenation) (Elliott et al. 2013).

Advantages: Produces high-quality hydrocarbons, works with existing infrastructure and engines, lowers oxygen concentration, and can employ both food- and non-food-based feedstocks.

Challenges include the process's relatively high energy requirements, dependence on a steady supply of hydrogen, restriction to triglyceride-containing feedstocks, and potential competition for feedstocks from other sectors of the economy (Gómez-Castro et al. 2023).

(b) *Synthesis Through Fischer-Tropsch (FT)*:

Wide variety of feedstocks are usable, including syngas (carbon monoxide and hydrogen) derived from biomass. Transform syngas by catalytic processes into liquid hydrocarbons (Fischer-Tropsch synthesis). To make bio-jet fuel, more refinement of the generated hydrocarbons is required (Zhai et al. 2021).

Advantages: Has the potential for large-scale production, can employ a variety of feedstocks, gives flexibility in feedstock selection, can use biomass gasification, and generates high-quality hydrocarbons.

Challenges include complex processes needing exact catalysts and reactor conditions, expensive initial and ongoing expenses, and occasionally problematic carbon efficiency (Aghbashlo et al. 2021).

(c) *Alcohol-to-Jet Conversion (ATJ)*:

Uses alcohols generated from biomass as feedstocks, like as ethanol, to make feed.

Alcohols are catalytically dehydrated, and then hydroprocessed to create hydrocarbons that are appropriate for use as aviation fuel (Goh et al. 2022).

Benefits include the ability to use a range of feedstocks, such as agricultural residues and cellulosic biomass, compatibility with current infrastructure, the potential for integration with current ethanol production plants, and lower oxygen content.

Challenges include the need for more processing stages compared to direct hydrocarbon pathways, competition from alternative bio-based alcohol applications, and potential technical difficulties in producing high yields and the appropriate fuel characteristics (Schubert 2020).

There are a number of things to take into account while contrasting these pathways (Gómez-Castro et al. 2023):

Feedstock Availability: The feedstock compatibilities and needs differ for each method. The choice of pathway may be affected by the accessibility of acceptable feedstocks in a particular area.

Process Efficiency: Taking energy input and yields into account, different conversion paths have different overall conversion efficiencies for producing bio-jet fuel.

Fuel Quality: Energy content, combustion characteristics, and compatibility with aviation engines are crucial factors to take into account.

Environmental Impact: Different environmental factors, such as greenhouse gas emissions, land use, and water use, are associated with each pathway.

Economic Viability: The economic viability of each pathway is mostly based on capital and operating costs as well as prospective revenue sources.

The final decision of a bio-jet fuel pathway is influenced by a number of factors, including technological viability, economic viability, environmental sustainability, and political backing in a particular setting. The decision-making process is also influenced by elements including the availability of feedstock, technical development, and changing legislation (Martinez-Valencia et al. 2021).

12.14 Regulations, Guidelines, and Accounting Standards

Regulations, rules, and accounting standards have a significant impact on how bio-jet fuel is developed, produced, and used. They guarantee the bio-jet fuel sector's openness, environmental sustainability, and safety (Gray et al. 2021).

1. *Rules and Requirements*:

A crucial standard created by ASTM International expressly for aviation biofuels is ASTM D7566. It outlines the standards and specifications for synthetic blending materials used in the manufacture of aircraft turbine fuels, including bio-jet fuels (López-Gómez et al. 2023). It guarantees that bio-jet fuels meet the requirements for quality for use in aviation. The International Civil Aviation Organization (ICAO) established Annex 16, which details environmental protection guidelines for aircraft engine emissions (Korkut and Fowler 2021). This includes rules governing the use of alternative fuels, such as bio-jet fuels, in an effort to lessen the impact of aviation on climate change.

Regional Rules: Different nations and regions have their own rules governing the creation, distribution, and usage of bio-jet fuel. These rules may cover topics like safety requirements, greenhouse gas emissions, and sustainable feedstocks (Liu et al. 2021).

2. *Sustainable Development Principles*:

An organization called the Roundtable on Sustainable Biomaterials (RSB) creates and executes sustainability standards and certification programs for bio-based goods, including bio-jet fuels (Muijden et al. 2020). Aspects of biofuel production that are economically, socially, and environmentally sustainable are covered by their certification. European Union Directive on Renewable Energy (RED II): Aviation biofuels must meet the sustainability standards specified by RED II in the European Union (Chiaramonti and Goumas 2019). In addition to requiring that biofuels be produced without having a detrimental social or environmental impact, it also establishes requirements for greenhouse gas emissions reductions.

3. *Standards for Carbon Accounting*:

Life cycle assessment (LCA) is a methodology used to evaluate a product's or processes environmental effects over the course of its full life cycle. This involves calculating the whole carbon footprint of several bio-jet fuel production methods, from feedstock generation to end consumption (Algren et al. 2021).

Carbon Intensity: The quantity of greenhouse gas emissions generated per unit of energy or product is referred to as carbon intensity. It is used to meet regulatory criteria for emissions reduction as well as to compare the environmental effects of various fuels, including bio-jet fuels (Mandegari et al. 2023).

Carbon Offsetting: Under certain rules and voluntary initiatives, expenditures in projects that lessen or eliminate an equivalent quantity of carbon dioxide from the environment are required to counterbalance the emissions caused by the use of bio-jet fuels (Bergero et al. 2023).

4. *Reporting and Openness*:

Transparency Reporting: Producers and suppliers of bio-jet fuels frequently have to provide information on the sources of their feedstocks, their manufacturing procedures, their efforts to reduce emissions, and other pertinent details (Tiwari et al. 2023). This encourages supply chain transparency for bio-jet fuel.

Environmental Impact Reporting: To help assure compliance with sustainability standards, governments and regulatory authorities may compel companies producing bio-jet fuel to report on the environmental impact of their operations (Kumar et al. 2023).

It is crucial to remember that rules, policies, and accounting standards for bio-jet fuels are dynamic and subject to change as the sector develops and new scientific knowledge is discovered (Alam and Dwivedi 2019). To ensure compliance and encourage sustainable bio-jet fuel production and use, industry stakeholders, including as producers, regulators, and researchers, should keep up with the most recent advancements.

12.15 Conclusion

The life cycle assessment (LCA) of bio-jet fuel has shed important light on its sustainability and environmental impact when compared to conventional jet fuels derived from fossil fuels. Reduced Greenhouse Gas Emissions: When compared to conventional jet fuel, bio-jet fuel shows a considerable reduction in greenhouse gas emissions. Lower net carbon dioxide emissions are produced as a result of the production of feedstock, its processing, and ultimate combustion. This decrease aids in reducing emissions and achieving emission reduction goals. The selection of the feedstock is important since it affects how environmentally friendly bio-jet fuel will be in general. The results of the LCA are more positively impacted by sustainable feedstocks that require little modification to the land's use, use little water, and produce a lot of energy. Utilizing waste products or specially bred energy crops can further improve the environmental advantages. Despite the potential environmental advantages of producing bio-jet fuel, some feedstocks, particularly those grown on a large scale, have the potential to alter land use, cause deforestation, and reduce biodiversity. Prioritizing feedstocks that reduce these adverse effects and using sustainable land management techniques are essential. The LCA highlights the significance of utilizing resources effectively during the production, processing, and conversion of feedstock. The manufacturing of bio-jet fuel can be made more sustainably overall by optimizing the use of energy, water, and chemical inputs. Bio-jet fuel's environmental performance can be greatly enhanced by ongoing improvements in biofuel production methods. By improving conversion effectiveness, feedstock processing, and waste utilization, bio-jet fuel can become even more competitive with fossil-based jet fuels while reducing its environmental impact. Positive policies, incentives, and market demand are key factors in the adoption of bio-jet fuel. Governments and industry participants ought to work together to establish a supportive environment that promotes investment in biofuels' development and commercialization. Collaboration is necessary because life cycle assessment is a dynamic instrument that calls for cooperation between academics, decision-makers, businesspeople, and environmentalists. To guarantee that the assessment appropriately reflects the evolving realities of bio-jet fuel generation and its implications, the LCA data and techniques are routinely updated. The life cycle assessment of bio-jet fuel concludes by highlighting its potential to lower greenhouse gas emissions, promote sustainable resource use, and support an aviation industry that is more environmentally conscious. However, in order to maximize the environmental advantages of bio-jet fuel while minimizing any potential downsides, the choice of feedstock, technological improvements, responsible land management, and supportive regulations will all play crucial roles. For bio-jet fuel to remain a competitive substitute for traditional aviation fuels over the long-term, more research, development, and cooperation are required.

References

Aghbashlo M, Peng W, Tabatabaei M, Kalogirou SA, Soltanian S, Hosseinzadeh-Bandbafha H et al (2021) Machine learning technology in biodiesel research: a review. Prog Energy Combust Sci 85:100904

Ahlström J, Jafri Y, Wetterlund E, Furusjö E (2023) Sustainable aviation fuels–options for negative emissions and high carbon efficiency. Int J Greenhouse Gas Control 125:103886

Ahmed M, Alam MN, Abdullah A, Ahmad Z (2023) Bio-jet fuel: an overview of various feedstock and production routes. In AIP conference proceedings, vol. 2785, no. 1. AIP Publishing, July

Alam A, Dwivedi P (2019) Modeling site suitability and production potential of carinata-based sustainable jet fuel in the southeastern United States. J Clean Prod 239:117817

Algren M, Fisher W, Landis AE (2021) Machine learning in life cycle assessment. In: Data science applied to sustainability analysis. Elsevier, pp 167–190

Ali AR, Cerdas F, Ryberg M, Herrmann C (2023) Framework for developing technology roadmap based on absolute environmental sustainability assessment: a case study on electric vehicles. Procedia CIRP 116:360–365

Andooz A, Eqbalpour M, Kowsari E, Ramakrishna S, Cheshmeh ZA (2023) A comprehensive review on pyrolysis from the circular economy point of view and its environmental and social effects. J Clean Prod 388:136021

Balogu SA, Ghazali I, Mohammed AT, Hermansyah D, Amanah A, Kurnia MT (2022) Renewable aviation fuel: review of bio-jet fuel for aviation industry. Eng Sci Lett 1(1):7–11

Barbhuiya S, Das BB (2023) Life cycle assessment of construction materials: methodologies, applications and future directions for sustainable decision-making. Case Studies Constr Mater 19:e02326

Bergero C, Gosnell G, Gielen D, Kang S, Bazilian M, Davis SJ (2023) Pathways to net-zero emissions from aviation. Nat Sustainability 6(4):404–414

Böckin D, Goffetti G, Baumann H, Tillman AM, Zobel T (2022) Business model life cycle assessment: a method for analysing the environmental performance of business. Sustainable Prod Consumption 32:112–124

Braungart M, McDonough W, Bollinger A (2007) Cradle-to-cradle design: creating healthy emissions–a strategy for eco-effective product and system design. J Clean Prod 15(13–14):1337–1348

Cainelli G, D'Amato A, Mazzanti M (2020) Resource efficient eco-innovations for a circular economy: Evidence from EU firms. Res Policy 49(1):103827

Cervi WR, Lamparelli RAC, Gallo BC, de Oliveira Bordonal R, Seabra JEA, Junginger M, van der Hilst F (2021) Mapping the environmental and techno-economic potential of biojet fuel production from biomass residues in Brazil. Biofuels Bioprod Biorefin 15(1):282–304

Cervi WR, Lamparelli RAC, Seabra JEA, Junginger M, de Jong S, van der Hilst F (2020) Spatial modeling of techno-economic potential of biojet fuel production in Brazil. GCB Bioenergy 12(2):136–157

Chiaramonti D, Goumas T (2019) Impacts on industrial-scale market deployment of advanced biofuels and recycled carbon fuels from the EU renewable energy directive II. Appl Energy 251:113351

Colbertaldo P, Parolin F, Campanari S (2023) A comprehensive multi-node multi-vector multi-sector modelling framework to investigate integrated energy systems and assess decarbonisation needs. Energy Convers Manag 291:117168

Colombo SM, Roy K, Mraz J, Wan AH, Davies SJ, Tibbetts SM et al (2023) Towards achieving circularity and sustainability in feeds for farmed blue foods. Rev Aquac 15(3):1115–1141

Cooreman-Algoed M, Boone L, Uitterhaegen E, Taelman SE, De Soete W, Dewulf J (2023) Environmental life cycle assessment of nutraceuticals: a case study on methylcobalamin in different packaging types. Sci Total Environ 893:164780

Corma A, Huber GW, Sauvanaud L, O'connor P (2007) Processing biomass-derived oxygenates in the oil refinery: catalytic cracking (FCC) reaction pathways and role of catalyst. J Catal 247(2):307–327

Cro S, Morris TP, Kahan BC, Cornelius VR, Carpenter JR (2020) A four-step strategy for handling missing outcome data in randomised trials affected by a pandemic. BMC Med Res Methodol 20(1):1–12

Csikós N, Tóth G (2023) Concepts of agricultural marginal lands and their utilisation: a review. Agric Syst 204:103560

Dahal K, Brynolf S, Xisto C, Hansson J, Grahn M, Grönstedt T, Lehtveer M (2021) Techno-economic review of alternative fuels and propulsion systems for the aviation sector. Renew Sust Energ Rev 151:111564

De Wolf C, Cordella M, Dodd N, Byers B, Donatello S (2023) Whole life cycle environmental impact assessment of buildings: developing software tool and database support for the EU framework level (s). Resour Conserv Recycl 188:106642

Donnelly GE, Blanco C, Spanbauer C, Stienecker SL (2023) The effects of item dirtiness on disposal decisions. J Assoc Consum Res 8(3):000–000

Ebhodaghe SO, Imanah OE, Ndibe H (2022) Biofuels from microalgae biomass: a review of conversion processes and procedures. Arab J Chem 15(2):103591

Elliott DC, Hart TR, Schmidt AJ, Neuenschwander GG, Rotness LJ, Olarte MV et al (2013) Process development for hydrothermal liquefaction of algae feedstocks in a continuous-flow reactor. Algal Res 2(4):445–454

Eloka-Eboka AC, Bwapwa JK, Maroa S (2019) Biomass for CO2 sequestration. Encyclopedia of renewable and sustainable materials, 1:4252

Emmanouilidou E, Mitkidou S, Agapiou A, Kokkinos NC (2023) Solid waste biomass as a potential feedstock for producing sustainable aviation fuel: a systematic review. Renew Energy 206(C):897–907

Eswary Devi T, Parthiban R, Arun J, Gopinath KP (2022) Processing of marine microalgae biomass via hydrothermal liquefaction for bio-oil production: study on algae cultivation, harvesting, and process parameters. Biomass Convers Biorefinery 13(6):1–14

Fortier MOP, Roberts GW, Stagg-Williams SM, Sturm BS (2014) Life cycle assessment of bio-jet fuel from hydrothermal liquefaction of microalgae. Appl Energy 122:73–82

Gaurav G, Kumar V, Gupta S, Meena ML, Dangayach GS, Jindal MK (2023) Life cycle assessment of extraction of edible oil from mustard seeds: a case study of an oil industry. Materials Today: Proceedings

Geleynse S, Brandt K, Garcia-Perez M, Wolcott M, Zhang X (2018) The alcohol-to-jet conversion pathway for drop-in biofuels: techno-economic evaluation. ChemSusChem 11(21):3728–3741

Ghoroghi A, Rezgui Y, Petri I, Beach T (2022) Advances in application of machine learning to life cycle assessment: a literature review. Int J Life Cycle Assess 27(3):433–456

Gibbs NK, Angus C, Dixon S, Parry CDH, Meier PS (2023) Stakeholder engagement in the development of public health economic models: an application to modelling of minimum unit pricing of alcohol in South Africa. Appl Health Econ Health Policy 21(3):395–403

Goh BHH, Chong CT, Ong HC, Seljak T, Katrašnik T, Józsa V et al (2022) Recent advancements in catalytic conversion pathways for synthetic jet fuel produced from bioresources. Energy Convers Manag 251:114974

Gómez-Castro FI, Gutiérrez-Antonio C, Romero-Izquierdo AG, May-Vázquez MM, Hernández S (2023) Intensified technologies for the production of triglyceride-based biofuels: current status and future trends. Renew Sust Energ Rev 184:113580

Gray N, McDonagh S, O'Shea R, Smyth B, Murphy JD (2021) Decarbonising ships, planes and trucks: an analysis of suitable low-carbon fuels for the maritime, aviation and haulage sectors. Advances Appl Energy 1:100008

Haldar A, Sucharita S, Dash DP, Sethi N, Padhan PC (2023) The effects of ICT, electricity consumption, innovation and renewable power generation on economic growth: an income level analysis for the emerging economies. J Clean Prod 384:135607

Hari TK, Yaakob Z, Binitha NN (2015) Aviation biofuel from renewable resources: routes, opportunities and challenges. Renew Sust Energ Rev 42:1234–1244

Henchion MM, Shirsath AP (2022) Developing and implementing a transdisciplinary framework for future pathways in the circular bioeconomy: the case of the red meat industry. J Clean Prod 380:134845

Hiloidhari M, Sharno MA, Baruah DC, Bezbaruah AN (2023) Green and sustainable biomass supply chain for environmental, social and economic benefits. Biomass Bioenergy 175:106893

Huber GW, Corma A (2007) Synergies between bio-and oil refineries for the production of fuels from biomass. Angew Chem Int Ed 46(38):7184–7201
Hussin F, Hazani NN, Khalil M, Aroua MK (2023) Environmental life cycle assessment of biomass conversion using hydrothermal technology: a review. Fuel Process Technol 246:107747
Ikiz Kaya D, Pintossi N, Dane G (2021) An empirical analysis of driving factors and policy enablers of heritage adaptive reuse within the circular economy framework. Sustainability 13(5):2479
Inayat A, Rocha-Meneses L, Ghenai C, Abdallah M, Shanableh A, Al-Ali K et al (2022) Co-pyrolysis for bio-oil production via fixed bed reactor using date seeds and plastic waste as biomass. Case Stud Therm Eng 31:101841
Jacob-Lopes E, Dias RR, Zepka LQ (eds) (2023) Microalgae-based systems: process integration and process intensification approaches. Walter de Gruyter GmbH & Co KG
Jennings R, Henderson AD, Phelps A, Janda KM, van den Berg AE (2023) Five US dietary patterns and their relationship to land use, water use, and greenhouse gas emissions: implications for future food security. Nutrients 15(1):215
Joensuu T, Leino R, Heinonen J, Saari A (2022) Developing buildings' life cycle assessment in circular economy-comparing methods for assessing carbon footprint of reusable components. Sustain Cities Soc 77:103499
Julio AAV, Batlle EAO, Trindade AB, Nebra SA, Reyes AMM, Palacio JCE (2021) Energy, exergy, exergoeconomic, and environmental assessment of different technologies in the production of bio-jet fuel by palm oil biorefineries. Energy Convers Manag 243:114393
Kara S, Hauschild M, Sutherland J, McAloone T (2022) Closed-loop systems to circular economy: a pathway to environmental sustainability? CIRP Ann 71(2):505–528
Karunanidhi R (2015) The aviation fuel and the passenger Aircraft for the Future-bio Fuel, synthetic fuel (Doctoral dissertation, Master Thesis. Department of Automotive and Aeronautical Engineering)
Khoo KS, Ahmad I, Chew KW, Iwamoto K, Bhatnagar A, Show PL (2023) Enhanced microalgal lipid production for biofuel using different strategies including genetic modification of microalgae: a review. Prog Energy Combust Sci 96:101071
Klankermayer J, Wesselbaum S, Beydoun K, Leitner W (2016) Selective catalytic synthesis using the combination of carbon dioxide and hydrogen: catalytic chess at the interface of energy and chemistry. Angew Chem Int Ed 55(26):7296–7343
Korkut E, Fowler LB (2021) Regulatory and policy analysis of production, development and use of sustainable aviation fuels in the United States. Front Energy Res 9:750514
Kumar T, Basakran GK, Marsuki MZ, Wash AM, Mohsin R, Majid ZA, Ghafir MFA (2023) Exploring socioeconomic and political feasibility of aviation biofuel production and usage in Malaysia: a thematic analysis approach using expert opinion from aviation industry. Green Energy Intell Transp 2:100111
Labeckas G, Slavinskas S (2021) Comparative evaluation of the combustion process and emissions of a diesel engine operating on the cetane improver 2-Ethylhexyl nitrate doped rapeseed oil and aviation JP-8 fuel. Energy Convers Manage X 11:100106
Lahijani P, Mohammadi M, Mohamed AR, Ismail F, Lee KT, Amini G (2022) Upgrading biomass-derived pyrolysis bio-oil to bio-jet fuel through catalytic cracking and hydrodeoxygenation: a review of recent progress. Energy Convers Manag 268:115956
Le Foll R, Lechevalier V, Hamon P, Guérin-Dubiard C, Lambert X, Deglaire A, Nau F (2023) Beef protein ingredients from fat rendering process are promising functional ingredients. Food Chem 433:137298
Li C, Stomph TJ, Makowski D, Li H, Zhang C, Zhang F, van der Werf W (2023) The productive performance of intercropping. Proc Natl Acad Sci 120(2):e2201886120
Liu H, Zhang C, Tian H, Li L, Wang X, Qiu T (2021) Environmental and techno-economic analyses of bio-jet fuel produced from jatropha and castor oilseeds in China. Int J Life Cycle Assess 26:1071–1084
Lo SLY, How BS, Leong WD, Teng SY, Rhamdhani MA, Sunarso J (2021) Techno-economic analysis for biomass supply chain: a state-of-the-art review. Renew Sust Energ Rev 135:110164

López Gómez M, Posada J, Silva V, Martínez L, Mayorga A, Álvarez O (2023) Diagnosis of challenges and uncertainties for implementation of sustainable aviation fuel (SAF) in Colombia, and recommendations to move forward. Energies 16(15):5667

Lorenzo JA (2021) A multilateral track for sustainable development along the belt and road: aligning country ownership with international rule of law, KFG Working Paper Series 47:1–69

Lucchetti MG, Paolotti L, Rocchi L, Boggia A (2019) The role of environmental evaluation within circular economy: an application of life cycle assessment (LCA) method in the detergents sector. Rigas Tehniskas Universitates Zinatniskie Raksti 23(2):238–257

Makepa DC, Chihobo CH, Ruziwa WR, Musademba D (2023) A systematic review of the techno-economic assessment and biomass supply chain uncertainties of biofuels production from fast pyrolysis of lignocellulosic biomass. Fuel Commun 14:100086

Mandegari M, Ebadian M, Saddler JJ (2023) The need for effective life cycle assessment (LCA) to enhance the effectiveness of policies such as low carbon fuel standards (LCFS's). Energy Policy 181:113723

Manikandan S, Vickram S, Sirohi R, Subbaiya R, Krishnan RY, Karmegam N et al (2023) Critical review of biochemical pathways to transformation of waste and biomass into bioenergy. Bioresour Technol 372:128679

Martinez-Valencia L, Garcia-Perez M, Wolcott MP (2021) Supply chain configuration of sustainable aviation fuel: review, challenges, and pathways for including environmental and social benefits. Renew Sust Energ Rev 152:111680

McCabe C, Paulden M, Awotwe I, Sutton A, Hall P (2020) One-way sensitivity analysis for probabilistic cost-effectiveness analysis: conditional expected incremental net benefit. PharmacoEconomics 38:135–141

McKechnie J, Colombo S, Chen J, Mabee W, MacLean HL (2011) Forest bioenergy or forest carbon? Assessing trade-offs in greenhouse gas mitigation with wood-based fuels. Environ Sci Technol 45(2):789–795

McManus MC, Taylor CM (2015) The changing nature of life cycle assessment. Biomass Bioenergy 82:13–26

Melero JA, Iglesias J, Garcia A (2012) Biomass as renewable feedstock in standard refinery units. Feasibility, opportunities and challenges. Energy Environ Sci 5(6):7393–7420

Michaga MFR, Michailos S, Akram M, Cardozo E, Hughes KJ, Ingham D, Pourkashanian M (2022) Bioenergy with carbon capture and storage (BECCS) potential in jet fuel production from forestry residues: a combined techno-economic and life cycle assessment approach. Energy Convers Manag 255:115346

Mishra RK, Kumar P, Mohanty K (2022) Hydrothermal liquefaction of biomass for bio-crude production: a review on feedstocks, chemical compositions, operating parameters, reaction kinetics, techno-economic study, and life cycle assessment. Fuel 316:123377

Misra P, Alvarez-Majmutov A, Chen J (2023) Isomerization catalysts and technologies for biorefining: opportunities for producing sustainable aviation fuels. Fuel 351:128994

Motalebi S, Barnes T, Lu L, Leibowicz BD, Niet T (2023) The role of U.S.-Canada electricity trade in North American decarbonization pathways. Energ Strat Rev 41(May):100827, 1–22. https://doi.org/10.1016/j.esr.2022.100827

Muijden JV, Stepchuk I, de Boer AI, Kogenhop O, Rademaker ER, van der Sman ES et al (2020) Final results alternative energy and propulsion technology literature study: deliverable D1. 1 of the Transcend project

Mungodla SG, Linganiso LZ, Mlambo S, Motaung T (2019) Economic and technical feasibility studies: technologies for second generation biofuels. J Eng Des Technol 17(4):670–704

Nagy Z, Felkner J, Beck AL, Reeves DC, Richter S, Shastry V et al (2022) IMPACT: integrated multi-domain emission pathways for cities under land-use policy, Technology Adoption, Climate Change And Grid Decarbonization. arXiv preprint arXiv:2202.07458

O'Donoghue C, Buckley C, Chyzheuskaya A, Green S, Howley P, Hynes S et al (2021) The spatial impact of rural economic change on river water quality. Land Use Policy 103:105322

Okolie JA, Awotoye D, Tabat ME, Okoye PU, Epelle EI, Ogbaga CC et al (2023) Multi criteria decision analysis for the evaluation and screening of sustainable aviation fuel production pathways. Iscience
Ong HC, Tiong YW, Goh BHH, Gan YY, Mofijur M, Fattah IR et al (2021) Recent advances in biodiesel production from agricultural products and microalgae using ionic liquids: opportunities and challenges. Energy Convers Manag 228:113647
Osman AI, Zhang Y, Lai ZY, Rashwan AK, Farghali M, Ahmed AA et al (2023) Machine learning and computational chemistry to improve biochar fertilizers: a review. Environ Chem Lett 21:1–86
Patón-Romero JD, Baldassarre MT, Rodriguez M, Piattini M (2019) Application of ISO 14000 to information technology governance and management. Comput Stand Interfaces 65:180–202
Peters MA, Alves CT, Onwudili JA (2023) A review of current and emerging production technologies for biomass-derived sustainable aviation fuels. Energies 16(16):6100
Presbitero A, Melnikov VR, Krzhizhanovskaya VV, Sloot PM (2021) A unifying model to estimate the effect of heat stress in the human innate immunity during physical activities. Sci Rep 11(1):16688
Pudel F, Wiesen S (2019) Vegetable oil-biorefinery. Biorefineries 166:69–98
Rajak J (2021) A preliminary review on impact of climate change and our environment with reference to global warming. Int J Environ Sci 10:11–14
Ramos A, Monteiro E, Rouboa A (2022) Biomass pre-treatment techniques for the production of biofuels using thermal conversion methods–a review. Energy Convers Manag 270:116271
Ravindran MX, Asikin-Mijan N, Ong HC, Derawi D, Yusof MR, Mastuli MS et al (2022) Feasibility of advancing the production of bio-jet fuel via microwave reactor under low reaction temperature. J Anal Appl Pyrolysis 168:105772
Rebitzer G, Ekvall T, Frischknecht R, Hunkeler D, Norris G, Rydberg T et al (2004) Life cycle assessment: part 1: framework, goal and scope definition, inventory analysis, and applications. Environ Int 30(5):701–720
Ribul M, Lanot A, Pisapia CT, Purnell P, McQueen-Mason SJ, Baurley S (2021) Mechanical, chemical, biological: moving towards closed-loop bio-based recycling in a circular economy of sustainable textiles. J Clean Prod 326:129325
Rioux B, Combes J, Woolley JM, Rodrigues NDN, Mention MM, Stavros VG, Allais F (2022) From biomass-derived p-hydroxycinnamic acids to novel sustainable and non-toxic phenolics-based UV-filters: a multidisciplinary journey. Front Chem 10:886367
Rissman J, Bataille C, Masanet E, Aden N, Morrow WR III, Zhou N et al (2020) Technologies and policies to decarbonize global industry: review and assessment of mitigation drivers through 2070. Appl Energy 266:114848
Rumizen MA (2021) Qualification of alternative jet fuels. Front Energy Res 9:760713
Saavedra-Rubio K, Thonemann N, Crenna E, Lemoine B, Caliandro P, Laurent A (2022) Stepwise guidance for data collection in the life cycle inventory (LCI) phase: building technology-related LCI blocks. J Clean Prod:132903
Sarkar S, Datta D, Deepak KS, Mondal BK, Das B (2023) Comprehensive investigation of various re-refining technologies of used lubricating oil: a review. J Mater Cycles Waste Manag:1–31
Schubert T (2020) Production routes of advanced renewable C1 to C4 alcohols as biofuel components–a review. Biofuels Bioprod Biorefin 14(4):845–878
Seber G, Escobar N, Valin H, Malina R (2022) Uncertainty in life cycle greenhouse gas emissions of sustainable aviation fuels from vegetable oils. Renew Sust Energ Rev 170:112945
Sharew S, Montastruc L, Yimam A, Negny S, Ferrasse JH (2022) Alternative energy potential and conversion efficiency of biomass into target biofuels: a case study in ethiopian sugar industry-Wonji-Shoa. Biomass 2(4):279–298
Sharma P, Bano A, Singh SP, Atkinson JD, Lam SS, Iqbal HM, Tong YW (2022) Biotransformation of food waste into biogas and hydrogen fuel–a review. Int J Hydrog Energy 52(C):46–60
Sharno MA, Hiloidhari M (2022) Life cycle carbon, energy, water, and landuse footprints of agricultural residue-based biojet fuel in Bangladesh. Bioresour Technol Rep 20:101234

Singh S, Pant KK, Krishania M (2023) Current perspective for bio-oil production from agricultural residues in commercialization aspect: a review. J Anal Appl Pyrolysis 175:106160
Teng Y, Li CZ, Shen GQ, Yang Q, Peng Z (2023) The impact of life cycle assessment database selection on embodied carbon estimation of buildings. Build Environ 243:110648
Thabrew L, Wiek A, Ries R (2009) Environmental decision making in multi-stakeholder contexts: applicability of life cycle thinking in development planning and implementation. J Clean Prod 17(1):67–76
Tien NH, Phu PP, Chi DTP (2019) The role of international marketing in international business strategy. Int J Res Marketing Manage Sales 1(2):134–138
Tiwari R, Mishra R, Choubey A, Kumar S, Atabani AE, Badruddin IA, Khan TY (2023) Environmental and economic issues for renewable production of bio-jet fuel: a global prospective. Fuel 332:125978
Torkayesh AE, Rajaeifar MA, Rostom M, Malmir B, Yazdani M, Suh S, Heidrich O (2022) Integrating life cycle assessment and multi criteria decision making for sustainable waste management: key issues and recommendations for future studies. Renew Sust Energ Rev 168:112819
Wait AD (2021) The importance of data reliability and usability when assessing impacts of marine mineral oil spills. Toxics 9(11):302
Wang WC, Tao L (2016) Bio-jet fuel conversion technologies. Renew Sust Energ Rev 53:801–822
Wei H, Liu W, Chen X, Yang Q, Li J, Chen H (2019) Renewable bio-jet fuel production for aviation: a review. Fuel 254:115599
Why ESK, Ong HC, Lee HV, Gan YY, Chen WH, Chong CT (2019) Renewable aviation fuel by advanced hydroprocessing of biomass: challenges and perspective. Energy Convers Manag 199:112015
Wood ML, Mann AW, Kraus AL (2021) Characterizing undetected stellar companions with combined data sets. Astron J 162(4):128
Xu D, Wang Y, Lin G, Guo S, Wang S, Wu Z (2019) Co-hydrothermal liquefaction of microalgae and sewage sludge in subcritical water: ash effects on bio-oil production. Renew Energy 138:1143–1151
Zein SH, Antony A (2022) Techno-economic analysis and feasibility of industrial-scale activated carbon production from agricultural pea waste using microwave-assisted pyrolysis: a circular economy approach. PRO 10(9):1702
Zeng J, Xu J, Li W, Dai X, Zhou J, Shan Y et al (2022) Evaluating trade-off and synergies of ecosystem services values of a representative resources-based urban ecosystem: a coupled Modeling framework applied to Panzhihua City, China. Remote Sens 14(20):5282
Zhai P, Li Y, Wang M, Liu J, Cao Z, Zhang J et al (2021) Development of direct conversion of syngas to unsaturated hydrocarbons based on Fischer-Tropsch route. Chem 7(11):3027–3051

Kolibri 360 GmbH
Ferdinand-Jühlke-Straße 7
99095 Erfurt, Deutschland
produktsicherheit@kolibri360.de